Fundamental Symmetries

ETTORE MAJORANA
INTERNATIONAL SCIENCE SERIES
Series Editor:
Antonino Zichichi
European Physical Society
Geneva, Switzerland

(PHYSICAL SCIENCES)

Recent volumes in the series:

Volume 21 **ELECTROWEAK EFFECTS AT HIGH ENERGIES**
Edited by Harvey B. Newman

Volume 22 **LASER PHOTOBIOLOGY AND PHOTOMEDICINE**
Edited by S. Martellucci and A. N. Chester

Volume 23 **FUNDAMENTAL INTERACTIONS IN LOW-ENERGY SYSTEMS**
Edited by P. Dalpiaz, G. Fiorentini, and G. Torelli

Volume 24 **DATA ANALYSIS IN ASTRONOMY**
Edited by V. Di Gesù, L. Scarsi, P. Crane, J. H. Friedman , and S. Levialdi

Volume 25 **FRONTIERS IN NUCLEAR DYNAMICS**
Edited by R. A. Broglia and C. H. Dasso

Volume 26 **TOKAMAK START-UP: Problems and Scenarios Related to the Transient Phases of a Thermonuclear Fusion Reactor**
Edited by Heinz Knoepfel

Volume 27 **DATA ANALYSIS IN ASTRONOMY II**
Edited by V. Di Gesù, L. Scarsi, P. Crane, J. H. Friedman, and S. Levialdi

Volume 28 **THE RESPONSE OF NUCLEI UNDER EXTREME CONDITIONS**
Edited by R. A. Broglia

Volume 29 **NEW MECHANISMS FOR FUTURE ACCELERATORS**
Edited by M. Puglisi, S. Stipcich, and G. Torelli

Volume 30 **SPECTROSCOPY OF SOLID-STATE LASER-TYPE MATERIALS**
Edited by Baldassare Di Bartolo

Volume 31 **FUNDAMENTAL SYMMETRIES**
Edited by P. Bloch, P. Pavlopoulos, and R. Klapisch

A Continuation Order Plan is available for this series. A continuation order will bring delivery of each new volume immediately upon publication. Volumes are billed only upon actual shipment. For further information please contact the publisher.

Fundamental Symmetries

Edited by

P. Bloch

P. Pavlopoulos

and

R. Klapisch

CERN
Geneva, Switzerland

Plenum Press • New York and London

Library of Congress Cataloging in Publication Data

Fundamental symmetries.

(Ettore Majorana international science series. Physical sciences; v. 31)
Proceedings of the first course of the International School of Physics with Low-Energy Antiprotons, Fundamental Low Energy, held September 26–October 3, 1986, at Erice, Trapani, Sicily, Italy.
Bibliography: p.
Includes index.
1. Symmetry (Physics)—Congresses. 2. Antiprotons—Congresses. I. Bloch, P. II. Pavlopoulos, P. III. Klapisch, R. IV. International School of Physics "Ettore Majorana." V. Series.
QC793.3.S9F86 1987 530.1′42 87-20228
ISBN 0-306-42673-0

Proceedings of the first course of the International School of Physics with Low-Energy Antiprotons, Fundamental Low Energy, held September 27–October 3, 1986, at Erice, Trapani, Sicily, Italy

A Division of Plenum Publishing Corporation
233 Spring Street, New York, N.Y. 10013

Printed in the United States of America

PREFACE

The first course of the International School on Physics with Low Energy Antiprotons was held in Erice, Sicily at the Ettore Majorana Centre for Scientific Culture, from September 26 to October 3, 1986.

The purpose of this School is to review the physics accessible to experiments using low energy antiprotons, in view of the new era of the CERN LEAR ring opened by the upgrade of the antiproton source at CERN (ACOL). In 1986 the first course covered topics related to fundamental symmetries.

These Proceedings contain both the tutorial lectures and the various contributions presented during the School by the participants. The contributions have been organized in six sections.

The first section is devoted to gravitation, a particularly "hot" topic in view of recent speculations about deviations from Newton's and Einstein's theories.

Section II covers various problems related to the matter-antimatter symmetries such as comparison of the proton and antiproton, inertial masses or spectroscopy of antihydrogen or other antiprotonic atoms.

CP and CPT violations in weak interaction are presented in Section III.

The test of symmetries in atomic physics experiments and the strong CP problem are covered in Section IV.

Section V groups contributions related to high precision measurements of simple systems like protonium, muonium or the anomalous moment of the muon.

The last section is devoted to the experimental challenge of polarizing antiproton beams.

We thank the lecturers and the participants of the School for their efforts and their contributions and hope that they found this School as fruitful as we did.

We are particularly grateful to Mrs. Anne Marie Bugge for her friendly and efficient help in the organization of the School as well as for the editing of these Proceedings. We should also like to thank Dr. Alberto Gabriele and his staff of the Ettore Majorana Centre who made the running of the School very easy and our stay extremely pleasant.

Ph. Bloch, P. Pavlopoulos, R. Klapisch

CONTENTS

I GRAVITATION

II MATTER AND ANTIMATTER

Mass measurements for antiprotons

Cooling and decelerations of antiprotons

Atomic physics experiments with antiprotons

Charge measurements

III CP AND CPT VIOLATION IN WEAK INTERACTION

IV TIME REVERSAL INVARIANCE AND SYMMETRIES IN STRONG INTERACTION

V SIMPLE SYSTEMS AND QED

VI POLARIZATION OF ANTIPROTONS

GRAVITY*

J.S. Bell

CERN, Geneva, Switzerland
Geneva, Switzerland

LECTURE No. 1: OLD THEORIES OF GRAVITATION

NEWTON'S LAW

Newton's law of gravitation is expresssed in Eq. (1.1)-

$$V = -\ m_1 m_2 G / r \tag{1.1}$$

where m_1 and m_2 are the masses of the two objects and G is Newton's constant. This formula can be reexpressed in terms of the so-called Planck mass, m_p:

$$V = hc/r\ (m_1/m_p)\ (m_2/m_p) \tag{1.2}$$

where h is Planck's constant divided by 2π. It is important to note that m is the <u>inertial</u> mass and not something specially introduced for gravity.

The values of Newton's constant and of the Planck mass are listed below:

$$G = 6.672(4)\ 10^{-8}\ \mathrm{cm}^3\ \mathrm{g}^{-1}\ \mathrm{sec}^{-2}$$
$$m_p = 2.172\ 10^{-5}\ \mathrm{g} = 1.22\ 10^{19}\ \mathrm{Gev}$$

The Compton wave length associated to the Planck mass is:

$$l_p = hc/(m_p c^2) = 1.62\ 10^{-33}\ \mathrm{cm}.$$

For two protons the gravitational interaction energy is therefore given by:

$$V = 6\ 10^{-39}\ hc/r \tag{1.3}$$

*These are verbatim lecture notes by Ph. Bloch, R. Klapisch and P. Pavlopoulos. They have not been edited by J.S. Bell.

This can be compared with the Coulomb interaction:

$$V_C = + e^2/r = (1/137)\, hc/r \tag{1.4}$$

where the coefficient 6 10^{-39} is replaced by 1/137. So we are concerned with an extremely small interaction on the scale of those that we normally meet in elementary particle physics.

Table 1.1 presents some characteristics of some of the elementary objects of the theory of gravitation, namely the Sun, the Earth and the Moon: their masses, densities and radii. We should especially look at the 4th quantity in this table, $Gm/(Rc^2)$, where R is the radius and m the mass of the object concerned. This quantity measures the importance of the relativistic effects in gravitational phenomena about these bodies. You can see that it is rather tiny: One part in a million for the Sun and seven parts in 10^9 for the Earth. For a satellite going very closely around the body in question, this quantity is the ratio of its potential energy to its rest energy; it is also the square of the ratio of the velocity of this satellite to the velocity of light. It measures also the ratio of the gravitational energy of the source to its ordinary rest energy.

Table 1.1

Sun ⊙ — $1.50\ 10^{13}$ cm — Earth ⊕ — $3.84\ 10^{10}$ cm — Moon ☾

mass m	$1.99\ 10^{33}$ g	$5.98\ 10^{27}$ g	$7.35\ 10^{25}$ g
density g cm^{-3}	1.409	5.52	3.34
radius cm R	$6.96\ 10^{10}$	$6.37\ 10^{8}$	$1.74\ 10^{8}$
$G\,m/R\,c^2$	$2.1\ 10^{-6}$	$7.0\ 10^{-9}$	$3.1\ 10^{-11}$

We see that relativistic effects are very small in the phenomena we are discussing. Nevertheless they can be seen, and even if they could not be seen, it would be important to make Newton's theory relativistic, because we believe in relativity. When we say relativity, we are talking about special relativity (there will be no mention of general relativity until rather near the end of this first lecture).

RELATIVISTIC THEORY OF GRAVITATION

In special relativity, the mass is associated with the internal energy of the body as shown in Einstein's famous formula

$$E = mc^2 \tag{1.5}$$

The energy is also given by the integral of an energy density T_{00} :

$$E = \int d^3 x\ T_{00} \tag{1.6}$$

where T_{00} is one of the 16 quantities $T_{\mu\nu}$ which form the energy momentum tensor. $T_{\mu\nu}$ is a symetric tensor ($T_{\mu\nu} = T_{\nu\mu}$).

One must therefore expect that a relativistic theory of gravitation will have some fields coupled to this tensor $T_{\mu\nu}$. There are two obvious ways to do it: One is to introduce a tensor field $h_{\mu\nu}$ and to contract it with $T_{\mu\nu}$. This leads us to a theory which ressembles Einstein's theory. One can also use a scalar field h and couple it to the trace of the energy momentum tensor, h $T_{\mu\mu}$, a theory which was put forward by Nordström in 1913 after Einstein had put forward the first ideas of general relativity, but before he had completed it. These terms are the interaction terms in the Lagrangian. We have, however, to find a complete Lagrangian by adding all the Lagrangian terms for the fields and all the Lagrangian terms for matter.

In the following we will follow the Poincaré-Minkowski notations, using the the "zero" component of the four-vector, x_0, related to the fourth component by $x_4 = ix_0$ such that the scalar product of two vectors is:

$$\begin{aligned} x_\mu y_\mu &= x_1y_1 + x_2y_2 + x_3y_3 + x_4y_4 \\ &= x_1x_1 + x_2y_2 + x_3y_3 - x_0y_0 \end{aligned}$$

THE FIELD TERMS IN THE LAGRANGIAN

1. Case of a scalar field

We will first examine the case of a scalar field. We will assume that the theory follows an action principle: the action is the integral of the Lagrangian

$$A = \int d^4 x\, L \tag{1.7}$$

Setting its variation to zero

$$\delta A = 0 \tag{1.8}$$

defines the classical equations of motion, or even the Heisenberg's equations of motion of quantum mechanics. The standard Lagrangian for a scalar field is given by

$$L = -1/2\ \partial_\mu\varphi\ \partial_\mu\varphi - \mu^2/2\ \varphi + \sqrt{4\pi G}\ \varphi\ T_{\mu\mu} \tag{1.9}$$

where we have set h and c to 1 and ∂_μ is the differential operator $\partial_\mu = \partial/\partial x_\mu$. The second term in Eq. (1.9) is a mass term and the third term is an interaction term where we have related the coupling constant to Newton's constant G. The equation of motion which follows from such a Lagrangian is:

$$\partial_\mu\partial_\mu\ \varphi\ - \mu^2\ \varphi\ = \sqrt{4\pi G}\,T_{\mu\mu} \tag{1.10}$$

It is the Klein-Gordon equation augmented by a source term.

If we take a static source, the energy density is a delta function, the coefficient of the delta function being the mass of this source. The trace is given by:

$$T_{\mu\mu} = -m_2\,\delta^3(r)\ ;\quad T_{00} = m_2\,\delta^3(r) \tag{1.11}$$

the minus sign in $T_{\mu\mu}$ being due to the fact that the T_{44} component has a sign opposite to the T_{00} term. If we have a static source, we have a static field and therefore the derivatives relative to time disapear from Eq. (1.10) which becomes:

$$\nabla^2\varphi - \mu^2\varphi = \sqrt{4\pi G}\, m_2\,\delta^3(r) \tag{1.12}$$

The familiar solution of this equation is:

$$\varphi = -\sqrt{G/4\pi}\,(m_2/r)\,e^{-\mu r} \tag{1.13}$$

If one considers now a static particle in the neighbourhood of that static source, we can calculate its interaction energy. T_{00} for the static particle is equal to:

$$T_{00} = -m_1\,\delta^3(r)$$

and the interaction energy is given by the integral:

$$\int d^3x\,\sqrt{4\pi G}\,\varphi\, T_{00} = G\,(m_1 m_2/r)\ e^{-\mu r} \tag{1.14}$$

We obtain Newton's formula augmented by a range factor.

2. Case of a vector field

There are two reasons for reminding you of some aspects of vector field theory. One is that it is easier than tensor field theory and important points can be made with fewer indices. The second reason is that now people do speculate that the gravitational field may not be entirely described by Einstein's tensor but may also have a vector component (and perhaps a scalar component too).

We have more freedom in constructing the Lagrangian than in the scalar field case:

$$L = -1/2\,\partial_\mu A_\nu\,\partial_\mu A_\nu - \mu^2/2\,A_\nu A_\nu + b/2\,\partial_\mu A_\nu\,\partial_\nu A_\mu + f\,j_\mu A_\mu\sqrt{4\pi} \tag{1.15}$$

The first two terms in this Lagrangian are entirely analogous to what we had for the scalar theory: we have the derivative of the field squared and a mass term. The index ν that we did not have for the scalar field is summed over from 1 to 4, as if we had four different scalar fields. The third term is another Lorentz invariant quantity, where the indices are associated in a different way. The coefficient b in this term is not arbitrary and there are good reasons for giving it the value 1. To see these reasons, let us consider fields which do not vary with x_1, x_2 and x_3 and are functions only of the time. The Lagrangian simplifies then to:

$$L = 1/2\dot{\vec{A}}^2 - 1/2\dot{A}_0^2 - \mu^2/2\vec{A}^2 + \mu^2/2A_0^2 + b/2\dot{A}_0^2 \tag{1.16}$$

Comparing to Eq. (1.9), we see that both the first two terms in A_0^2 have other signs than in the scalar case.

For $b < 1$ the term in A_0^2 corresponds to a negative energy and correspondingly our theory becomes unstable, since a system of finite energy can raise its energy by radiating negative energy, which does not sound good. (Going to quantum field theory, we could get rid of that difficulty of negative energy in a certain sense by trading it against negative probability: instead of a positive probability of negative energy, we could have a negative probability of positive energy. It is not much better! In fact I, would say it is much worse, since although negative energies are embarassing, negative probabilities are nonsense!)

This difficulty disapears if b = 1. The mass term has still the wrong sign, but that is not so bad because you find that, when the kinetic energy term is absent, the component A_0 does not go anywhere: the embarrasing term remains in the neighborhood of the source.

If we take $b > 1$ to get the kinetic energy positive, the mass term still has the wrong sign and we get tachyons, i.e. particles faster than the light.

Therefore the best way of getting rid of both difficulties is to set b = 1. This choice gives the standard Lagrangian for a vector field:

$$L = - 1/4 \, (\partial_\mu A_\nu - \partial_\nu A_\mu)^2 - \mu^2/2 \, A_\nu{}^2 + f \, j_\mu A_\mu \sqrt{4\pi} \tag{1.17}$$

When the mass μ is 0, and if we do not look at the source term, this Lagrangian is gauge invariant: this means that if we replace A_μ by $A_\mu - \partial_\mu x$, where x is a scalar, the Lagrangian does not change. Gauge invariance is not an optionnal luxury. Without gauge invariance, the theory is a disaster!

We have now written the Lagrangian for the case of massive particles. Let us consider the case of massless particles.

The equations of motions follow from the Lagrangian:

$$\partial_\mu (\partial_\mu A_\nu - \partial_\nu A_\mu) - \mu^2 A_\nu = - \sqrt{4\pi} \, f \, j_\nu \tag{1.18}$$

If we multiply the left hand side of this equation by ∂_ν and contract over μ we obtain:

$$- \mu^2 \partial_\nu A_\nu = - \sqrt{4\pi} \, f \, \partial_\nu j_\nu \tag{1.19}$$

The 4-divergence of the vector field is directly related to the 4-divergence of the current. This is what I meant when I said that the embarassing component of the vector meson field does not propagate when b = 1, because this term is directly related to the divergence of the source.

Using Eq. (1.18) we can rewrite Eq. (1.19) as:

$$\partial_\mu \partial_\mu A_\nu - \mu^2 A_\nu = - \sqrt{4\pi} \, f \, (j_\nu - \partial_\nu \partial_\mu j_\mu / \mu^2) \tag{1.20}$$

We get back the Klein-Gordon equation with an effective source which contains both the current and the divergence of the current. Since μ appears now at the denominator of the extra term, it is not possible to take μ to 0 unless $\partial_\nu j_\nu = 0$, i.e. unless we have a conserved current. A massless theory needs a conserved current.

The fact that the limit $\mu \to 0$ is delicate, is very important for the hyperphoton theory: several times people have speculated that the hypercharge, which distinguishes K meson from K meson, should be coupled to something analogous to a photon. However, hypercharge is not conserved, so we are faced here with a case where $\partial_\nu j_\nu$ is not zero and therefore we can not have $\mu = 0$. The range λ associated with the hyperphoton, which is the inverse of μ, must not be infinite. By looking at experimental bounds for the reaction $K \to \pi + \gamma$, one can put a restriction relating the range λ to the coupling constant. This has been evaluated by Fischbach et al. who quote:

$$(f^2/G \, m_p{}^2) \cdot \lambda \leq 5 \tag{1.21}$$

where

> λ is the range associated to hyperphoton, expressed in meter;
> f is the coupling constant of the hyperphoton, and
> m_p is the proton mass.

Even if we are considering coupling constants which are comparable with gravitational strength, the range λ must be rather small compared to macroscopic objects.

If we again consider static situation where $\partial/\partial t = 0$, and where only the 4th component of the current is significant

$$j_0 = Q_2\, \delta^3(r)$$

we obtain :

$$(\nabla^2 - \mu^2)A_0 = -\sqrt{4\pi}\, f\, Q_2\, \delta^3(r) \tag{1.22}$$

A_0 is the only non vanishing component and is proportional to the inverse distance from the source. For a slow particle, which has its own charge density $j_0 = Q_1\delta^3(r)$, we can calculate the interaction energy:

$$E = f^2\, Q_1 Q_2 / r_{12}\, e^{-\mu r} \tag{1.23}$$

Q_1 and Q_2 here are some kind of charge related to what some people have called the graviphoton, a vector particle which might or might not be associated with gravitation.

3. Case of the tensor field

We all know that it must be the main component of gravitation since the tensor field is Einstein's field and it works extremely well.

For the tensor field case, the possibilities for making a Lagrangian are still richer than in the vector case and we see below a Lagrangian in which there are many terms:

$$\begin{aligned} L = 1/4(&\partial_\lambda h_{\mu\nu}\partial_\lambda h_{\mu\nu} - 2\partial_\lambda h_{\mu\lambda}\partial_\sigma h_{\mu\sigma} + 2\partial_\lambda h_{\mu\lambda} - \partial_\mu h_{\sigma\sigma}\partial_\mu h_{\lambda\lambda} \\ &+ \mu^2 (h_{\mu\nu}\, h_{\mu\nu} - 1/2\ h_{\mu\mu} h_{\nu\nu}) - \sqrt{8\pi G}\, h_{\mu\nu}\, T_{\mu\nu} \end{aligned} \tag{1.24}$$

If you started from the beginning, you could put arbitrary coefficients in front of all these terms, so you would have a lot of parameters to play with. However, again you find that nearly all of these Lagrangians are unacceptable. If you again require that there is no negative energy in the classical theory or no negative probabilities (which one describes as no "ghost") in the quantum theory, and no tachyons, you find that the coefficients which were arbitrary are restricted to precisely the values written in Eq. (1.24).

Let us ignore for the moment the factor 1/2 in the 6th term. So you get the Lagrangian (1.24) and it looks as if you were well started. Indeed, as long as you put the mass $\mu \neq 0$, it is a perfectly good theory. But of course we know that the real gravitation has $\mu = 0$ (or at least extremely tiny). Therefore we are specially interested in the limit $\mu = 0$. And now there is a shock as compared to the vector case: in this theory the field h does not have a limit when μ goes to 0. You cannot approach an infinite range theory of gravity in this way. What you find is that, if you want a theory in which this limit exists, not only must you conserve the tensor $T_{\mu\nu}$ (i.e. $\partial_\mu T_{\mu\nu} = 0$) just as we had to conserve the current in the vector theory, but you must also put in a very special form to the mass term, where there is a factor 1/2 in it. But putting in this factor 1/2, you are in contradiction with the previous arguments dictating this coefficient to be 1! We have a very curious situation, which was discovered by Veltman and Van Dam, that the theory which is sensible for finite mass does not have a limit and the theory which does have a limit is not sensible for finite mass.

You could ask for what kind of sense does it make to put $\mu = 0$ and in fact it does make sense. Although there is a scalar ghost in the theory with the factor 1/2, you find that in the limit μ goes to 0, this ghost is always and unavoidably accompanied by the zero helicity component of the tensor of the gravitationnal field, which is a spin 2 object and which has various helicity components. These two, always going together, are such that the negative probability is always associated with a positive probability. This gives to the theory a sensible interpretation. By these means the Einstein's theory (I call it the Einstein's theory but at the moment you only see that it ressembles Einstein's theory in having a tensor field) has a sort of stability: when you have decided that you are going to work with zero mass, you can not have a little mass without adding non-sense; you are stuck with zero mass.

The fact that you can live when $\mu = 0$ with what was a ghost before μ was 0, must reopen the question: what about the other ghosts that we eliminated by requiring a conserved tensor? And in fact you do find again that, if you decide to permit these other ghosts for $\mu \neq 0$, you can cancel the negative probabilities against positive probabilities when going to $\mu = 0$, so that the Lagrangian becomes more free for zero masses than for finite masses. But what you discover is that this freedom is a familiar one: it corresponds to the freedom that you have in a zero mass gauge invariant theory to add gauge fixing terms, forcing the field into one gauge rather than another. So it is not a freedom to make different physics, but a freedom to write the same theory in a richer variety of ways.

I spoke here of gauge, and you are familiar with gauge in the vector case. The tensor theory which I described is also invariant in the limit μ goes to 0 under the gauge transformations:

$$h_{\mu\nu} \rightarrow h_{\mu\nu} + \partial_\mu \xi_\nu + \partial_\nu \xi_\mu \tag{1.25}$$

which is an obvious generalisation of the vector one.

The field equation which follows from the Lagrangian (1.24) is written below :

$$\partial_\sigma \partial_\sigma h_{\mu\nu} - 2h_{\mu\sigma,\sigma\nu} = - \lambda \overline{T}_{\mu\nu} \tag{1.26}$$

where $\overline{T}_{\mu\nu} = T_{\mu\nu} - 1/2\, \delta_{\mu\nu} T_{\lambda\lambda}$, etc.

and $\lambda = \sqrt{8\pi G}$

The easiest way to solve that equation is to first neglect the term $h_{\mu\sigma,\sigma\nu}$. That spoils the gauge invariance, and this means that the solution of the remaining equation is unique. When you find this solution, you discover that the term that you put equal to zero is in fact zero. That is to say that neglecting this term is an admissible gauge (just as we are used to use the Lorentz gauge where $\partial_\mu A_\mu = 0$ when we are doing the theory of electromagnetism).

When you go to a time independent situation, Eq. (1.26) becomes

$$\nabla^2 h_{\mu\nu} = - \lambda (T_{\mu\nu} - 1/2\, \delta_{\mu\nu} T_{\sigma\sigma}) \tag{1.27}$$

Taking a source concentrated at a point with a mass m_2, we find the non vanishing components of the tensor field

$$h_{00} = h_{11} = h_{22} = h_{33} = \lambda m_2/(8\pi r) \tag{1.28}$$

The interaction energy with a static test particle is obtained by integrating the field of the source over the energy density of the test particle, and we come back to Newton's law:

$$- \lambda^2 \int T_{00}(1)\, h_{00}(2) \; = G\, m_1 m_2 / r \tag{1.29}$$

THE RED SHIFT

The tensor and scalar theories, but not the vector theory, produce a red shift in the sense that periodic objects deeper in the gravitational potential oscillate more slowly and in particular spectral lines are shifted into the red.

Such an object has a Lagrangian or a Hamiltonian of its own, and on to that I have added an interaction term ($h_{\mu\nu}\; T_{\mu\nu}$ or $h\; T_{\mu\mu}$) involving the field h which I now treat as an external field. So not only do these fields exerce forces on the object, but also they change its internal dynamics: I am adding something to the Hamiltonian. If I express these T terms in function of the internal coordinates and momenta of the object, you see that I have added something to the internal Hamiltonian of the object. In fact the terms I put have changed it in a simple way. The easiest way to see it is in quantum mechanics, because it happens that we are more familiar with perturbation theory in quantum mechanics than in classical mechanics. So imagine I calculate the perturbations, due to these terms, of the energy of the stationary states. I have to calculate the expectation value of the $h \cdot T$ terms integrated over the object in question

$$\delta E_n = \int \langle n \mid h_{\mu\nu} T_{\mu\nu} \text{ or } h\; T_{\mu\mu} \mid n \rangle \tag{1.30}$$

Let us suppose that the object is small and h or $h_{\mu\nu}$ approximately constant. I can take the h outside the integral and integrate over T_{00}.

$$\delta E_n = (h_{00} \text{ or } h) \int \langle n \mid T_{00} \mid n \rangle = (h_{00} \text{ or } h)\, E_n \tag{1.31}$$

The second part is nothing but the energy E_n: I have got that the shift in the energy E_n is porportionnal to E_n multiplied by the field h or h_{00}, whether I was dealing with a scalar or a tensor theory. I have used here a theorem due to Van Laue, that the expectation value in this stationary state of a conserved tensor comes from the T_{00} component, and the expectation of the other components which complicate this expression are just 0. So you see that the tensor and scalar theory are coming together. This means also that transition frequencies which, in the classical limit are the frequencies of periodic motions, are changed in proportion to what they were before:

$$\delta \omega_{nm} = (h_{00} \text{ or } h)\, (E_n - E_m) \tag{1.32}$$

If I translate these h into a more meaningful language, I find that the frequency is multiplied from its unperturbed value by 1+ the potential of the object in the gravitational field divided by its rest energy:

$$\omega_{nm} \rightarrow (1 + V/mc^2)\, (E_n - E_m) \tag{1.33}$$

That is the standard red shift, which first was appeared in Einstein's theory. Here it comes out of a theory which is not Einstein's, but it comes out nevertheless and (a point that I want to emphasize) not by any philosophizing about space and time, but by seing what happens if you add a term to the Lagrangian.

THE CASE OF MOVING OBJECTS

What happens when the test particle which goes around some heavy source moves and moves fast?

$T_{\mu\nu}$ is a tensor; the only tensor referring to a particle considered as a whole is its 4 momentum; therefore the obvious tensor is:

$$T_{\mu\nu} = p_\mu p_\nu \, \delta^3(r)/p_0 \tag{1.34}$$

You can understand the $1/p_0$ factor if you think of this object existing in a little sphere; when you make it move, that sphere will Fitzgerald contract, so that the density will be over a smaller volume. The δ function of Eq. (1.34) has, in a certain sense, a constant volume and we must Fitzgerald contract it.

For the scalar field, I am interested in its trace. Since we have

$$p^2 = m^2$$

the trace is just

$$T_{\mu\mu} = - \delta^3(r) \, m^2 / p_0 \tag{1.35}$$

Finally, in the vector case I have the current:

$$j_\mu = Q \, \delta^3(r) \; p_\mu / p_0 \tag{1.36}$$

I have to calculate the interaction energy of these terms with the field of a static source. I am considering a body moving rapidly around a heavy static object. In the scalar case, we obtain:

$$\text{scalar case:} \quad V = - G \, (m_1 m_2/r) \; e^{-\mu r} \; m_1/p_0 \tag{1.37}$$

which is the same as the static case Eq. (1.14) multiplied by m/p_0. You see that this interaction energy becomes rapidly less effective as the particle moves faster. The scalar gravitational field, which contributes to the fall of apples, does not contribute to the deflection of light.

In the vector field case, the interaction energy does not fall off with velocity

$$\text{vector case:} \quad V = f^2 \, Q_1 Q_2 / r \; e^{-\mu r} \tag{1.38}$$

because the relevant component is p_0 (the static source has only a zero component in its field), and we therefore get p_0 divided by p_0. This is a bit more substantial at high energy, but nevertheless it is overcome by the increase in relativistic mass.

Finally, in the tensor field case we get

$$\text{tensor case:} \quad V = - G \, (m_1 m_2 / r) \; (p_0{}^2 + p^2) / m_1 p_0 \tag{1.39}$$

the $p_0{}^2$ coming from the h_{00} component and the p^2 term from the other diagonal elements of the field (see Eq. (1.28)). It just increases with high energy in such a way to counteract the relativistic mass increase.

In conclusion, only the tensor field deflects and retards fast particles, photons and light pulses. In a theory of gravity which mixes up the three cases, one of the things you will have to worry about is that you may get the deflection of light wrong unless you take some precautions in adding up your admixtures.

CONNECTION WITH GENERAL RELATIVITY

Tensor theory comes increasingly to ressemble Einstein's theory on general relativity. I have said nothing about space and time, nothing about geometry and physics,nothing about invariance under general transformation of coordinates and yet you see, coming out of the theory, all the following goodies:

- light deflection

- delay of light passing near massive bodies (That is not one of the classical tests of general relativity, but it should have been; it was discovered more recently by Shapiro but probably, if Einstein had imagined that it could be measurable, he would have discovered it himself.)

- red shift

And indeed all theese goodies come out with the right magnitude.

What does not come out is the motion of the perihelion of Mercury. The theory that I have presented to you certainly does imply that the perihelion of Mercury moves, but it will not give the right answer to it. And in fact, to get the right answer is more delicate because, whereas the previous points are first order effects, proportional to Newton's constant G, most of the perihelion of Mercury is a second order effect, and it needs to take into account that the gravitational energy of the gravity field is itself a source of gravitation.

When you have a massive body, in first approximation all the energy is concentrated in the body; the body creates a gravitational field, the gravitational field has energy. Some of the source is now distributed and there is a corresponding modification of Coulomb's law which is important for the perihelion of Mercury.

You might say: "This tensor theory is not so good!", and you will discover at the same time that this theory is not self-consistent because I told you it only makes sense for a conserved current. It is conserved approximately in so far as I consider the matter left to itself, but now I am putting this matter in a gravitational field, and the gravitational field exchanges energy with the object and the original term $T_{\mu\nu}$ of matter is conserved only to an approximation in which the gravitational field is neglected. To make my theory sensible and the equations self-consistent I have to take into account that the gravitational field has energy which must be accounted for in the conservation law. So I have to consider a Lagrangian which in the equation of motion, will produce not only a term $T_{\mu\nu}$ of matter but a term $T_{\mu\nu}$ of matter + $T_{\mu\nu}$ of gravity. The latter term is likely to be of second order in the field strength. To get that out of the Lagrangian you have to add to it a term cubic in the field strength:

$$h_{\mu\nu}\, T_{\mu\nu} \rightarrow h_{\mu\nu}\, T_{\mu\nu} + 0(h^3) + 0(h^4) + \ldots \qquad (1.40)$$

If you add these extra terms in a minimal way, what you come to is Einstein's equation:

$$R_{\mu\nu} - 1/2\, g_{\mu\nu}\, R = T_{\mu\nu} \qquad (1.41)$$

written here with the extra terms collected on the left. Therefore you find that Einstein's equation is the universal low energy limit of sensible tensor theories. That is the main moral of the lecture I gave to-day, because you will see, when I will describe newer theories, that the way in which Einstein's theory is incorporated is like the way I developed here to-day: you do not start with philosophy and geometry; you start with a Lagrangian and you find that there is a spin 2, zero mass gravity field in it, and that it satisfies Einstein's equation in the low energy limit. This is a developement which is, I think, close to others that you could find in papers from Thirring or Weinberg, or in the article of Boulware and Deser[1], which is a sort of definitive paper on this topic.

GENERAL COVARIANCE AND EQUIVALENCE PRINCIPLES

So far I have said nothing about changing coordinates, but we know that Einstein's central technical idea was precisely that the theory should have the same form when written in arbitrary coordinate systems. How does that emerge from our approach?

We have seen that our approach has a gauge invariance (see Eq. (1.25). This can be seen as a first approximation to the way in which the tensor changes under general coordinate transformation. If you make a small change in coordinate:

$$x_\mu \rightarrow x_\mu + \xi_\mu \qquad (1.42)$$

and if you ask how does a tensor transform under such a coordinate change, you find that Eq. (1.25) is just the first term. Then there are terms which are themselves proportionnal to h, and so on. And you find moreover that if you extend the theory as I said, you must extend the definition of this gauge invariance to include the additional terms in the Lagrangian. And the transformation which the theory does respect turns out to be precisely that of general covariance, so the theory does have the invariance of Einstein's theory.

You remenber that in special relativity the quadratic form

$$ds^2 = \delta_{\mu\nu}\, dx^\mu\, dx^\nu$$

was especially important, giving you directly the trajectory of light. If you choose to write a special relativistic theory in terms of curvilinear coordinates, this simple expression ds^2 becomes in general, by linear transformation,

$$g_{\mu\nu}\, dx^\mu\, dx^\nu$$

and the tensor $g_{\mu\nu}$ appears all over the place in your theory. In a tensor theory you find that the combination

$$\delta_{\mu\nu} + \sqrt{4\pi G}\, h_{\mu\nu}$$

appears just in those places where $g_{\mu\nu}$ appears in special relativity as a consequence of co-ordinate transformations. You see that, if you call this quantity $g_{\mu\nu}$, you have the general covariance, that is that the difference between one set of coordinates and another can be regarded as beeing in a different gravitational field. This leads you to the famous Equivalence Principle that "gravity can be transformed away, locally, for all physics". You certainly cannot find a change of coordinates which reduces an arbitrary $g_{\mu\nu}$ everywhere to $\delta_{\mu\nu}$. But locally you can find a transformation which makes $g_{\mu\nu}$ and its first derivative equal to $\delta_{\mu\nu}$ so that the gravity can be transformed away locally and you are left with the physics of special relativity free of gravity. That is the famous Einstein's lift; if you go into a falling lift and use the frame of coordinates attched to the lift, there is no gravity, and that is so for all physics.

I want now to expand a little bit on the Equivalence Principle. There are many Equivalence Principles in the litterature; I have in Fig. 1.1 a classification taken from the book of Clifford Will. What I call the Equivalence Principle, he calls Strong Equivalence Principle: the whole of physics in a gravitational field can be obtained from gravitation-free physics by going to the accelerated coordinate system. Then there are all kinds of specialization in which you ask that only some restricted domain of physics should be reproduced in this way. Let us list some of them:

- the Einstein's Equivalence Principle (EEP): all physics but gravitation physics respects this principle. As long as you consider objects which do not have appreciable gravitational interactions, they will respect this principle that an external gravitaional

Theory and experiment in gravitational physics

CLIFFORD M. WILL

Department of Physics, Washington University, St. Louis

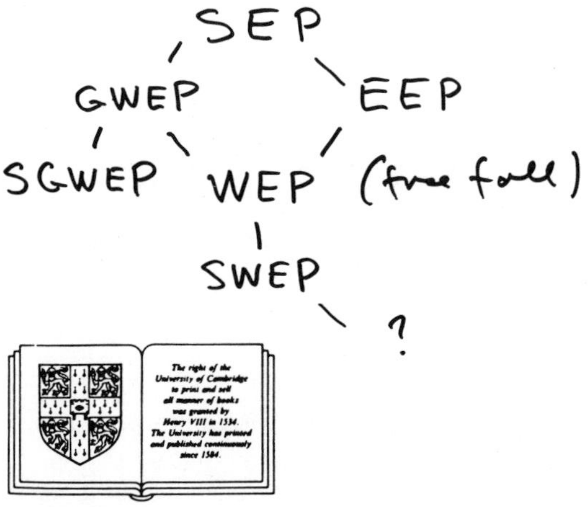

CAMBRIDGE UNIVERSITY PRESS

Cambridge 1981

London New York New Rochelle

Melbourne Sydney

Fig. 1.1

field can be transformed away. So you would worry about it if yo come to discuss the Moon, where an appreciable part of its energy is gravitational, but you are not worried about these distinctions when discussing apples;

- the Weak Equivalence Principle (WEP) in which you forget about most of physics and just consider free fall;

- the Gravitational Weak Equivalence Principle (GWEP) which is about how objects fall even when they have appreciable gravitational self energy.

Then I have introduced myself some specializations:

- the Slow Weak Equivalence Principle (SWEP) applied only to slowly moving objects, i.e. appples but not light or photons. A scalar theory respects this principle, but photons do not fall in a scalar theory;

- the Slow Gravitationnal Weak Equivalence Principle (SGWEP).

No matter what experiment you are doing, you should be able to devise an Equivalence Principle which is tested by your experiment and by nobody else's!

LECTURE No. 2: NEW THEORIES OF GRAVITY

ANTIGRAVITY

Once upon a time some people suggested serious reasons for thinking about antigravity: How is it that the world we are living in is matter rather than antimatter? It could be that matter and antimatter have separated in the world because they repelled.

In a celebrated paper in 1958 Morrison[2)] made the best case he could for something like antigravity but of course he realized that there were difficulties with the idea: difficulty with CPT, which I will not discuss, and difficulties with the equivalence principle and energy conservation.

The equivalence principle of Einstein says that, if you are doing experiments in a free-fall lift, you should not see any gravitational field; you should see both particles and antiparticles floating before you and therefore both falling with respect to a fixed laboratory. But the equivalent principle may be wrong. It would be rather a coincidence that Einstein has found the principle so fruitful, but that would not be the only time in physics that something good comes out of something wrong. For example, the Dirac equation, displayed in this lecture hall, was discovered as a result of a mistake. Dirac thought that a good wave equation should be of first order in the operator d/dt and set out to make one with conspicuous success. And now he has his name blessed in this marvellous lectur theatre. But he was wrong, and now the Klein-Gordon equation which has $(d/dt)^2$ operator, has an honoured place in theoretical physics. Maybe Einstein was just lucky with the equivalence principle?

Then, there is the question of energy conservation. To see the problem there, imagine that you have a balance and that on each side there are identical boxes. One is empty and the other contains an atom and an antiatom. Since the antiatom is accelerated upwards rather than downwards, the balance will be in equilibrium. Now, imagine that the atom is excited and that the excited atom weighs a little more because its energy is bigger. Therefore you can gain energy by tipping the balance down on the side of the box with the atom and antiatom and up on the side of the empty box. But you can imagine that the photon moves from the atom to the antiatom and now it is the antiatom which is the stronger and which is pulling upwards by hypothesis. So, by moving the balance back again you extract energy from the system. As the photon often floats between atom and antiatom you can continue indefinitely to extract energy from the system. Well, maybe the energy conservation is wrong and that would be extremely useful. But if you think that energy conservation is right, then you must admit that exchangeable energy does not antigravitate when it is attached to an antiparticle. If it gravitates, it continues to gravitate even if it moves to an antiparticle. We can make the same kind of argument for kinetic energy. If the objects are moving and if you have some kind of wall which keeps them apart but which allows them to exchange energy, you can also convince yourself that any kinetic energy of an antiatom must gravitate rather than antigravitate. At best one can go ahead on the hypothesis that there is some heart of an antiparticle which does not fall with the usual acceleration.

It was pointed out by Schiff[3)] that the matter we have all around us is not so pure. Actually it is a mixture of matter and antimatter because we have vacuum polarization. The nucleus of an atom emits virtual photons which polarize the vacuum; in an atom there are e^+e^- pairs, with some probability, and indeed you expect also to have $q\bar{q}$ pairs with some probability, in fact with greater probability. However, we are somehow more happy with electrodynamics than with QCD, which Schiff did not even know about. Therefore Schiff addressed himself to calculating the influence of the diagram in Fig. 2.1 on the discussion. What he said was the following: Suppose the rest masses of electrons and positrons cancel out as regards gravity (not their complete energy but their rest energy which I am considering). Then you can find that there is a contribution in the energy which is not matched by contribution in the weight. The pair does not weigh (due to the cancellation between particle and antiparticle) but contributes to the energy of the system, and Schiff calculated it by perturbation theory. If this contribution is very big, you will have a problem with the Eötvös experiment which says that weight goes very closely with inertial mass, i.e. internal energy.

Schiff calculated actually a lot of things to interpret the Eötvös experiment (listed in Table 2.1). His formula for the contribution of e^+e^- pairs to energy depends on the number of protons and on the nuclear radius. Also shown in Table 2.1 are the results of his calculations for three different elements. What is listed is the relative perturbation to the mass multiplied by 10^8, because this is bound by something of order 1 according to Eötvös experiment. Contributions to the mass which do not contribute to the weight, cannot make this bigger by about unity. The biggest contribution is the nuclear binding energy (10^6). What is more important is that the differences in nuclear binding energy between the various elements are of the order 10^5. So, nuclear binding energy must weigh. Finally, you see the e^+e^- pair rest mass contribution is just too big to be left out of the weight of the object.

Table 2.1 - Different contributions to the atomic mass

	δM	$10^8\ \delta M/M$		
		Magnalium	Copper	Platinum
Electron mass	$m Z$	26 400	25 200	21 900
Electron binding	$- 15.6\ Z^{7/3}\ eV/c^2$	25	68	223
Nuclear electrostatic	$.6\ Z(Z - 1)\ e^2/R$	1.44×10^5	2.42×10^5	3.97×10^5
Nuclear binding	$- B(Z, A)$	8.00×10^5	9.23×10^5	8.38×10^5
e^+e^- rest mass	$\frac{3m}{8\pi}\left(\frac{Z}{137}\right)^2\times(\ell n \frac{1}{mR} + .338)$	12	25	53
$\bar{q}q$ pairs	?	?	?	?

Of course one could say that it is all speculative, that we have never seen a virtual electron-positron pair. We did, however, see the Lamb shift. To the Lamb shift there is an important contribution from vacuum polarization. It is rather reasonable to regard vacuum polarization as real effect. So, I think this argument of Schiff is a serious one and should make people who speculate about antigravity find some way around it.

In the theories which I will present you later one does not say anything so crude as that perhaps gravity works upwards instead of downwards for antiparticles. One says: "Perhaps gravity is not entirely described by the tensor field which works equally for particles and antiparticles; perhaps there is a vector component which changes its sign between particles and antiparticles." With that hypothesis the Table 2.1 is even more relevant, because not only is the vector field going to overlook these pairs when the weight has to be calculated, but also it overlooks nuclear binding energy. The typical vector field does not distinguish if nucleons are free or assembled and does not change when they are bound. Therefore, if that contribution is more than very small, it is going to be a gross violation of Eötvös' experiment. So, a vector contribution to gravity, which changes sign between particles and antiparticles has to be less than about 10^{-5} the sum of tensor and scalar contributions.

Finally there is another argument to this discussion which is due to Good [4]. He observed that the mass difference between K_L and K_S is an extremely sensitive measure of any perturbation on the system that would introduce something like a difference on mass, weight or potential energy between K^0 and $\bar{K}^0$. Let us write the mass matrix of the neutral kaon system

$$\begin{pmatrix} M_k + V(1 + \delta) & i\,W \\ i\,W & M_k + V(1 - \delta) \end{pmatrix} \tag{2.1}$$

where V is the gravitational potential energy. It is the total energy which occurs in the propagation. Suppose that, in addition a fraction of this potential energy, δ, changes its sign when going from kaons to antikaons (or that there is a little fraction δ of antigravity). That would look like a violation of CP and CPT and indeed in the early days people said that maybe that is how CP is violated, i.e. there is some vector field contribution which changes its sign between kaons and antikaons. The δ would mix together K_1 and K_2 to an apparent CP violation. This type of mixing is

$$\frac{V\,\delta}{|M_L - M_S|} \simeq \varepsilon \tag{2.2}$$

where we know that $\varepsilon \approx 10^{-3}$, $|M_L - M_S| \approx 10^{-5}$ eV. So, you see that we are down to very, very tiny values of δ. This is very important also for another reason: Equations (2.1) and (2.2) are valid in the rest system of the neutral kaon and you have to transform into that system. The importance of a field in a rest system relative to the laboratory system depends on its tensor character: a scalar field does not change, but a vector field acquires an extra factor γ (experiments are done up to γ of order 100) and a tensor field goes with γ^2. So, if you take that into account and consider that the field may be scalar, vector and tensor, you can improve the limits. You also have to know what is contributing to the potential energy of the kaon and that depends on what you think about the range of gravity. Certainly the Earth has a gravitational field extending to its surface, certainly the Sun has a gravitational field extending to us; it is extremely likely that the Galaxy has a gravitational field which keeps the outer parts circulating. You may also imagine that the Universe has a gravitational potential. In Table 2.2 you can find the gravitational energy coming from the Earth, Sun, etc. for a K-meson. It is a remarkable coincidence that if you calculate the universal gravitational energy of a particle, you find it about equal to its rest energy! The limits that you get on the size of δ are also presented in Table 2.2. If the external field is a vector field (and that is the one that could change between particles and antiparticles according to orthodox field theory), you see that your limit is a very tiny fraction of the gravitational potential, 3×10^{-10}.

Table 2.2 - The limits on the parameter δ of different tensor field contributions

	Earth	Sun	Galaxy	Universe
Potential energy V	.4 eV	6 eV	300 eV	5 - 500 MeV
Scalar field; $\delta \leqslant$	3×10^{-8}	2×10^{-9}	4×10^{-11}	$4(10^{-15} - 10^{-17})$
Vextor field; $\delta \leqslant$ $\gamma \sim 10^2$	3×10^{-10}	2×10^{-11}	4×10^{-13}	$4(10^{-17} - 10^{-19})$
Tensor field; $\delta \leqslant$ $\gamma^2 \sim 10^4$	3×10^{-12}	2×10^{-13}	4×10^{-15}	$4(10^{-19} - 10^{-21})$

You can, however, increase this limit by remembering that, if you are dealing with a force that has a finite range, small compared to the radius of the Earth (which is quite possible), then it would be less effective; it would not be the whole Earth which contributes to the potential but only the fragment of it within the range, giving a factor

$$[(\lambda^3/\lambda)/(R_\oplus^3/R_\oplus)]^{-1} = (\lambda/R_\oplus)^{-2} \tag{2.3}$$

with $\lambda \ll R_\oplus$ and $R_\oplus \approx 6 \times 10^6$ m. This means that by going to shorter ranges you can tolerate relatively stronger forces. For example, for a vector field with a range of 200 m, you could have a basic δ of .3. One has seen in an experiment on the K-meson system something like a γ-dependence which could be due to a δ rather smaller than .3, maybe ten times smaller, but all other experiments do not show this effect. In their paper the authors find that a simple coupling of the kind we are talking of, which splits the mass of the K_L and the K_S, would not explain the details of what they see, i.e. the increasing rather than decreasing magnitude of CP violation as a function of energy. Nevertheless, in subsequent papers they have continued to speak as if this hypothesis of the so-called hyperphoton - hypercharge discriminating component in gravity - was relevant for their experiments and that is quite mysterious to me.

So that concludes the discussion on antigravity. Some experiments were started to see whether positrons fall upwards, but as far as I know, they never got to the end. Some results were obtained in the initial experiments on electrons and they were quite interesting results, but I will not talk about them.

NEW THEORIES

The idea that particles and antiparticles fall rather differently was revived in 1979. Scherk pushed the idea very strongly together with others like Fuji, Zachos, Fayet, Macrae and Riegert, Goldman and Nieto, Bars and Visser, Barbieri. In particular Goldmann, Hughes and Nieto are people keen to see experiments done and in their paper[5)] one can find all relevant references. The new ideas which were stated, are the ideas of supersymmetry, of Kaluza-Klein theory and the string theory. I will say a few words about what these ideas are and how they bear in this discussion. The essential idea is that in these theories a gravity field does not come alone, it comes accompanied by vector and scalar fields. Some of these fields may interact with gravitational strength and they may have very tiny masses and therefore modify the long-range forces.

1. Supersymmetry

First of all we discuss the notion of supersymmetry. The discovery that made supersymmetry was that you can have symmetry operators which commute with the Hamiltonian and lead to a different spin. This was a big discovery of the time and surprised most of us, that such symmetry operators could exist. Also, as with any other symmetry you can imagine making it local, i.e. making different transformations at different points in space as one can make gauge transformations in the electromagnetic theory or isospin transformations in the Yang-Mills theory. Then you are obliged to introduce gauge bosons. The gauge bosons of supersymmetry can now include a graviton, the gauge boson for a supergravity theory. In such a theory there is a gravitational field coupled à la Einstein to a tensor field and accompanied by another particle called gravitino ($S = {}^3/2$). Gravitino, as any half integer field, cannot carry a long range force. But an extended supergravity where there is more than one of these operators could enhance this picture: the graviton comes in it together with a set of particles which include not only the gravitino but also a spin 1 particle, a vector field, called the graviphoton. So, there is now something which distinguishes particles from antiparticles and it may be the kind of thing we are discussing. One has to look into the details of the theory to see how these things are coupled, but one can guess just by dimensions how they do it:

$$\alpha = \sqrt{8\pi G}\,(h_{\mu\nu}T_{\mu\nu}) + \sqrt{4\pi G}\,(A_\mu \bar{\Sigma}\, m^0\, \psi\, i\, \gamma_\mu\, \psi) \qquad (2.4)$$

The first term has to be a graviton and has to be coupled to the energy momentum tensor with the coupling constant $\sqrt{8\pi G}$, if it is going to reproduce gravitational phenomena. The second term has also to have G in its coupling constant but the dimensions of such a scalar source must be different and must have a mass in addition. A mass which appears explicitly in the coupling constant, must be a mass which appeared already in the Lagrangian. Therefore it must be the mass of the spinor sources of the the field. Since it is a parameter appearing in the Lagrangian, it is called "bare mass". That is tremendously embarrassing because the bare mass m^0 is something that we do not measure directly. We measure a particle which is dressed by interactions with other particles. In old fashioned theories the bare mass was infinite. One had to cancel infinities that come in radiative corrections so that the observed mass should be finite. But the fact that supersymmetry is broken, makes the relation of m^0 to the ordinary masses (which appear from the energy momentum tensor) perfectly obscure. So really, although the theory is nice and can be constructed like that, you do not know what to do with it when you apply it. However, if you suppose that the m^0 is somehow not absolutely different from the mass of the first term, you can see that the two terms have a competitive strength and so this will lead to an appreciable change in gravitation between particles and antiparticles.

This problem was looked at especially by Scherk and he guessed the masses that you will take in the coupling strength of these vector fields. For the lepton, there is not much doubt about that; the corrections may be infinite but they are not very big and therefore we will take $m^0(e) = .5$ MeV. For the quarks the so-called "current masses" are very tiny things and they could be taken perhaps as the masses to be used in our formula. For the strange quark he took it about equal to the mass of the kaon meson. If you do like that, you could guess also what is the mass of the graviphoton, which as gauge particle is initially massless and acquires mass by spontaneous symmetry breaking. Its mass becomes proportional to the vacuum expectation value of some Higgs field and since this particle is somehow gravitational, the coupling is G.

$$m(A) \simeq \sqrt{4\pi G}\, m^0(X) \cdot \langle X \rangle_0 \sim 10^{-9} \text{ GeV} \tag{2.5}$$

Therefore the mass becomes very small compared to ordinary masses. That mass corresponds to a range of the corresponding vector contribution to gravity of about 1 km. Then, if you write down the interaction between two slow particles

$$V = \frac{G}{r}\,(m_1 m_2 \pm m_1^{\,0} m_2^{\,0}) \tag{2.6}$$

it will involve the Newton interaction and also the product of the two bare masses. There is a change of the sign of this vector contribution if you deal with particles or antiparticles. In this theory, because vector forces are repulsive rather than attractive between light objects, a particle will fall towards the Earth less quickly than an antiparticle. Scherk estimated what is the fraction of antigravity at the surface of the Earth on ordinary matter by the term

$$\delta g/g = (\lambda^3/\lambda^2)\,(R_\oplus^2/R_\oplus^3)\,\left(\frac{m_1^{\,0} m_\oplus^{\,0}}{m_1 m_\oplus}\right) \tag{2.7}$$

Due to the short range, only a fragment of the Earth is contributing and the change of the gravitational field is only 10^{-7}, in full agreement with the Eötvös experiment.

For the K-mesons you are concerned with the potential rather than with the field and the dependence on the radius of the Earth to the range of the force has two parts instead of one and because the K-meson is mostly strange quark, you do not have such a suppression, when you go from the real masses to the bare masses. Therefore you can count a factor of

10^{-9} which is somewhat bigger than the limit of 3×10^{-10}. But you can easily suppress this discrepancy by assuming a "generation independence" of the bare masses. That means that whatever the charge, the source of this vector field is not going to change from one generation to the other and you can totally suppress the effect on the K-meson.

2. The KALUZA - KLEIN Theory

A second idea is the idea of Kaluza and Klein that there are more dimensions than we know, but those which we do not know are very restricted. In these new dimensions you cannot really move over a long distance but only over a short distance and over probably a periodic space. We can imagine that space has five dimensions and the relation between energy and mass has the relation

$$0 = p_1^{\,2} + p_2^{\,2} + p_3^{\,2} + p_4^{\,2} + p_5^{\,2} + \mu^2 \tag{2.8}$$

with $p_4^{\,2} = - p_0^{\,2}$ and $p_5^{\,2}$ is now as a mass term. Let us put $\mu = 0$ and consider only massive objects in this space. Let somehow the fifth dimension be compactified, i.e. the fifth dimension does not go from $- \infty$ to $+ \infty$ but goes around a circle:

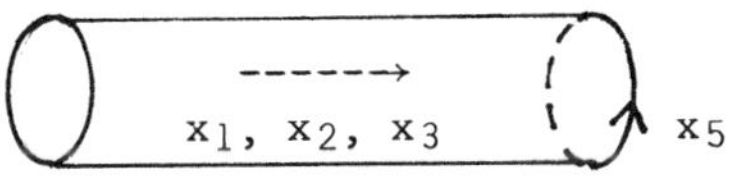

Then you can see that p_5 can be quantized as $p_5 = \pm\, n m_0$, because things have to be periodic in this dimension. It will be an integer multiple of some fundamental unit related to the radius of the compactification. The sign could be both positive and negative, and you can think it stands for particles and antiparticles. As regards the four-dimensional motion equation (2.8) now reads

$$p_0^{\,2} = p_1^{\,2} + p_2^{\,2} + p_3^{\,2} + (n m_0)^2 \tag{2.9}$$

The kinetic energy in the internal space has become a mass term. At the same time the tensor of gravity

$$h_{\mu\nu} \equiv [h_{\mu\nu},\, h_{5\mu} = h_{\mu 5},\, h_{55}]_{1....4} \tag{2.10}$$

if you have a gravity theory in this five-dimensional space, is going to be quantized. When you reclassify with respect to the transformations in the four-dimensional space, you see yet a tensor which, because there is a 5 in the index, becomes different: $h_{5\mu}$ becomes a scalar. Hence you have an even richer gravity coming out of that kind of thinking.

Let us now calculate the interaction energy between a couple of matter particles due to the exchange of this combined vector, scalar tensor interactions, when the particles are slow.

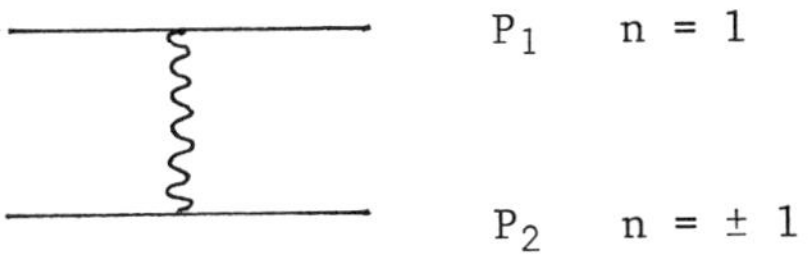

Fig. 2.1

Following the diagram in Fig. 2.1 we write

$$\begin{aligned} p_{10}\, p_{20}\, V &= (p_{1\mu}\, p_{1\nu})(p_{2\mu}\, p_{2\nu})\, G/r \\ &= (p_{1\mu}\, p_{2\mu})^2 \qquad \mu,\nu = 1....5 \qquad (2.10) \\ &= (p_{1\mu}\, p_{2\mu} \pm m_{10}\, m_{20})^2 \qquad \mu = 1....4 \\ &\simeq (-\, p_{10}\, p_{20} \pm m_{10}\, m_{20})^2 \qquad \text{(slow particles)} \end{aligned}$$

you find a factor which is the product of the two energies plus or minus the product of two masses. You see that these two masses, if they were ordinary masses, would perfectly cancel for slow particles when they have the plus sign. That is to say, particles would not fall at all and anti-particles would fall to the Earth. In this spirit you could know what "bare masses" would probably mean as far out as you consider theories without interactions. But once you consider renormalizations, etc., you have no idea whatsoever what they mean. The other problem is that the tensor, vector, scalar contributions have a completely different velocity dependence and we do not deal with static objects. The quarks, even in stationary nuclei, move.

STRING THEORY

The last idea is the String Theory. Many theoretical physicists are working with theses theories in the hope they will provide the whole theory of electroweak, strong and gravitational forces. The idea of the string is that instead of the concept of point particle, one has an extended object, in this case a string which can be open with two ends or closed with no ends, has a tension so it can be stretched, it can vibrate once it is stretched and it can rotate. So it has a lot of excitations and when these excitations are quantized, they correspond to particles of different masses. The theory gives immediately after beginning an infinite spectrum of particle masses. The string tension which is the only parameter of the theory has to be made equal to something like the Planck mass, because the gravitation constant which is also dimensional has to come out of this theory and if the only scale is the string tension, then we have to make it the Planck mass so that $G = m_p^{-2}$.

There is also something called the superstring, which incorporates fermionic degrees of freedom and once we require that, there are no ghosts in the theory; we find that it only works at 10 dimensions. Then the masses of the particles are either zero or large compared to the Planck mass because the Planck mass is the only scale. The ordinary particles have to be found in this massless sector. As far as these particles are concerned, maybe they can be described by an ordinary field theory, which is a D = 10 supergravity theory with Yang-Mills and fermion degrees of freedom. If you want to reinterpret the gravity field $h_{\mu\nu}$ which comes out of the theory in 10 dimensions as the thing which has to do with real space and time, you can hope (as you hope for a closed Universe in general relativity) for a spontaneous compactification in which six of the 10 dimensions curl up into small spaces, leaving only four for the free motion.

There is the hope to develop the string theory not in flat space but in curved space so to look like Einstein's theory from the beginning. Once you have the compactification, you find again particles that are rather massless and particles with masses large compared to m_p. They have to acquire masses, small compared to the Planck mass scale by spontaneous symmetry breaking.

In the heterotic string theory which is the nicest and most promising version of string theory, the sector where the masses are small compared to the Planck mass is thought to be described by a N = 1 supersymmetry, which is of course broken. That is a bit disappointing from the point of view of extra long-range forces. The symmetry group which comes out in practical limit is

$$\underbrace{(SU(3) \times SU(2) \times U(1)}_{\text{standard model}} \times U(1) \tag{2.11}$$

where the extra U(1) is related to a new extra $\bar{Z}^0$, a heavy boson of the weak interactions (like the Z^0 that we know). The theory does not include any new long-range carrier and for the long-range forces we have only the graviton and the photon. However, there is very much guesswork in translating the basic string Lagrangian into laboratory physics.

In these theories there is a so-called "shadow matter" sector, that is to say as well as the ordinary particles there is a whole world of extraordinary particles, which interact with the world that we know only by gravitation. This is studied now and people think that it could be the dark matter, the missing matter in the Universe.

CONCLUSION

In summary, new long-range forces are not needed by fashionable theories. However, some theories could make space for them if needed. If you ask these days what happens at the surface of the Earth, maybe in addition to the famous tensor contribution to the gravitational acceleration g, there is also a vector and a scalar contribution

$$g = T \mp V + S \tag{2.12}$$

The scalar contribution is rather well controlled if it is coupled to the trace of the energy-momentum tensor. If it is coupled to the energy-momentum tensor, then T and S both see the binding energy as well as the free energy of neutrons and protons in nuclei and both make a red shift, the gravitational red shift. The vector contribution V does not see the binding energy, nor does it give a red shift. That means that the limit on V, quite apart from theoretical reasons, but experimentally from the red shift, which works in one part to 10^4, is suggested to be less than $10^{-4}(T + S)$ and Eötvös' experiment suggests it to be even less than $10^{-5}(T + S)$.

So, most theorists are not expecting new things to be found by experiments on antigravity. But of course you can very well say: this is just the place to surprise theorists, therefore, go there!

LECTURE No. 3: OLD AND NEW EXPERIMENTS IN GRAVITY

To-day I will try to review the experimental situation in gravity, and I will do that in three boxes. The two big predictions of Newtonian gravity concern the inverse square law of forces; firstly, the fact that the attraction goes with r^{-2} and secondly, the fact that the force is given by the inertial masses and not by any new property of the body (that is to say the forces are independent of the composition of the bodies). Finally I will consider more specifically relativistic effects.

THE INVERSE SQUARE LAW

One way to parametrize a possible departure from the inverse square law is to put an r dependence in Eq. (3.1)

$$F = \frac{m_1 m_2}{r^2} G(r) \tag{3.1}$$

Another way is to put in a short-range component $e^{-r/\lambda}$ with some strength α in the following Eq. (3.2)

$$V = G_\infty(1 + \alpha\, e^{-r/\lambda}) \frac{m_1 m_2}{r^2} \tag{3.2}$$

Of course, these two things are related to one another. The latter is an increasingly popular way of presenting what one knows about the validity of the inverse square law. If you calculate the effective Newtonian constant $G(a)/G(b)$, it is clear that, for given distances a and b, it would be unity if λ is either very large or very small. But for two given objects, systems with scales a and b, the rate $G(a)/G(b)$ is a test for the presence of such a force with λ in the range characterized by a and b and from a given system you will get some limits or possibly some result on the value of α. The first example of this was when Newton found out that the inverse square law was valid for the fall of the Moon compared with the fall of an apple. This was the start of gravity theory. That is to say that α must be small compared with 1 or that λ is either very small compared to the radius of the Earth (such a small-range force would see only a very small part of the Earth and be rather ineffective, contribute very little to the acceleration due to gravity) or λ must be very large compared with the distance between the Earth and the Moon (so gravity has the strength $1 + \alpha$ on the whole way out and is constant over that distance). This would also be a test of the second question: independence of composition (unless the Moon were made of apples! You do find in the litterature on this subject hypotheses of that kind that, because of some strange coincidence, some effects which would be very embarrassing are not actually there).

Figure 3.1 shows a way of presenting the limits of what we know on the independence of Newton's constant on distance. This diagram has evolved through many hands and the version seen here is taken from a review by A. De Rujula (see also there for all the references not included in my lecture). What is plotted along the bottom is the $\log_{10}$ of the range λ which you can investigate in meters. What is plotted in ordinate is the $\log_{10}$ of α. The curves say that in various ranges we know that α must be less than some number; e.g. laboratory experiments tell us that in the range of a few cm α cannot be much bigger than 10^{-4} whereas at somewhat bigger distances you might have bigger tolerances. I will come back to what α might be.

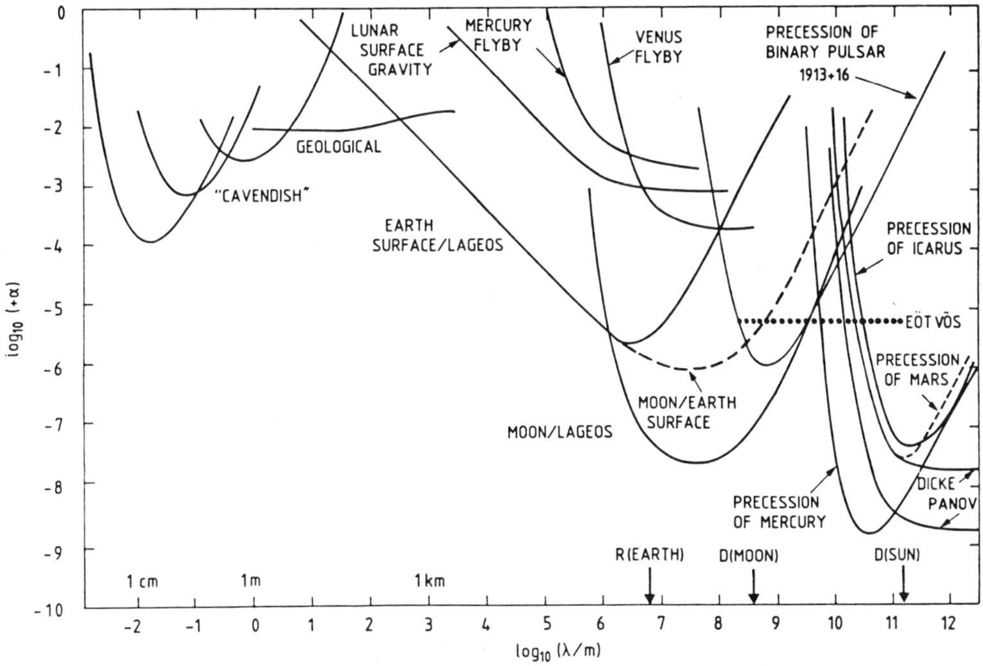

Fig. 3.1

For large ranges there is a limit coming from the precession of Mercury. This is a very delicate test of the inverse square law because it is characteristic of the inverse square law that the elliptical orbits are fixed in space: they repeat one another and any deviation from the inverse square law results in a precession of the orbit where the direction of the ellipse rotates. Of course, there is a general relativistic contribution to this effect which one has to subtract. One is looking for something other than the general relativistic effect which I will come to later and that I am here taking for granted. If you are asking what room is there for an additional component of the force which simply does not go with the inverse square law but for example falls off exponentially, you see that the possibilities for such an extra force are extremely minute, 10^{-8} or 10^{-9} if that force has a range about the distance from the Sun to Mercury. But if that force has a much smaller range, this system becomes irrelevant and if the force has a much bigger range, this system also is not very important, as I explained.

Also shown on Fig. 3.1 are curves from Dicke and Panov on their experiments on the composition dependence. They had an Eötvös apparatus sensitive to the Sun's gravity and found the force largely independent of the composition of the object being attracted to the Sun. That does not immediately give us something about the range but if you make a particular hypothesis about what the anomalous force might be, then you can say something about its range; if such a force does not reach from the Sun to us, it could be very strong but if it does reach from the Sun to us, the Dicke and Panow experiment limits its strength. You see therefore that they limit it quite severely provided it is long range.

As you see, there is a gap in the diagram where the limit is not very good and that is because we do not have astronomical systems intermediate in scale between the solar system and the Earth-Moon system. So, the next good limit comes from the Earth-Moon system. The dotted line is the modern version of Newton's comparison of the apple and the Moon falling towards the Earth. Now it is done by very careful observations of the motion of the Moon and very careful measurements of the strength of gravity at the Earth's surface. This latter is by no means trivial because any point at the surface of the Earth is peculiar: there are mountains nearby, there are lakes nearby and so on. That would not matter if you could measure everywhere because, by Gauss' theorem, only the integral would be significant. But you do not measure everywhere and therefore you have to correct as best you can and reduce the observations to what you would get for an ideally shaped ellipsoidal Earth. Without going into details, once this is done, the strength of gravity for the Moon and at the Earth surface agrees to something like one part in a million and so it is quite a tight limit on the possibilities of any anomalous forces which would have a range of order of the Earth-Moon distance. On this kind of scale the situation has also been greatly improved by the observation of the artificial satellite LAGEOS which goes around between us and the Moon and which has been especially designed for gravitational prospecting, specifically to look in detail at the Earth's gravitational field to see what there is under the surface. It is tracked very carefully and is very suited for our purpose. We can get two limits from it: one from the comparison of the attraction of the Moon and of LAGEOS towards the Earth and the other from comparison of the gravity at the Earth surface and the attraction of LAGEOS towards the Earth. You see that over a large region this is the most important check on the gravitational attraction.

Then there are less important limits given by space probes flying in the direction of Mercury and Venus. There is a big gap in range because we do not have systems intermediate in size between the Earth-Moon system and the laboratory. Of course you can measure gravity in the neighbourhood of a mountain but you do not know very well what there is inside the mountain so the analysis of the experiment is difficult.

In the short-range region the best limits come from experiments in which you go underground. As you go underground, part of the Earth, instead of pulling down, is pulling up, and the extent to which it is pulling up can be measured and it gives you a measure of the Newtonian constant G for the scale of distances involved. Out of these experiments comes a limit which is not very precise, at the level of 10^{-2} of the strength of gravity. That should not be interpreted to mean that you should only measure things at the level of 10^{-2} to obtain this limit because we are here discussing the part which relates to the force between two particles. To see the extent to which it contributes to the fall of bodies, it has to be integrated over the Earth and, if it has a short range, it is suppressed very effectively so that an anomaly of 10^{-2} level in the two particle force, translates to an anomaly of 10^{-5} or 10^{-6} level in the strength in which things fall to the Earth. Therefore the limits depend on quite accurate measurements of the acceleration of gravity on the Earth's surface and as you go down in mines.

The interesting thing is that the experiments in mines seem to give a somewhat higher value for Newton's constant than those that are done in the laboratory. The results shown in Fig. 3.2 come from the work of Stacey et al. who have done the most interesting experiments in this area. The data are compared with the experiments of the "Cavendish type" which involve room-size objects. The measurements done at depth of the order of 1 km seem, with large fluctuation, to be consistently higher. The dominant factor in the error bars is the uncertainty about the density of the Earth around the experiment. There could be something here and therefore it has been analysed in great detail.

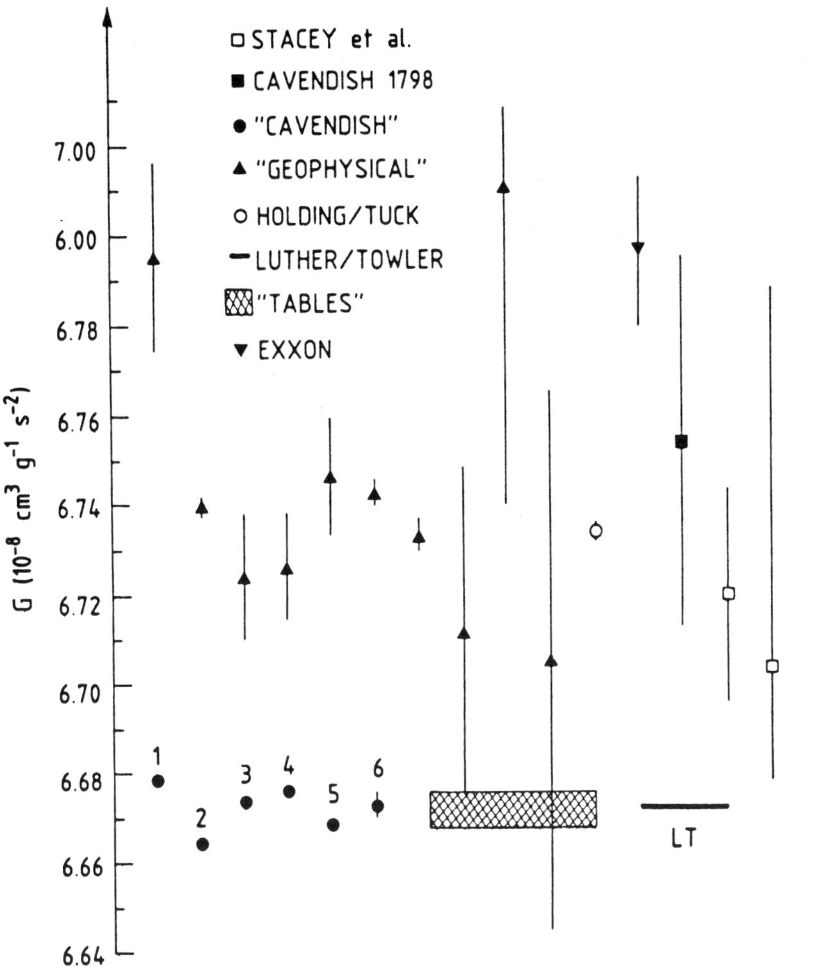

Fig. 3.2

Figure 3.3 shows again the diagram of the most important limits on the possibility of a short range force for negative and for positive α. The most important is that for the geological experiments there is now a hint of not only a limit but an actual result that there is a deviation at the level of 10^{-2}. Figure 3.4 shows the same results expanded. The shaded area are the limits coming from satellites, LAGEOS and the apples (or laboratory experiments), but in a range which could be anything from a few metres to 1 km there is a suggestion that there is a positive anomaly and the error is about half the distance away from zero. This is done at the Hilton mine in Queensland. One can fit that data to the kind of form which I presented in which you consider an additional short-range piece in Newton's law. Figure 3.5 shows the kind of fits you get. The various lines through the points are for various ranges and you see that you do not distinguish very much the range. The range is hardly determined from that data and the strength of a hypothetical component of the force is 0.008, i.e. 1% of the strength of gravity between two particles, only 10^{-6} of the rate fall at the Earth surface. It is a repulsive force and that might be due to a vector component in the gravitational force because a vector field gives you repulsion between two like objects, and one assumes that the Earth is like the material of the gravimeter for that purpose.

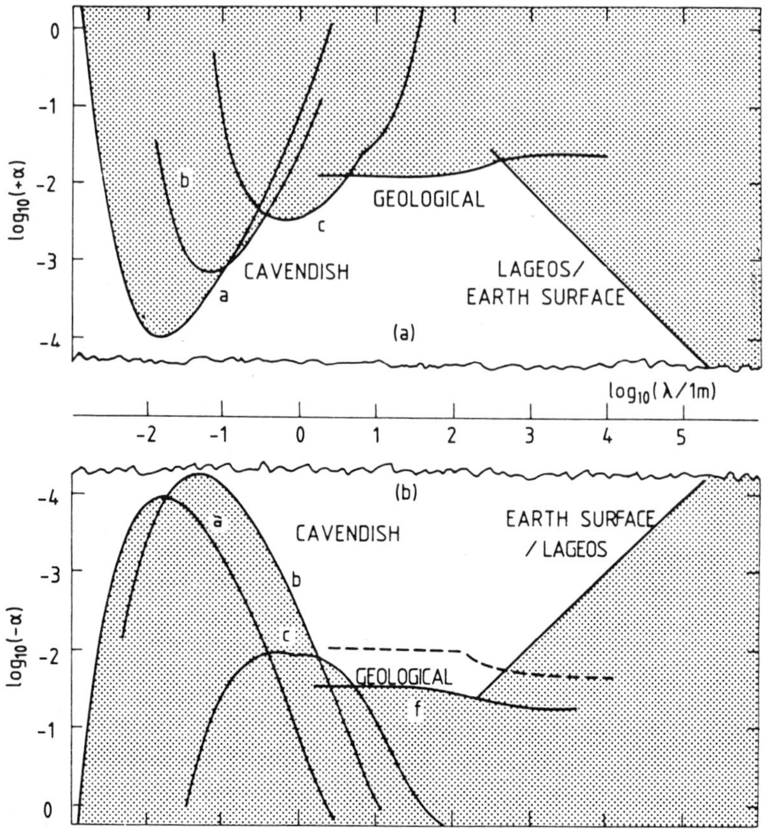

0
-1
-2
-3
-4
log10(+α)
b
c
a
GEOLOGICAL
CAVENDISH
LAGEOS/
EARTH SURFACE
(a)
log10(λ/1m)
-2
-1
0
1
2
3
4
5
(b)
-4
-3
-2
-1
0
log10(-α)
a
b
c
f
CAVENDISH
EARTH SURFACE
/ LAGEOS
GEOLOGICAL

Fig. 3.3

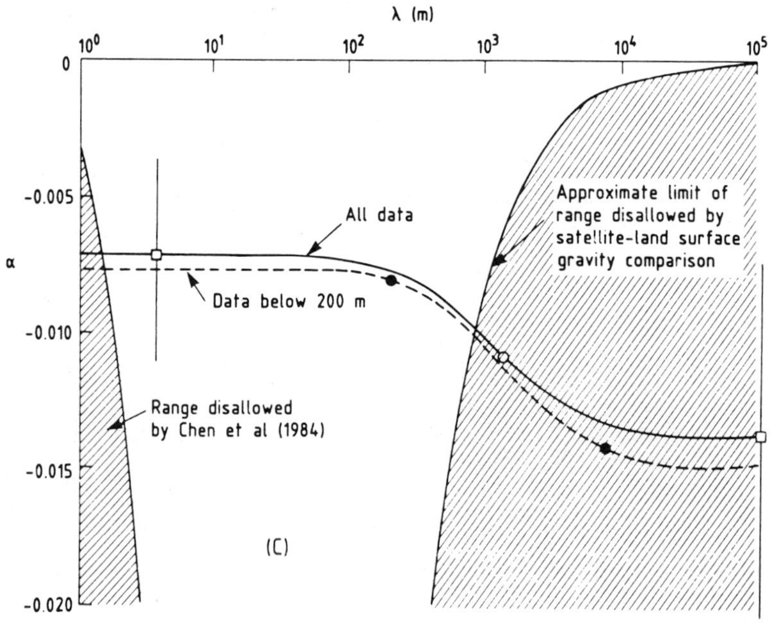

λ (m)
10^0
10^1
10^2
10^3
10^4
10^5
0
-0.005
-0.010
-0.015
-0.020
α
All data
Data below 200 m
Approximate limit of
range disallowed by
satellite-land surface
gravity comparison
Range disallowed
by Chen et al (1984)
(C)

Fig. 3.4

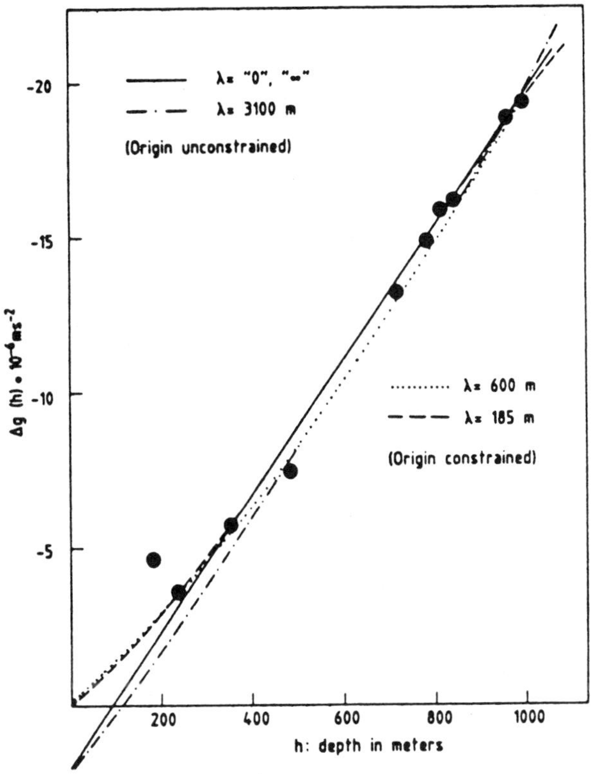

Fig. 3.5

Figure 3.6 shows the results of the fits for various ranges. The data points are omitted. The interesting thing about this figure is that, as you project it above the Earth, you find something which is much more sensitive to the range of the interaction. This gravitational anomaly, as you go down (it is an anomaly if you use the laboratory value of the Newtonian constant to calculate), builds up because the laboratory value is the wrong value. There is a short-range repulsion which has fallen off for the distances relevant in the mines and if you calculate with it, you get an increasing anomaly as you go down. But as you go up, what you get above the surface of the Earth does not depend on any assumption about Newton's constant because there is nothing between you and the Earth. And so what you get when you go up, is not depending on that error you have made in your calculation when assuming a value for Newton's constant and the result is that things are very sensitive to the range. There is some data above the ground, in chimney stacks; Mount Isa chimney and Tarong power station chimney. Unfortunately as you see, they do not lie along the curves, and those from Mount Isa even go in a surprising direction. So the authors Stacey et al. have to assume that there is something funny under the ground near this chimney so that as you go up, and get away from whatever this object is, the strength of gravity varies more quickly than usual. But even the ones which are reasonable, from the Tarong chimney, are scattered so that one cannot draw conclusions but clearly it is important to do experiments above the ground. Here I mention some which were the first I ever heard of:

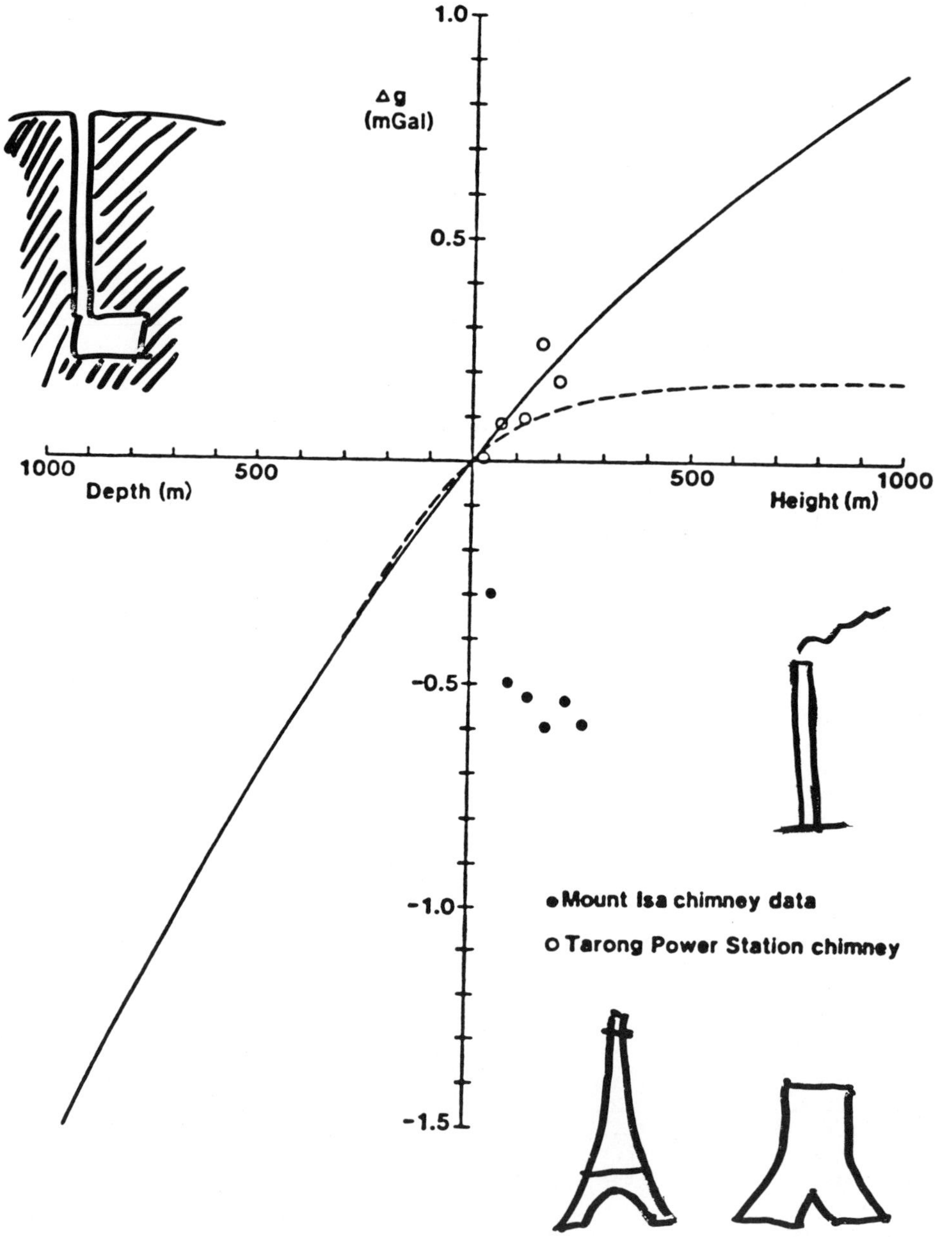

Fig. 3.6

P. Kabir told me a few years ago that he was on his way to the Eiffel Tower to borrow a gravimeter from the Bureau of Standards in Paris, in order to do experiments at various heights in the Tower. But nothing came of that because the people who own the Tower were not sufficiently co-operative and would not let him in when the tourists were not there, and the tourists shook the Tower so much that he could not do anything. He transferred his activities to the high-rise building at Fermilab and there, he tells me, he has been gathering data and has a computer program to analyse them, but he is unwilling to say anything about the result.

The trouble with experiments above the Earth is indeed to get a sufficiently stable object and also that, if you are going to move an instrument to different heights, it has to be a portable instrument. This is not so good and so it is best, instead of moving the instrument away from the Earth, to move the Earth away from the instrument! That can be done. Such an experiment is now under way at the Splityard Creek hydroelectric lake in Queensland (Fig. 3.7). What one has there, is a hydroelectric lake whose surface varies and one can mount the instrument on an electricity pylon which is reasonably solid. The instrument is a balance of which one weightarm is at the high level and the other one is below the variable level of the water, so you get the double effect: the effect between the variable level of the water reinforces between the two objects. You have about 50 m of variation and are thus sensitive to forces in the 20 to 50 m range. This is probably not enough and you would need experiments like that sensitive to much bigger ranges, but this one is the most interesting thing that I know of in this area, in the attempt to get above the Earth and to learn something about the range of the anomalous force.

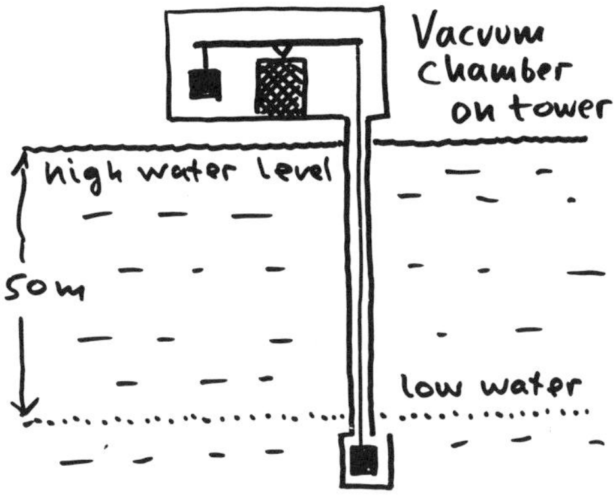

Fig. 3.7

The anomaly still needs a lot of work and I am sure you will hear a lot of that. For the moment one can say it is possible that the Newtonian interaction does have a short-range part at a strength of about 10^{-2} and it is repulsive, which might mean a vector component in the gravitational field. This is what I have to say about experiments on range, on the inverse square law.

THE COMPOSITION INDEPENDENCE

Now I come to the following question: Could the gravitational force depend on the objects involved other than through their inertial masses? Table 3.1 shows the history of the subject. Already Newton swung pendula with various materials on the bottom and found that there was independence to the material to one part in one thousand. Then there were a series of experiments. Of particular importance is the one by Eötvös, Pekar and Fekete which was reported in 1922 but which had gone on for a period of 20 years since about 1900. They brought that possible dependence to the level of one part in 10^{-8}. I will have more to say on that in a moment. Then there are the experiments of Roll, Krotkov and Dicke and Braginsky and Panov and their apparatus is sensitive to the Sun's gravitational field rather than the Earth's, i.e for the long-range part of the gravitational field they brought these limits down still further. Finally comes something particularly interesting: a proposal written in 1978 to consider objects of different compositions freely falling in a satellite, and they were hopeful to get to the level of 10^{-15} or 10^{-16}. At that time there was no reason to suspect that there would be an anomaly, but a subsequent reanalysis of the data suggests that there is an anomaly and that it is connected to the Earth's gravitational field so that these people now plan to run their apparatus on the Earth, in the neighbourhood of a mountain or something like that, and see if the gravitational field of the mountain is in some way anomalous.

Table 3.1

Experiment	Reference	Method	Substances tested	Limit on $\lvert\eta\rvert$
Newton	Newton (1686)	Pendula	Various	10^{-3}
Bessel	Bessel (1832)	Pendula	Various	2×10^{-5}
Eötvös	Eötvös, Pekar, and Fekete (1922)	Torsion balance	Various	5×10^{-9}
Potter	Potter (1923)	Pendula	Various	2×10^{-5}
Renner	Renner (1935)	Torsion balance	Various	2×10^{-9}
Princeton	Roll, Krotkov, and Dicke (1964)	Torsion balance	Aluminum and gold	10^{-11}
Moscow	Braginsky and Panov (1972)	Torsion balance	Aluminum and platinum	10^{-12}
Munich	Koester (1976)	Free fall	Neutrons	3×10^{-4}
Stanford	Worden (1978)	Magnetic suspension	Niobium, Earth	10^{-4}
Boulder	Keiser and Faller (1979)	Flotation on water	Copper, tungsten	4×10^{-11}
Orbital[a]	Worden (1978)	Free fall in orbit	Various	$10^{-15} - 10^{-18}$[a]

[a] Experiments yet to be performed.

So, what I want to do now, is to go in more detail into the Eötvös experiment. The "Eötvös Balance" consists essentially of a horizontal beam suspended on a wire and with weights of different materials hanging from it. The force on these weights is due to gravity combined with the centrifugal force due to the Earth rotation. Let us call f_i the force on weight i (i = 1,2) (Fig. 3.8). If these forces were parallel, they would have no tendency to turn the balance and to send a torsion in the wire. But if they are not balanced because the weights see the Earth's gravitational attraction and the centrifugal force of the Earth in different proportion because of a violation of the composition independence, then there will be a torque. The measurement of this torque is a measurement of the effect. For equilibrium the force F supplied by the wire must be the sum of the forces acting on the weights, and the torque T supplied by the wire must be the sum of the torques

$$- \vec{F} = \Sigma \vec{f}_n \tag{3.3}$$

$$- \vec{T} = \Sigma \vec{r}_n \times \vec{f}_n \tag{3.4}$$

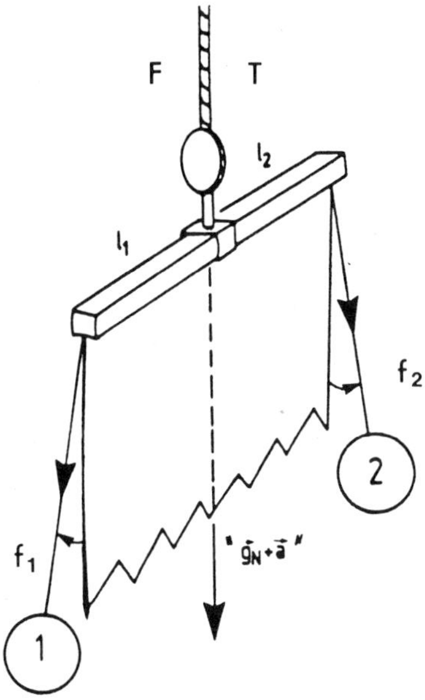

Fig. 3.8

It will be sufficient to consider the component of the torque along the force F since in a thin wire both the force and the torque will be along the wire. Let us therefore write the product

$$\begin{aligned}\vec{T} \cdot \vec{F} &= \Sigma\Sigma\, r_n \cdot \vec{f}_n \times \vec{f}_m \\ &= \tfrac{1}{2}\Sigma\Sigma(\vec{r}_n - \vec{r}_m) \cdot \vec{f}_n \times \vec{f}_m\end{aligned} \tag{3.5}$$

You see that what you need to turn this balance, is a lack of parallelism between the different forces acting on different weights. Let us suppose that this force is given by the expected force (the inertial mass times the acceleration due to gravity towards the centre of the Earth and the centrifugal acceleration away from the Earth's axis) plus an anomalous force which I will put at some multiple of mg:

$$f_1 = m_1(\vec{g} + \vec{c}) + m_1 g\, \vec{\delta}_1 \tag{3.6}$$

$$f_2 = m_2(\vec{g} + \vec{c}) + m_2 g\, \vec{\delta}_2 \tag{3.7}$$

One obtains in the leading order approximation

$$(m1 + m2)\, \vec{T} \cdot \hat{g} = m1m2(\vec{r}1 - \vec{r}2) \cdot (\vec{\delta}1 - \vec{\delta}2) \times (\vec{g} + \vec{c}) \tag{3.8}$$

The torque is given by the distance between the two weights dot produced with the difference between the two vector anomalies times the total acceleration due to gravity and the centrifugal force (g + c). If the anomalous forces just went with the masses, the result would be zero. What you have here is that the deflection of the balance is a measure of the failure of mass proportionality of the forces. Eötvös himself supposed that any anomaly would be parallel to g (he was interested in possible anomalies in what we call the gravitational force which is long range and points to the centre of the Earth). In this case, Eq. (3.8) becomes

$$\frac{m_1 + m_2}{m_1 m_2}\, g \cdot \vec{T} = (\vec{r}_1 - \vec{r}_2) \cdot (\delta_1 - \delta_2)\hat{g} \times \hat{c} \tag{3.9}$$

Something to note especially is that expression (3.8) is much more sensitive to horizontal than to vertical components. The vertical components which Eötvös assumed, do not see the main term g in the acceleration because the cross product is then zero, and the effect remaining depends on the little component c. g is about 600 times c at temperate latitude and so anything which pulls sideways is more effective by a factor of 600. This is important in the analysis of the experiment that I am going to describe.

Figure 3.9 shows Eötvös' famous balance. There is a triple metallic surround everywhere to insulate it very well thermally and to avoid convection currents. One weight is attached directly to the beam of the balance and the other hangs down. This difference of heights of the two weights was not an advantage in this experiment but the apparatus was originally designed for something else. A telescope observes a scale, through a little mirror attached to the beam, and measures sensitively any deflection in the balance. This is a telescope that you look through with a mirror inside and the eyepiece actually pointing towards you. That is important because it meant that the experimenter was seated just in the position to do maximum damage to the balance by the influence of his own gravitational field. This was subsequently a mystery, why it did not matter to Eötvös. It is assumed that in fact he set up his equipment, went away, let it settle down and observed quickly before he started himself to pull the balance out of equilibrium.

Figure 3.10 shows his results but plotted in a way that would have astonished him. Whereas he reported his experiment saying there was nothing to be seen at the level of 10^{-8}, Fischbach, Sudarsky, Szater, Talmadge and Aronson (FSSTA) recntly looked at his data again. They had reasons to hope that they would find some positive results here, and they did! If the experimental points, which look like a random distribution, are ploted against the specific baryon number of the falling material, one finds that there is a systematic variation which is quite striking. At the level of 10^{-8} it does look as if there is an anomaly in the gravitational field - a component that is sensitive to baryon number rather than to rest energy or inertial mass. Figure 3.11 shows the same material presented a little differently. A. De Rujula has fitted the data with a possible dependence on baryon number and charge number. You see that, if you plot the results against charge number, the points are again randomized. Data only tolerate a very small and insignificant dependence on charge number but a rather significant dependence on baryon number.

The force is attractive, that is to say, baryons attract one another independently of their mass to a very tiny extent - five parts in a million. De Rujula's hypothesis was that that could be a long-range attractive force and he looked to see what were other limits for such a long-range attractive force and what he found is shown on Fig. 3.1. Indeed, there is a gap in the limits coming from other places. There is a possibility to have a new force of the 10^{-5} strength, if its range is of order of $4 \cdot 10^6$ km. So, he has a very specific contribution to make: there is in gravity a component that sees baryon number and not rest mass, and that has a very well defined range in between the Earth-Moon distance and the Earth-Sun distance. Presumably that gap could be filled by sending artificial satellites around the Sun closer than the natural satellites that we can observe accurately, and then I expect that this hypothesis would be excluded. (De Rujula also expects this.)

The hypothesis of a very weak attractive force was not the one which appealed to the authors of the reanalysis. What they wanted was a repulsive short-range force - about 200 m - which was not too different from the one that Stacey et al. had found in the mines. Their analysis was

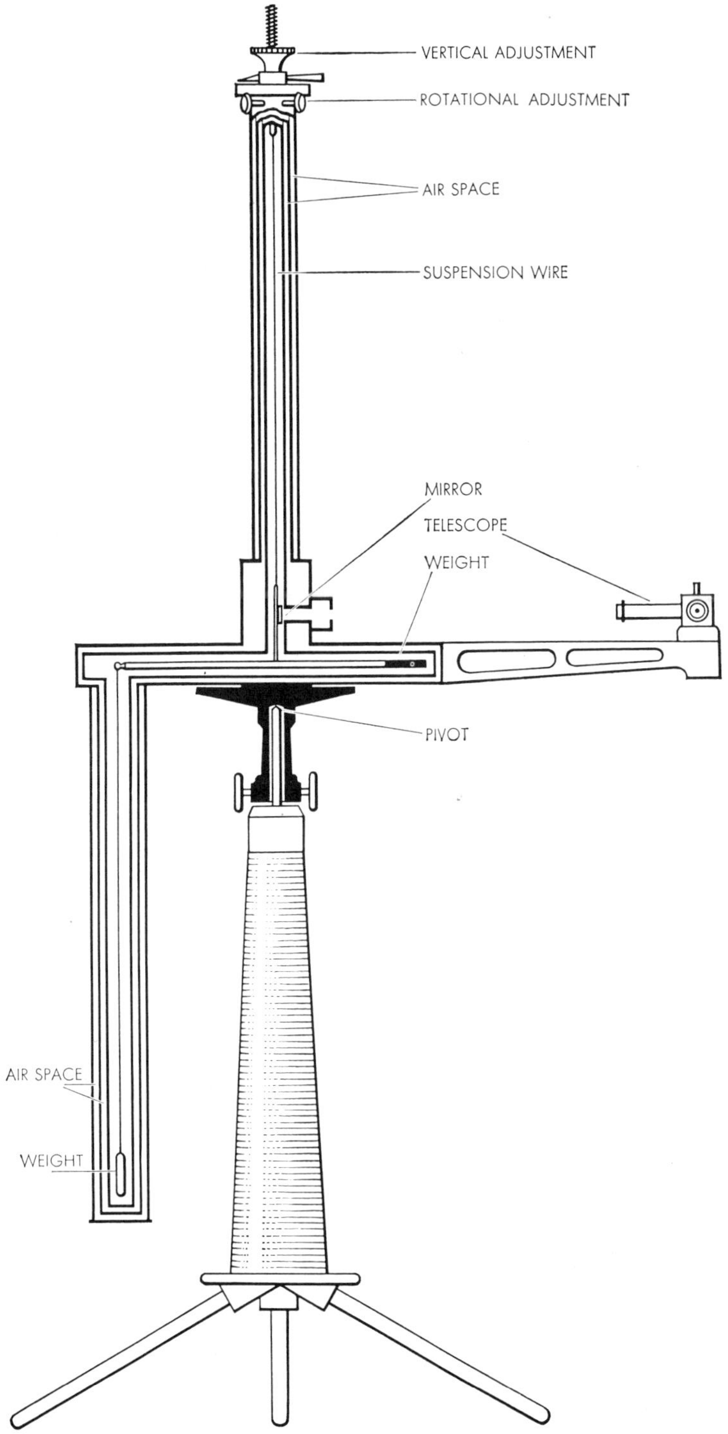
VERTICAL ADJUSTMENT
ROTATIONAL ADJUSTMENT
AIR SPACE
SUSPENSION WIRE
MIRROR
TELESCOPE
WEIGHT
PIVOT
AIR SPACE
WEIGHT

Fig. 3.9

soon the subject of a lot of criticism and also of self-criticism. First of all the sign was wrong; it was also 15 times too strong. So, when one corrected the sign mistake that these authors made, it was a force - 15 times that of Stacey et al.'s. So the comparison was not so good. The second criticism that emerged was that if the terrain had been ideal, perfectly flat terrain, for example the surface of a very deep ocean on a calm day, then any extra short-range force would be vertical, at right angle to the surface, because being short range, it is concerned only with local things, not with where the centre of the Earth lies. Moreover, this flat terrain is not at right angle to a line drawn to the centre, but at right angle to the effective total acceleration which contains the centrifugal force as well as the gravitation. That is to say, the centrifugal force distorts the surface and makes that effective flatness which is not at right angles to g. δ is along g + c and the unperturbed forces acting on the weights are also along g + c; all the forces remain parallel and the balance is not deflected. So, in ideal conditions, the short-range forces which they were putting forward simply would not have deflected the balance and would not have accounted for the results. Moreover, for a short-range force - remember that the sideways pull in the Eötvös experiment is very important - there is a factor of 600 leverage as compared with any vertical pull. If your forces are short range and if there are buildings within range, these are going to be significant. When FSSTA reanalyzed things in this perspective, they found that the buildings were in the wrong direction to change the sign on their force. However, the basements were excavated and represented a negative source of their field. Finally they found - lo and behold! - that they did indeed need a repulsive field, in the same way as Stacey.... Moreover, if one sticks to the range of 200 m which is just a representative number not very well defined, they found that a force about three times that of Stacey and with the same sign could do the job. Of course, they put that forward with all kinds of necessary reservations, saying that more experiments would be required and indeed there are many experiments now under way to see if it is really like that.

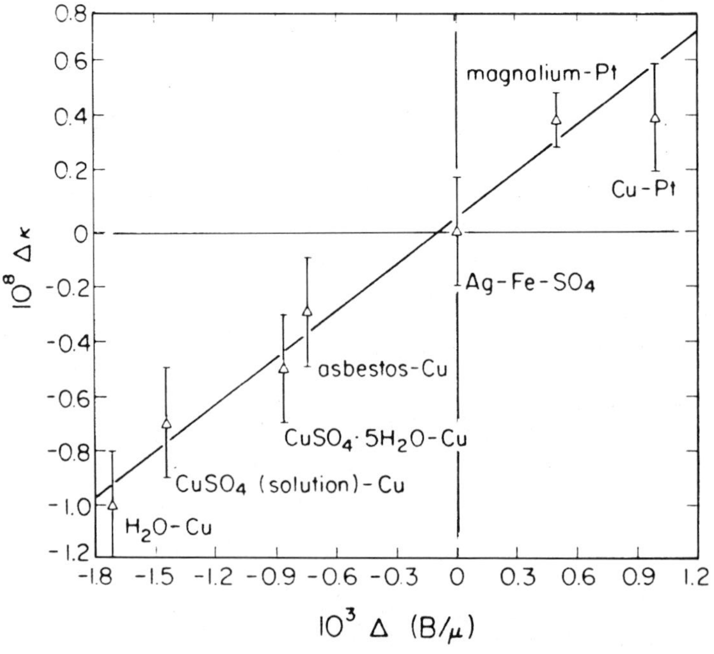

Fig. 3.10

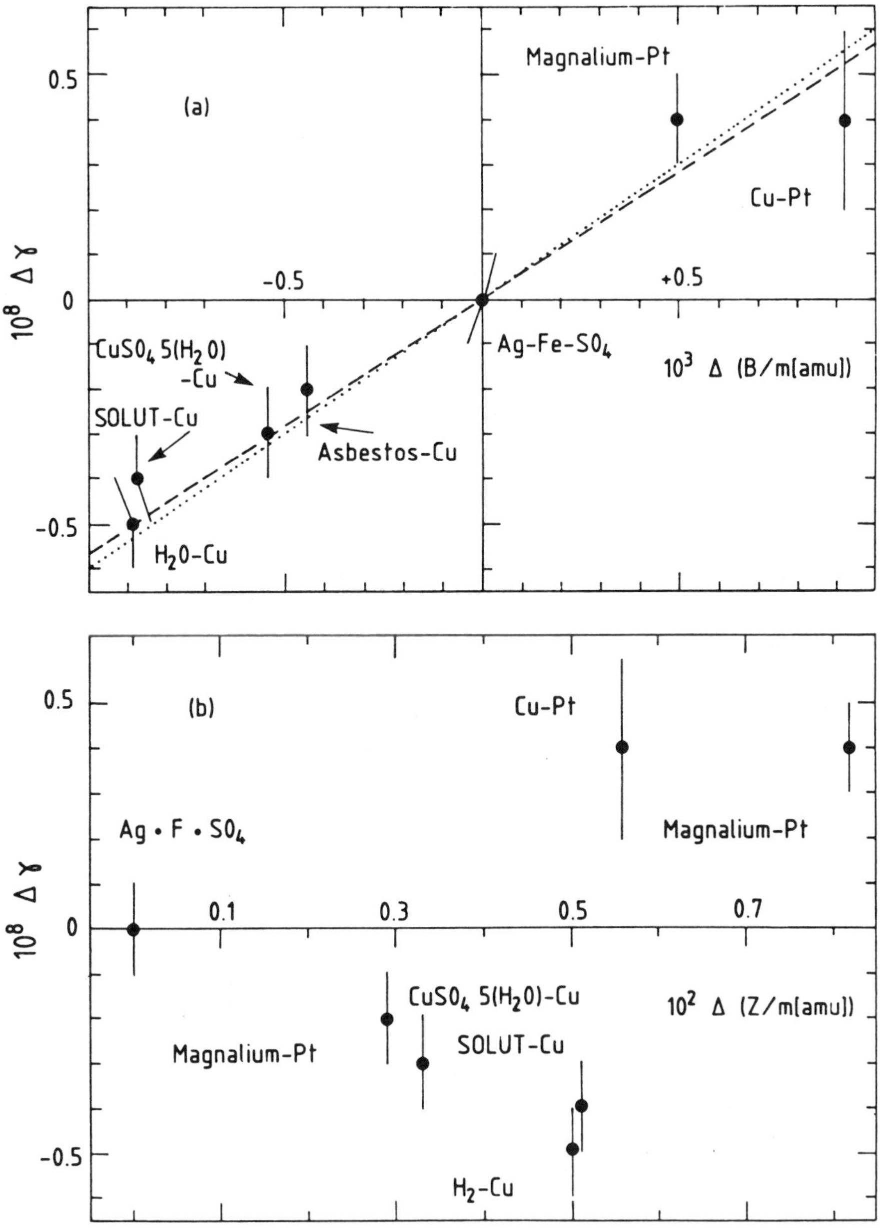

Fig. 3.11

Clearly it is important to repeat this experiment and in the neighbourhood of large objects sideways of relatively known composition. There are experiments going on in the neighbourhood of mountains or going into tunnels in mountains, and I suppose it will not be long before we have results from them.

Before leaving the question of the composition dependence I want to mention one more thing - the lunar laser ranging experiment. It is a kind of Galileo experiment rather than an Eötvös experiment because it concerns bodies in something like free fall. What is done is to use lasers bouncing off reflectors, placed on the Moon by astronauts, to keep very careful track of the motion of the Moon. One can detect if there is any systematic difference in the acceleration of the Moon and Earth towards the Sun. That would result in a perturbation of the motion which you can analyze and extract from all the other complications of the Moon's motion. That is to say Galileo's experiment not with weights from the tower of Pisa but with the Earth and the Moon falling towards the Sun. It is found that the difference is extremely tiny; the accelerations are the same to within $7 \cdot 10^{-12}$. It is not as good as Dicke's experiment but it is in that area. What is distinctive about these objects, however, is that they have important gravitational self energies, 3 parts in 10^{11} for the Moon and 7 parts in 10^{9} for the Earth. So you see that, if gravitational self energy did not gravitate, you would have a substantial effect in that experiment. This test is therefore an extension of the weak equivalence principle, that gravitational self energy also gravitates. If we knew what the Earth an the Moon were made of, for example if the Moon were made of cheese, then this would also be a test of the independence of composition in the more ordinary sense. I suppose the Moon is somewhat differently composed from the Earth, maybe less iron, so again it is in the line of the Eötvös experiment.

RELATIVISTIC EFFECTS

I come now to the relativistic effects where I will skip all the ones which are more or less ripe and where the tests have greatly improved in recent years with the advent of radioastronomy and radar ranging, and I will just discuss two. I will first talk about the motion of the perihelion of Mercury which was the first great success of Einstein's theory. Table 3.2 shows the status of the data as of the review article of Will in 1981. He reported radar ranging experiments from 1976 to 1978 to keep very careful track of the motion of Mercury. The results and the theoretical expectations are presented as a fraction of Einstein's prediction. You see it is looking quite good. But in 1974 there was a terrible stir in this business because there is a contribution in the theoretical expression coming from the quadrupole moment of the Sun. If the Sun is not quite spherical, the Newtonian force law is not quite inverse square. There is also an inverse fourth power contribution. It was thought that the quadrupole moment of the Sun would be pretty small. On the basis of its apparent rotation and the nature of centrifugal forces, it was thought that the distortion coefficient (10^{6} times the quadrupole moment) would be at the level of 10^{-1} which would not be significant in comparing theory and experiment. However, Dicke and Goldenberg in 1974, looking at the Sun and interpreting what they saw, decided that it was 250 times that expectation. This would very substantially spoil the agreement between the theory and the experiment and in fact create room for Dicke's own theory of gravity. Other people found smaller values, for example a limit of less than 5, which would be much less serious and, as far as I know, although much discussion took place, it was not settled.

More recently there was a new element in the game in that one has seen oscillations of the Sun with periods of from 10 to 120 min. and these give you some clue to what is inside. You can make a model of the Sun including high rotation inside as distinct from outside. This has an effect on the various oscillation modes, on the way their frequencies are split, on the way they are coupled to one another and information on the quadrupole moment can be extracted. There are two different analysis of the same data, the authors having decided to publish separately although it seems to me they are not so different. They quote respectively values of 5.5 ± 1.3 and $3.6 \pm$ an error which the author was unwilling to specify for the distortion coefficient (Table 3.2). If one takes, say, the biggest of these numbers and puts it into the theoretical formula, you obtain a theoretical correction of 1.016. If you take the smallest of the experimental numbers, you see that there is the possibility of a discrepancy at the 1% level and perhaps that creates a little room for the people that want to modify at long range. (Gough, for example, has written a paper along these lines.)

Table 3.2

Mercury perihelion precession	
Radar ranging:	Einstein x
Shapiro et al (1976)	1.003 ± .005
Anderson et al. (1978)	1.007 ± .005
Theory	$1 + .0030\,\underbrace{(10^6 J_Z)}_{\text{expected } 10^{-1}}$ ↓
Sun quadrupole moment	
Dicke and Goldenberg (1974)	~ 25
Hill et al. (1974)	⩽ 5
from solar oscillations:	
Hill et al (1982) [1]	5.5 ± 1.3
Gough (1982) [2]	3.6 ± ?

1) Phys. Rev. Lett. 49 (1982) 1794
2) Nature 298 (1982) 334

Table 3.3

Experiment	Reference	Method	Limit on $\lvert\alpha\rvert$
Pound–Rebka–Snider	Pound and Rebka (1960) Pound and Snider (1965)	Fall of photons from Mössbauer emitters	10^{-2}
Brault	Brault (1962)	Solar spectral lines	5×10^{-2}
Jenkins	Jenkins (1969)	Crystal oscillator clocks on GEOS-1 satellite	9×10^{-2}
Snider	Snider (1972, 1974)	Solar spectral lines	6×10^{-2}
Jet-Lagged Clocks (A)	Hafele and Keating (1972a,b)	Cesium beam clocks on jet aircraft	10^{-1}
Jet-Lagged Clocks (B)	Alley (1979)	Rubidium clocks on jet aircraft	2×10^{-2}
Vessot-Levine Rocket Red-shift Experiment	Vessot and Levine (1979) Vessot et al. (1980)	Hydrogen maser on rocket	2×10^{-4}
Null Red-shift Experiment	Turneaure et al. (1983)	Hydrogen maser vs. SCSO	10^{-2}
Close Solar Probe[a]	Nordtvedt (1977)	Hydrogen maser or SCSO on satellite	10^{-6}[a]

[a] Experiments yet to be performed.

The other experiments involve the deflection of light past the Sun; the deflection of radio waves from quasars improved that. Formerly it was a 10% sort of accuracy and now it is a fraction of a percentage. There is the time delay of signals bounced off planets passing near the Sun, and that is at the level of a tenth of a percentage, and finally the thing which I will talk a little bit more about is the red shift.

The red shift is the fact that an atom deep in a gravitational potential emits redder lines than an atom higher up in a gravitational potential. Experiments with astronomical objects are not so clean because of various complications, and the laboratory experiments are in this respect the most interesting. Table 3.3 presents the experimental results on red shift. The first line is the famous experiment of Pound, Rebka and Snider on the tower of the Jefferson Physical Laboratory at Harvard using the Mossbauer effect and pinning down the red shift to the theoretical result to one part in 10^2. The last line is a proposal to send an artificial object close to the Sun, and maybe you can get an accuracy of 10^{-6} but that remains to be done.

The most impressive experiment which has been done is the one that involves sending a rocket up to 10 000 km that falls back into the ocean and I think it takes about 15 min. During that time you keep very careful check by sending radio signals on the running of a Maser clock on the rocket in comparison with a Maser clock on the Earth. One finds that the clock on the rocket (we must allow not only for the red shift but also for Doppler effects first and second order and so on) confirms the theory to two parts in 10 000. Now, that is an important limitation on any desire to tinker with the theory of gravity with forces that have ranges of this kind or shorter (in fact not so much shorter because it is an integrated effect). But it is important and it is also a kind of Eötvös' experiment in the sense that you can show just by looking at conservation of energy; any component in the gravitational field which depends on the energy of a body and which is also stronger for excited states, gives you a red shift because as you lift an excited and a non-excited object up, if the gravity pulls more strongly on the more excited object, it takes more work to get to the top, and so the photon that can be emitted from the top is bluer than the photon that can be emitted from the bottom. Indeed, this is not independent of the proportionality of the acceleration due to gravity of the inertial mass, it is really the same thing if you believe in the conservation of energy. A curious thing in Table 3.3 is the description of this experiment: "The fall of photons from Mossbauer emitters". I looked up the original papers to see what they were called and here is what I found: "The apparent weight of photons; effect of gravity on γ radiation". I do not think these titles are appropriate. Suppose that this experiment be done differently. Figure 3.12 is again the 74 ft tower of the Jefferson Physical Laboratory at Harvard. What I want to do is, instead of putting Fe57 sources in the top and bottom, I put atomic clocks. The effect is about one part in 10^{15} and I find that this means a 10 min. disagreement in these clocks after 10^{10} years. So you wait 10^{10} years and you compare them. Now of course you could use photons or radio waves but you do not have to. For example the man at the top could put his head out of the window and shout down: "What's the time?" But I guess this would not be "bon ton" at Harvard. Another proposal would be to use carrier pigeons. To increase reliability you could have a whole flight of carrier pigeons released. The carrier pigeons, although somewhat erratic, are taking a few seconds which would be unimportant compared to the 10 min. that have built up in the clock. So, you could do that experiment and report it to the Physical Review under the title "Apparent Weight of Carrier Pigeons".

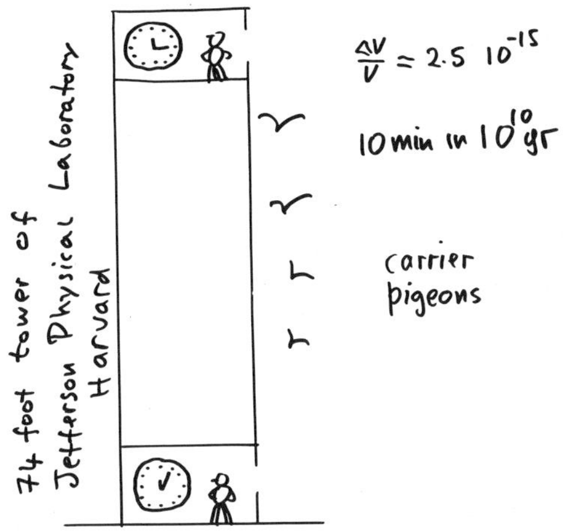

Fig. 3.12

REFERENCES

1. D.G. Boulware and S. Deser; Phys.Rev. D6 (1972) 3368.
2. P. Morrison; Am.J.Phys. 26 (1958) 358.
3. L.I. Schiff; Proc.Natl.Acad.Sci.USA 45 (1959) 69.
4. M.L. Good; Phys.Rev. 121 (1961) 311.
5. T. Goldmann, R.J. Hughes, M.M. Nieto; Phys.Lett. B171 (1986) 217.

THE GRAVITATIONAL PROPERTIES OF ANTIMATTER

Richard J. Hughes, T. Goldman and Michael Martin Nieto

Los Alamos National Laboratory

Los Alamos, New Mexico 87545

In classical gravitational physics a particle couples to the local gravitational potential with a strength known as its "gravitational mass" [1,2]. In principle, the gravitational mass is physically distinct from the inertial mass, which is a kinematic property of the particle. Together they determine the particle's gravitational acceleration. There would be no violation of CPT-symmetry if a particle and its antiparticle should fall with different accelerations in the same gravitational potential. (By "gravity" we mean all forces other than the strong, electromagnetic and weak ones of macroscopic range and gravitational strength.) Specifically, CPT-symmetry equates the gravitational acceleration of a particle towards a particular source with that of its antiparticle towards an "anti-source". That is, a proton falls towards the earth with the same acceleration that an antiproton has towards an "anti-earth". CPT does not tell us how an antiproton falls towards our earth. However, a different behavior of an antiproton from a proton in the earth's gravitational field would violate the weak equivalence principle[3] of classical physics. This principle may be expressed mathematically using Newton's inverse-square law,

$$m_I \, g = G \, M_0 m_G / r^2 \qquad (1)$$

for the acceleration, g, of an object of inertial mass m_I, gravitational mass m_G, in the gravitational field of an object of mass M_0. The principle states that,

$$m_I = m_G \quad . \qquad (2)$$

Although incorporated into general relativity, the weak equivalence principle is not an a priori concept, but has been distilled from the results of experiments performed over a 2,000 year period[4]. Indeed, this principle has never been tested for antimatter, and so there is a valid scientific question to be answered: what is the gravitational acceleration of antimatter?[5] (By "antimatter" and "antiparticle" we mean composite objects built out of antiquarks, and antileptons). Furthermore, a generic feature of modern quantum gravity theories is that matter and antimatter have different gravitational properties. In this paper we will argue that a determination of the gravitational acceleration of antimatter (towards the earth) is capable of imposing powerful constraints on such theories.

Various principles of classical physics fail when quantum effects are taken into account. For instance, Newton's first law, which might be re-expressed as "the universality of free-motion"[6], implies that the trajectories of freely-moving particles are determined kinematically in classical physics. This cannot be the case quantum-mechanically because the Heisenberg uncertainty relations involve the momentum, a dynamical quantity.

The classical gravitational analog of Newton's first law is the weak equivalence principle, also known as "the universality of free-fall". It implies that the trajectories of freely-falling classical bodies in a gravitational potential are determined kinematically. This also fails quantum-mechanically[7], as verified by the C-O-W experiment[8]. Furthermore, Wigner[9] has emphasized the incompatibility of general relativity, which embodies weak equivalence, and quantum mechanics. It is therefore not surprising that modern quantum gravity theories, motivated by renormalizability, include interactions of gravitational strength which violate the weak equivalence principle. In order to determine the status of the weak equivalence principle, one must investigate whether these interactions persist in the classical limit, and if so, with what strengths and ranges?[10]

As first noted by Zachos[11], in quantum gravity theories based on local supersymmetry, vector and scalar partners of the graviton appear naturally. Furthermore, vectors and scalars also appear in the reduction to four-dimensions of higher-dimensional gravity theories[12]. Although originally identified with the photon in this last context, it is now clear that the vector is more naturally associated with the graviton. The vector and scalar fields both couple directly to matter. This is quite different from metric theories of gravity, such as the Brans-

Dicke[13] or Hellings-Nordvedt theories[14], in which the new fields do not couple directly to matter. Indeed, the vector field of interest here is reminiscent of the Lee-Yang vector,[15] and will be presumed to couple to some linear combination of baryon and lepton numbers[16]. The new scalar will be somewhat similar to that of Nordström's second theory[17].

The common phenomenology of these quantum gravity theories is the existence of $J = 1$ and 0 partners of the graviton which couple with gravitational strength. The vector is termed the "graviphoton", and the scalar, the "graviscalar"[18]. Additional scalar[19] or vector[20] components of gravity have also been suggested in other contexts. The new feature here is the occurrence of both.

New classical effects of gravitational strength, associated with the graviphoton and graviscalar, will arise from the coherent sum over many sources. However, in the static limit of the unbroken theory, with matched couplings, there would be no corrections to Newtonian gravity for ordinary matter from the virtual exchange of the graviphoton and graviscalar[21]. On the other hand, if only the vector were present its coupling would have to be enormously suppressed relative to the graviton[22].

The usual theoretical expectation is that both the graviphoton and graviscalar acquire masses from symmetry breaking. Thus, at the phenomenological level, the observable classical effects of a broad class of quantum gravity theories consist of additional, finite-range (Yukawa) interaction potentials, with approximately gravitational strength. We may expect the ranges to be comparable, and the coupling strength difference to be small. In the linear approximation, the form of the total "gravitational" interaction energy between two massive fermionic objects, separated by a distance r, with four-velocities

$$u_i = \gamma_i(1, \vec{\beta}_i) \tag{3}$$

is then

$$I(r) = - G_\infty \frac{M_1 M_2}{\gamma_1 \gamma_2 r} \times [2(u_1 \cdot u_2)^2 - 1 \mp a(u_1 \cdot u_2) e^{-(r/v)} + b\, e^{-(r/s)}] \tag{4}$$

where a and b are the products (in units of $G_\infty M_1 M_2$) of the vector and scalar charges, and v and s are the inverse masses (in units of length) of the graviphoton and graviscalar, respectively. G_∞ is Newton's constant at infinite separation.

The -(+) sign in front of a in Eq. (4) is chosen for the interaction between matter and matter (antimatter). This arises from the well-known properties of vector boson exchange. The vector component is repulsive between matter and matter (so-called "null"[20] gravity) and attractive between matter and antimatter.

A general prediction of this type of theory is, then, that antimatter would experience a greater gravitational acceleration towards the Earth than matter. Note how different this is from older ideas about "antigravity"[23].

Indeed, there is a general rule of field theory that the exchange of an even-spin particle leads to an attractive force, while the exchange of an odd-spin one produces the rule: "like charges repel, opposites attract"[24]. It is clear that the notion of "antigravity" cannot be accommodated in this framework.

The question immediately arises as to the range of values to be expected for a and b in quantum gravity theories. One would naively expect $a \sim b \sim 1$ for each graviphoton and graviscalar in such theories, and for a simple reduction from 5 to 4 dimensions, there is just one vector and one scalar[21]. However, Scherk has explicitly observed that there could be more than one of each. In particular, we note that for N=8 supergravity, 28 vector and 35 scalar helicity states are present (for each of the two graviton helicity states), raising the possibility that the effective values of a and b are significantly larger than one. (If the scalar does not exist, then b=0.)

Unfortunately, there are no theoretical constraints for the values of v and s. In globally supersymmetric theories, for instance, massive superpartners of massless degrees of freedom may be very light for virtually any value of the supersymmetry breaking vacuum expectation value. Recently, Bars and Visser[25] have argued that the symmetry breaking scale must be related to a vacuum expectation value. This suggests that the weak symmetry breaking scale, Λ, or even the lightest fermion mass, m, may be relevant. Then

$$v^{-1},\ s^{-1} \sim \sqrt{\kappa} \times (m^2,\ \Lambda^2) \tag{5}$$

where κ is the gravitational coupling constant. From this we conclude that

$$10\text{cm} < v,s < 10^6\text{km} \tag{6}$$

The naive theoretical expectation, however, is that the graviphoton and graviscalar should have masses $\sim 10^{19}$GeV, although masses of $\sim 10^{-9}$eV may be possible in a geometric hierarchy scheme[26]. Meanwhile, in the absence of such an argument, we will adopt a phenomenological approach and turn to gravitational experiments to find bounds on the values of the parameters in Eq. (4).

One classical test would be to search for variations in Newton's constant as a function of the length scale on which it is measured. In fact, the Newtonian limit of gravity has only been tested to a high accuracy at laboratory distance scales, and in the solar system at distances of 10^6 to 10^{13} meters. Deviations from the inverse-square force law are not excluded at intermediate distances[27].

The intermediate region could be tested by experiments such as the Hills' Kepler-Orbit proposal[28]. A pair of large spheres, of say 1 meter diameter of dense material, could be placed in high earth orbit to minimize tidal forces, and gravitationally bound to each other. For a 10 meter separation, the period would be on the order of a few days. This would allow a very precise measurement of Newton's constant over a range of distances.

In geophysical experiments, Stacey and co-workers[29,30] have found anomalies which are consistent with deviations from Newtonian gravity on length scales between ~ 1 and $\sim 10^6$ meters. They analyzed their data using only one Yukawa term

$$I(r) = - \frac{G_\infty M_1 M_2}{r} [1 + \alpha e^{(-r/\lambda)}] \quad , \tag{7}$$

and found an effective repulsion with parameters[29,30]

$$1 \text{ m} \lesssim \lambda \lesssim 10^6 \text{m} \quad , \tag{8a}$$

$$\alpha = -0.010 \pm 0.005 \quad . \tag{8b}$$

Despite the large uncertainties in Eqs. (8), observation of a definite repulsive component is claimed. However, the measured data is not sufficiently precise to restrict the repulsion to a single Yukawa term. Indeed, the data is consistent with many functional forms[30].

In particular, if a form such as the static limit of Eq. (4) is used,

$$I(r) = - \frac{G_\infty M_1 M_2}{r} [1 + a e^{(-r/v)} + b e^{(-r/s)}] \quad , \tag{9}$$

the small effective coupling, α, may be produced by an approximate cancellation between the vector and scalar contributions. This can occur in two ways: there can be a small difference between the values of v and s or there can be quantum corrections which produce a small net difference between the values of a and b.

One could also look for a material dependence of Newton's constant, as did Eötvös, and indeed, Galileo. Recently, Fischbach, et al.[(31)] found anomalies in the data from the original Eötvös[(32)] experiment. (Although Dicke and Braginskii[(33)] verified the weak equivalence principle to a higher accuracy, their experiments were performed with reference to the sun. Therefore, their experiments could well have been unaffected by additional forces of limited range. On the other hand, Eötvös performed his experiment relative to the Earth.) The anomaly was apparently viewed by Eötvös as a systematic effect which was not understood. His quoted error is larger than the uncertainties of the individual points, and in fact is determined by the spread between the points. What Fischbach, et al. found was that the trend of variations is systematic with baryon number, a concept which had not even been invented at the time of Eötvös' experiment!

Although the interpretation of the results as evidence of a (fifth) hyperforce is now controversial, it prompted speculation. A purely theoretical problem with this hypothesis is that an extremely small coupling ($\sim 10^{-2} \times$ the gravitational coupling) must be introduced ad hoc. Such a small coupling is difficult to reconcile within the framework of grand unification. While this certainly does not rule out the hypothesis, a gravitational-strength interaction is definitely more natural, because it avoids the necessity of intrinsically small values of a and b.

Aside from the geophysical studies referred to earlier, what other experiments bear on the issue of a new force? Light deflection by the sun[(34)] does not provide any information, since the interaction(s) do not couple to photons. A variant of an argument due to Good[(35)], using K_s vacuum-regeneration, would apply if the new interaction coupled differently to strange particles, as Fishbach et al. originally speculated. A gravitational mass difference between K_0 and $\bar{K}_0$ would lead to an anomalous K_s-regeneration from a K_L beam. However, Macrae and Riegert[(21)] and Scherk[(18)] all argued that the new gravitational interactions must be family independent, thereby avoiding this problem. Finally, in a recent paper, Lusignoli and Pugliese[(36)] show that coupling to a non-conserved current (such as strangeness) produces a large branching ratio for the decay, $K^+ \rightarrow \pi^+$ plus nothing else observed, in conflict with experimental results.

In an astrophysical context, it could be significant that the graviphoton introduces a new velocity-dependent interaction as shown in Eq. (4). Matter on the surface of a pulsar of radius 10 km, with a period of a msec, has a speed which is a significant fraction of the velocity of light. The graviphoton could yield a significant new repulsive interaction for such high velocities. Since 10 km may well be within the range of the new interactions, they would have to be considered in discussing rapidly rotating pulsars(37) or black holes.

An exciting new possibility is to make a comparison between the gravitational interactions of matter, and of antimatter, with the earth. If the smallness of the observed effects in the matter interactions is due to a cancellation between the vector and scalar terms for matter, then the anomalous effects would add, not cancel, between matter and antimatter. Thus the attraction could be much larger for antimatter, as much as three times the normal gravitational effect, if $a \sim b \sim 1$. A measurement of the gravitational interaction between antimatter and the earth would then be a first-order test of quantum gravity theories, whereas Eötvös-experiments are second-order(18). Indeed, such second-order effects may be absent if the coefficients a and b in Eq. (4) are composition-independent.

An experiment (PS-200) has been recently approved at LEAR(38) to measure the gravitational interaction between matter (the earth) and antimatter(5,39). It takes advantage of the unique availability, at LEAR, of low energy antiprotons. These are to be ejected from LEAR and further decelerated and cooled to ultra-low velocities. They may then be directed up a drift tube for a precise (±0.3%) measurement, using extensions of the techniques pioneered by Witteborn and Fairbank(40). Eötvös-type experiments would be complementary to this experiment, but by no means a substitute for it.

Although we have phrased our discussion in the context of quantum gravity, a measurement of the gravitational acceleration of antimatter is a new, direct test of a fundamental principle (weak equivalence) which has implications beyond any particular class of theories. This principle has never before been tested with antimatter.

We would like to comment on an argument of Morrison(23) and one of Schiff(41) which severely constrained the "antigravity" notion. Although the models discussed in this paper do not embody this concept, it is worthwhile to see if these old arguments impose any constraints on them, since they do involve different gravitational properties of antimatter.

Morrison[23] constructed a gedanken experiment in which he proposed adiabatically lifting a particle-antiparticle pair in a static gravitational field, allowing them to annihilate, and transporting the produced photons down to the initial height. The resulting photon energy must be equal to the rest-energy of the initial pair, plus the energy expended in lifting them. With a conservative gravitational field the "antigravity" idea ran into serious difficulty with this requirement, because the weight of the pair was not equal to the weight of the photons. However, the models discussed here are Lagrangian based, will therefore embody energy-conservation, and so have no difficulty in accommodating Morrison's gedanken experiment.

Schiff[41] argued that virtual antimatter occurs in atoms, and so if "antigravity" existed, it would have been noticeable from the results of the Eötvös experiment. In the models which we have discussed here, the gravitational difference between matter and antimatter arises from the graviphoton, which couples to a conserved charge. Virtual effects cannot change the value of this charge for an atom, and so Schiff's argument imposes no additional constraint.

In summary, there are theoretical reasons to expect, and experimental suggestions of, non-Newtonian non-Einsteinian effects of gravitational strength. In modern quantum gravity theories, only the classical effects of these new interactions are observable at present energies. Typical quantum effects would be expected to be apparent only at the Planck mass scale, $\sim 10^{19}$ GeV. Thus, classical gravitational experiments of the kind we have described here are now at the forefront of modern particle physics. We emphasize that empirical knowledge of the gravitational behavior of antimatter is crucial for a complete understanding.

References

1. G. Burniston Brown, Am. J. Phys. 28, 475 (1960).
2. M. Jammer, "Concepts of Mass", Harvard, Cambridge (1961).
3. R. H. Dicke, "Proc. Int. School Phys. Enrico Fermi, Course 20, G. Polyani ed., Academic, New York (1962).
4. H. C. Ohanian, "Gravitation and Spacetime", Norton, New York (1976).
5. T. Goldman and M. M. Nieto, Phys. Lett. 112B, 437 (1982).

6. J. S. Bell (private communication).

7. D. Greenberger, Ann. Phys. 47, 116 (1968).

8. R. Colella, A. W. Overhauser and S. A. Werner, Phys. Rev. Lett. 34, 1472 (1975).

9. E. P. Wigner, Rev. Mod. Phys. 29, 255 (1957); Bull. Am. Phys. Soc. 24, 633 (1979).

10. T. Goldman, R. J. Hughes and M. M. Nieto, Phys. Lett. 171B, 217 (1968).

11. C. K. Zachos, Phys. Lett. 76B, 329 (1978); Ph.D. Thesis, Caltech (1979).

12. T. Kaluza, Sitz. Preus. Acad. Wiss. K1, 966 (1921); O. Klein, Zeit. Phys. 37, 895 (1926).

13. C. Brans and R. H. Dicke, Phys. Rev. 124, 925 (1981).

14. R. W. Hellings and K. Nordvedt, Phys. Rev. D7, 3593 (1973).

15. T. D. Lee and C. N. Yang, Phys. Rev. 98, 1501 (1955).

16. T. Goldman, R. J. Hughes, and M. M. Nieto, LA-UR-86-3617, Submitted to Phys. Rev. D.

17. G. Nordström, Ann. Phys. 42, 533 (1913), 43, 1101 (1914); A. L. Harvey, Am. J. Phys. 33, 449 (1965).

18. J. Scherk, La Recherche 8, 878 (1977); Phys. Lett. 88B, 265 (1979); in "Unification of the Fundamental Particle Interactions" eds., S. Ferrara, J. Ellis and P. van Nieuwenhuizen, Plenum, NY (1981).

19. Y. Fujii, Nature Phys. Sci. 234, 5 (1971); J. O'Hanlon, Phys. Rev. Lett. 29, 137 (1972); A. Zee, Phys. Rev. Lett. 42, 417 (1979).

20. P. Fayet, Phys. Lett. 171B, 261 (1986); L. B. Okun, Sov. Phys. JETP 56, 502 (1982).

21. K. I. Macrae and R. J. Riegert, Nucl. Phys. B244, 513 (1984).

22. R. H. Dicke, Phys. Rev. 126, 1580 (1962); L. B. Okun, Sov. J. Nucl. Phys. 10, 206 (1969).

23. P. Morrison, Am. J. Phys. 26, 358 (1956).

24. D. C. Peaslee, Science 124, 1292 (1956); K. Jagannathan and L. P. S. Singh, Phys. Rev. D33, 2475 (1986).

25. I. Bars and M. Visser, Phys. Rev. Lett. 57, 25 (1986).

26. S. Raby, private communication.

27. D. R. Mikkelsen and M. J. Newman, Phys. Rev. D16, 919 (1977); G. W. Gibbons and B. F. Whiting, Nature 291, 636 (1981); P. Hut, Phys. Lett. 99B, 174 (1981).

28. J. G. Hills, A. J. 92, 986 (1986).

29. F. D. Stacey, G. J. Tuck, S. C. Holding, A. R. Maher, and D. Morris, Phys. Rev. D23, 1683 (1981); F. D. Stacey and G. J. Tuck, Nature 292, 230 (1981); S. C. Holding and G. J. Tuck, Nature 307, 714 (1984); F. D. Stacey, p. 285 in Science Underground, M. M. Nieto, W. C. Haxton, C. M. Hoffman, E. W. Kolb, V. D. Sandberg, and J. W. Toevs, eds., (AIP, New York, 1983), AIP Conference Proceedings no. 96.20.

30. S. C. Holding, F. D. Stacey, and G. J. Tuck, Phys. Rev., D33, 3487 (1986).

31. E. Fischbach, D. Sudarsky, A. Szafer, C. Talmadge, and S. H. Aronson, Phys. Rev. Lett. 56, 3 (1986).

32. R. V. Eötvös, D. Pekár, and E. Fekete, Ann. Phys. (Leipzig) 68, 11 1922).

33. P. G. Roll, R. Krotkov and R. H. Dicke, Ann. Phys. (NY) 26, 442 (1964); R. H. Dicke, Sci. Am. 205, 84 (December 1961); V. V. Braginskii and V. I. Panov, Sov. Phys. JETP 34, 463 (1972).

34. F. W. Dyson, A. S. Eddington, and C. Davidson, Phil. Tran. Roy. Soc. A220, 291 (1919).

35. M. L. Good, Phys. Rev. 121, 311 (1961).

36. M. Lusignoli and A. Pugliese, Phys. Lett. 171B, 468 (1986).

37. T. Goldman, R. J. Hughes, and M. M. Nieto (in preparation).

38. N. Beverini et al., "A measurement of the gravitational acceleration of the antiproton", LANL report LA-UR-86-260 (1986).

39. J. H. Billen, K. R. Crandall, and T. P. Wangler, in Physics with Antiprotons at LEAR in the ACOL Era, U. Gastaldi, R. Klapisch, J. M. Richard, and J. Tran Thanh Van, eds. (Editions Frontieres, Gif sur Yvette, France, 1985), p. 107; T. Goldman and M. M. Nieto, ibid, p. 639; M. V. Hynes, ibid, p. 657.

40. F. C. Witteborn and W. M. Fairbank, Phys. Rev. Lett. 19, 1049 (1967); Rev. Sci. Inst. 48, 1 (1977); Nature 220, 436 (1968).

41. L. I. Schiff, Proc. Nat. Acad. Sci. 45, 69 (1959); Phys. Rev. Lett 1, 254 (1958).

PROPOSAL TO MEASURE POSSIBLE VIOLATIONS OF THE g-UNIVERSALITY AT THE 10^{-10} LEVEL OF ACCURACY ON THE EARTH'S SURFACE

E. Iacopini

Scuola Normale Superiore

Pisa, Italy

1. INTRODUCTION

Newton's theory of gravity assumes that the source of the gravitational force is the inertial mass. As a consequence of this hypothesis and of Newton's second law, the acceleration of a body in the presence of a gravitational field is assumed to be independent of its mass (g-universality).

The equivalence of the inertial and the gravitational mass has been taken as a principle in the general theory of relativity and in several other, alternative theories of gravitation. With this assumption one can in fact describe the gravitational effects on matter with a modification of the local space–time geometry. All these theories of gravitation must be reviewed if g-universality is violated. This is why the experimental check of the g-universality is of great interest in physics.

The first limit on $\Delta g/g$, after Galileo's pioneering experiment, was given by Newton. He measured the period of a simple pendulum made with bobs of different substances and he obtained $|\Delta g/g| < 10^{-3}$ for gold, silver, and lead. The same method was improved by Bessel[1] (in 1832), who obtained $|\Delta g/g| < 2 \times 10^{-5}$, and by H.H. Potter[2] (in 1923) who reached $|\Delta g/g| < 3 \times 10^{-6}$.

The precision of the measurement was greatly increased by von Eötvös[3] who, using a torsion balance and comparing the Earth's gravity attraction with the centrifugal force due to the Earth's daily rotation, tested the g-universality at the 3×10^{-9} level of accuracy. This limit has been pushed to 7×10^{-10} by Renner,[4] using the same method.

The long-range effects due to the Sun's attraction on a torsion balance were first measured by Dicke,[5] who obtained $|\Delta g/g| < 10^{-11}$ for gold and aluminium. More recently, the method has been improved by Braginskii,[6] who reached $|\Delta g/g| < 9 \times 10^{-13}$ for aluminium and platinum using an eightfold torsion balance.

Quite recently, a reanalysis of the Eötvös experiment[7] has raised the question of the possible existence of a 'fifth force' with a range of $\approx$ 10 m to 10^4 m. This force would manifest itself with an apparent violation of the g-universality due to the Earth's attraction of the order of $|\Delta g/g| \approx 10^{-9}$. Dicke's and Braginskii's results are not significant in this case since the Earth to Sun distance is much larger than the hypothetical fifth-force range.

In this context we proposed an experiment[8] which should detect possible differences of the order of $\Delta g < 10^{-10}g$ in the free-fall acceleration on the Earth's surface of two sense bodies of different material.

Assuming that the interaction potential between two point-like masses m_1 and m_2 can be expressed by the equation

$$V(r) = -(Gm_1m_2/r)\,[1 + \alpha_{12}\exp(-r/r_0)] \ , \tag{1}$$

with α_{12} depending on the materials, the experiment proposed will set the following limits on the difference $\Delta\alpha$ between the couplings of the sense bodies with the Earth, as a function of the range r_0:

$$\begin{aligned} \Delta\alpha &\approx (2/3)(\Delta g/g)(R_E/r_0) \qquad &&\text{for} \quad 10\ \text{m} < r_0 < R_E \\ \Delta\alpha &\approx \Delta g/g &&\text{for} \quad r_0 > R_E \ , \end{aligned} \tag{2}$$

where R_E is the Earth's radius.

2. THE METHOD

The torque on a mass distribution due to a homogeneous gravitational field is zero. As a consequence, no angular acceleration should appear during the free fall of a rigid body, no matter what its shape and density distribution are.

The method we propose is based on the measurement of the angular acceleration of a free-falling rigid disk which is made of two half-disks of different materials. If the free-fall accelerations of the two materials constituting the disk differ by Δg, the disk will experience a torque which will produce an angular acceleration[8]

$$\dot{\omega} = (4/3\pi)(\Delta g/R) \ , \tag{3}$$

where R is the radius of the disk (we have assumed that the two half-disks have the same mass). Therefore a value of $\dot{\omega} \neq 0$ is direct evidence of a $\Delta g \neq 0$ between the two half-disks.

To measure $\dot{\omega}$ we intend to observe the angular motion of the disk around its axis using a modified Michelson interferometer in which the two arms terminate in two cube-corner reflectors fixed on the disk (see Fig. 1). The light intensity from the interferometer reads

$$J = I_1 + I_2 + 2\sqrt{I_1I_2}\cos\Phi(t) \ , \tag{4}$$

where I_1 and I_2 are the light intensities in the two interferometer arms and $\Phi(t)$ is the phase difference between the two optical paths given by[8]

$$\begin{aligned} \Phi(t) &= \Phi_0 + (8\pi R/\lambda)\,\{[\omega_0 t + (1/2)\dot{\omega}t^2]\} \\ &= \Phi_0 + [(8\pi R/\lambda)\omega_0 t + (16/3)\,(\Delta g/\lambda)t^2] \ , \end{aligned} \tag{5}$$

where λ is the wavelength of light and ω_0 is the angular velocity of the disk at $t = 0$. The presence of a t-squared term in the time-dependence of $\Phi(t)$ is the signature of a possible g-universality violation.

To check for possible systematic errors, we intend to start with a homogeneous disk and then we will repeat the measurement with three disks of composition A–B, B–C, and A–C.

3. EXPERIMENTAL APPARATUS

The experimental apparatus is shown schematically in Fig. 1. The light source is a commercially-available, frequency-stabilized He-Ne laser ($\lambda = 632.8$ nm; power $\approx$ 1 mW). The laser light intensity is monitored by the photodetector D0, via the beam-splitter BS0. The beam-splitter BS1, the mirror M, and the cube-corner reflectors R1 and R2 are installed in a

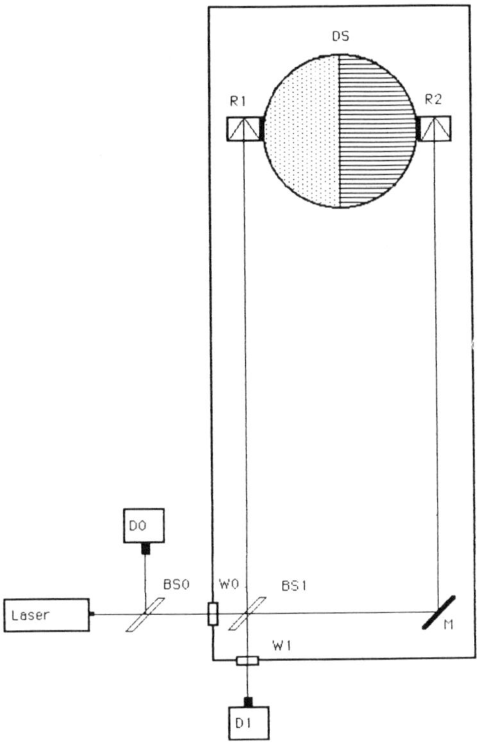

Fig. 1 Schematic view of the apparatus

cylindrical stainless-steel vessel 8 m in length and 270 mm in diameter. The vessel has its axis aligned with the vertical direction and it is evacuated down to 10^{-6} Torr.

The disk axis in the upper rest position is supported by a carriage. A measurement cycle is initiated by releasing the carriage which starts a free fall. After a short time interval the disk is also released and starts its own free fall. The time interval is chosen so as to guarantee an initial separation of a millimetre between the carriage and the disk. The carriage is guided by vertical rails on the first 30 cm of its path. After 5 m of free fall, when the disk-to-carriage separation is of the order of 15 cm, a progressive deceleration is produced on the carriage by braking pads rubbing on spring-loaded bars. A soft landing of the disk on the carriage is obtained at the beginning of this braking by a suitable damping catch on the carriage. The whole assembly (disk + carriage) comes to rest in ≈ 2.0 m, the maximum deceleration being ≈ 4g. A gear is used to pull the carriage with the disk on it to the top of the vessel. When the carriage is in the top position, an angular rotation ω_0 (≈ 10 mrad/s) is given to the disk by an externally-controlled gear. This rotation is necessary to bring the signal into the frequency region of a few kilohertz, where the laser noise is quite low. The signals from the PIN photodiodes D0 and D1 are sampled at a fixed frequency of ≈ 50 kHz by a 14-bit analog-to-digital converter (ADC) and the data are stored in a fast buffer memory which is read by a microcomputer at the end of each measurement for off-line analysis.

4. SENSITIVITY AND SYSTEMATICS

The sensitivity on $\Delta g/g$ obtained with a least-squares fit of the data with the theoretical signal form of Eq. (4) is given by

$$(\Delta g/g)_{sens} = 9\sqrt{5}/8\sigma(\lambda/gT^2)\,(1/\sqrt{N}) \ , \qquad (6)$$

where σ is the r.m.s. noise deviation on a single sample normalized to the peak cosine amplitude, T is the fall time (≈ 1 s), and N is the number of samples (≈ 50000).

Because of the laser-intensity fluctuations, the sensitivity is limited to[8]

$$(\Delta g/g)_{laser} \approx 5 \times 10^{-13} \;, \qquad (7)$$

the shot noise limit being more than one order of magnitude smaller.

The ADC quantization noise for a 14-bit conversion is[8]

$$(\Delta g/g)_{ADC} \approx 10^{-13} \;. \qquad (8)$$

The main limitations to the method come from systematic errors. It turns out that the disk mechanics set a limit to the sensitivity of the order of $\Delta g/g \approx 10^{-10}$.

In fact, the disk will not fall in a homogeneous gravitational field. Taking into account the dependence of g on the height, the disk will undergo an angular acceleration around its axis given by

$$\dot{\omega}_{inh} = (3/2)\,(g/R_E)\,(\Delta I/I)\,\sin 2\xi \;, \qquad (9)$$

where ΔI is the difference between the two principal moments of inertia in the vertical plane, I is the moment of inertia of the disk with respect to its axis, and ξ is the angle between the principal axis of the disk in the vertical plane and the direction of grad (g) evaluated at the centre-of-mass position of the disk. From the ratio

$$|\dot{\omega}_{inh}/\dot{\omega}| = (9\pi/8)\,(g/\Delta g)\,(R/R_E)\,(\Delta I/I) \;, \qquad (10)$$

it turns out that in order to detect a value of $\Delta g \approx 10^{-10} g$ we need to keep $(\Delta I/I) < 4 \times 10^{-3}$. We expect that this requirement will be fulfilled by accurate machining and assembly of the disk components. However, the principal moments of inertia of the disk in the vertical plane will also be measured by the torsion pendulum method so that corrections can be applied if required. More complex configurations with several cube-corner reflectors on the disk can also be envisaged in order to reduce higher multipole contributions.

REFERENCES

1. F.W. Bessel, Pogg. Ann. 25:401 (1832).
2. H.H. Potter, Proc. Roy. Soc. 104:588 (1923).
3. R. von Eötvös, D. Pekar, and E. Fekete, Ann. Phys. 68:11 (1922).
4. J. Renner, Mat. es termeszettudomanyi ertsito 53:542 (1935).
5. P.G. Roll, R. Krotkov, and R.H. Dicke, Ann. Phys. (NY) 26:442 (1964).
6. V.B. Braginskii and V.I. Panov, Sov. Phys.-JETP 34:463 (1972).
7. E. Fischbach, D. Sudarsky, A. Szafer, C. Talmadge, and S.H. Aronson, Phys. Rev. Lett. 56:3 (1986).
8. V. Cavasinni, E. Iacopini, E. Polacco, and G. Stefanini, Phys. Lett. 116A:157 (1986).

GRAVITATIONAL SPIN INTERACTIONS?

F.P. Calaprice

Department of Physics
Joseph Henry Laboratories
Princeton University, Princeton, New Jersey 08540

Some years ago H. Dass pointed out that essentially nothing is known about the possibility of an interaction between the spin **S** of an elementary particle and the field of a gravitating body of mass M (ref. 1). I will comment here briefly about the nature of such an interaction and the possibilities for measuring such terms by optical pumping NMR methods.

The phenomenological spin interaction energy introduced by Dass is given by the following equation.

$$U(r) = \alpha_1(GM/cr^3)\mathbf{S}\cdot\mathbf{r} + \alpha_2(GM/cr^2)\mathbf{S}\cdot\mathbf{v} + \alpha_3(GM/c^2r^3)\mathbf{S}\cdot(\mathbf{r} \times \mathbf{v})$$

The mass of the gravitating body, which in our case will be the earth, is denoted by M and the particle spin and its velocity and position relative to the body are given by **S**, **v**, and **r**, respectively. The coupling constants α_1, α_2, and α_3 are dimensionless and, according to Dass, they have the values 0, 0, and 2/3, respectively, from the strong Equivalence Principle, generalized to include intrinsic spin. Deviations from these values therefore indicate violation of this principle. In addition, the first term violates time reversal and parity invariance and the second term violates parity invariance.

There are no limits for the parameters α_i that follow from direct searches. However, an indirect limit has been obtained by Dass from a consideration of the hydrogen hyperfine structure with the result $\alpha_1 < 10^{10}$. A result attributed by Dass to B.K. Dennison on the polarization dependence of the bending of 13 cm radiowaves at the solar limb leads to the more restrictive limit of $\alpha_1 < 10^4$ (see note d of ref. 1). This is still very crude compared to the value of unity which is suggested by Dass to be in the interesting range.

The α_1 term is similar in form to the interaction between a magnetic dipole moment μ and an external magnetic field B. As is well known, such an interaction produces a precession of the spin about the magnetic field with an angular frequency given by

$$\omega_m = -\gamma B$$

where $\gamma = \mu/S$ is the gyromagnetic ratio. Similarly, the α_1 term produces a precession of the spin about **r** with a frequency of precession for a nuclear moment at the surface of the earth of roughly $\omega_1 \approx \alpha_1 x 10^{-7}$ rad/sec. This is a small rate and the rotation of the earth

($\approx 10^{-5}$Hz) must be accounted for at this level, but for simplicity we ignore the earth's rotation in the following discussion. The total precession rate of a spin in a combined magnetic and gravitational field is then the sum of the magnetic and gravitational frequencies, $\omega_{\mathbf{tot}} = \omega_{\mathbf{m}} + \omega_1$. From the above discussion we see that the presence of the gravitational term can be detected by comparing the total frequency of precession of the spin for two configurations in which the magnetic field is either parallel or anti-parallel to **r**. A difference in the frequencies implies a nonzero value for α_1, provided the magnetic field strengths are exactly the same.

Making certain that the magnetic fields are opposite in direction but exactly the same in magnitude is the main experimental difficulty for such an experiment. Here we point out that we may be able to get around this difficulty by using two nuclear systems with the same angular momentum but different magnetic dipole moments. The $11/2^-$ isomeric states ^{129m}Xe and ^{131m}Xe are examples. Suppose we place both isotopes in the same bottle, so that they experience the same magnetic field, and measure the precession frequencies for both species simultaneously, to avoid the problem of a time variation of the field. We measure the precession frequencies for both isotopes for the case where the magnetic field is up ($\Uparrow$), or parallel to **r**, and down ($\Downarrow$) or anti-parallel to **r**. Then we compute the ratio of the isotopic frequencies for each configuration. That is, for the parallel case,

$$R_{\Uparrow} = [\omega(129)/\omega(131)]_{\Uparrow} = [\gamma_m(129)B + \omega_1(129)]_{\Uparrow} / [\gamma_m(131)B + \omega_1(131)]_{\Uparrow}$$

and similarly for for the antiparallel case we compute $R_{\Downarrow}$,

$$R_{\Downarrow} = [\omega(129)/\omega(131)]_{\Downarrow} = [\gamma_m(129)B - \omega_1(129)]_{\Downarrow} / [\gamma_m(131)B - \omega_1(131)]_{\Downarrow}$$

where we note the relative sign change.

In the physically interesting case, the gravitational precessional frequency is much smaller than the typical magnetic precession frequency ($\omega_1 << \gamma B \sim$ Hz). If we also assume that ω_1 is the same for both isotopes (since they have the same angular momentum) and that the magnetic fields for "up" and "down" configurations are nearly the same, we find the following approximation for the ratio of ratios

$$R_{\Uparrow}/R_{\Downarrow} \approx 1 + 2[1-\gamma(129)/\gamma(131)]\ \omega_1/\omega_m(129)$$
$$\approx 1 + 0.2\omega_1/\omega_m(129),$$

where in the second line we have used the magnetic moments $\mu(129) = 0.891\ \mu_N$ and $\mu(131) = 0.994\ \mu_N$ (ref. 2).

A deviation of the above double ratio from unity is the signature for a non-zero value of α_1. This conclusion is independent of the assumptions used to obtain the approximate expression and in particular does not depend on exact reversal of the magnetic field, which affects only the correction to unity.

However, we do stress that there will be no deviation from unity in the double ratio if the gravitational precession frequencies $\omega_1(129)$ and $\omega_1(131)$ scale as the magnetic moments. Even though the spins of the two nuclei are the same, so that naively one might expect that

ω_1 is the same for both isotopes, there may be some dependence of the gravitational coupling on the nuclear structure. However, it seems that it would be surprising if the dependence of the "gravitational" dipole moment on nuclear structure were identical to the dependence of the magnetic dipole moment on nuclear structure.

Now we speculate on how well one might be able to measure the ratio of the two precession frequencies. We focus on the optical pumping method that has been used to polarize xenon isotopes (see ref. 2 and related paper in this conference) but other NMR methods can be used too, though perhaps with less sensitivity. First we note that the xenon isotopes can be highly polarized by spin exchange on alkali atoms that are polarized by optical pumping. The gamma ray anisotropy method has been used to observe the nuclear alignment and to observe NMR transitions. The rare gas atoms have long spin relaxation times, resulting in narrow NMR resonances (0.01Hz) and, therefore, high sensitivity to small frequency shifts.

The simultaneous orientation in a common cell of several isotopes has been achieved and observation of the nuclear orientation and NMR transitions of the isotopes was straightfoward using the gamma ray energy to distinguish the isotopes. Simultaneous measurement of the precession frequencies can be accomplished by measuring the free precession of the isotopes about a field normal to the initial polarization axis. The sensitivity to ω_1 is enhanced for low external magnetic field. Assuming an external field of 0.01G, a linewidth of 0.01Hz, and a count rate of 10^4counts/sec gives a sensitivity to frequency shifts of 10^{-7}Hz in 10 days of counting, implying a limit of $\approx 10^{-6}$Hz for the gravitational precession frequency, or equivalently a limit of $\approx$ 100 for α_1. This is more than a factor of 100 better than the current limit but not down to the level of unity for α_1. This limit is possible with slight extensions of currenlty used methods. Further developments may lead to improved sensitivites.

References

1. N.D. Hari Dass, Annals of Physics **107**, 337-359 (1977).
2. M. Kitano, M. Bourzutschky, F.P. Calaprice, J. Clayhold, W. Happer, and M. Musolf, Phys. Rev. **C34,** 1974 (1986).

PENNING TRAPS, MASSES AND ANTIPROTONS*

G. Gabrielse
Department of Physics, FM-15
University of Washington
Seattle, Washington 98195

Abstract

Penning traps are an important new tool for mass spectroscopy, aided by an invariance theorem which facilitates precise mass spectroscopy in an imperfect trap. The motions of particles in a Penning trap are discussed and the features which make it very attractive to do mass spectroscopy in a trap are illustrated. Careful attention is paid to the motivations and prospects for a measurement of the inertial mass of the antiproton. Prospects for such a measurement are now excellent since our TRAP Collaboration actually captured antiprotons in a Penning trap only 2 months ago. An overview of ways to cool particles within the trap is provided and brief speculations upon the possibility of producing antihydrogen in a trap are included.

* Invited Lecture at the International School of Physics with Low Energy Antiprotons: Fundamental Symmetries, Sept. 24 - Oct. 4, 1986, Erice, Italy.

A. Motion in a Penning Trap

Consider a negatively charged particle of charge -e in a homogeneous magnetic field $B\hat{z}$. The motion is familiar cyclotron motion about a field line with angular frequency

$$\omega_c = \frac{eB}{mc}. \tag{1}$$

Such a simple system would be very nice for mass spectroscopy, since ω_c for two different ions could be measured in the same magnetic field. Although the particle is confined to a magnetic field line, however, it is free to leave the trap along the axis of the magnetic field. Also, it is not immediately clear how to measure ω_c.

Actual measurements can be performed within a slightly more complicated structure called a Penning trap, which I will describe as simply as possible, leaving details to a recent review.[1] An electrostatic quadrupole potential is added to the magnetic field using electrodes such as those shown in Fig. 1. The particles are now prevented from leaving the trap along the axis of the magnetic field because they are repelled by negatively charged electrodes above and below. The axial potential well comes at the expense of a radial potential hill which exerts an outward radial force on the particle and which must be overcome by the strong magnetic field. The added quadrupole potential modifies the simple cyclotron motion in 3 ways.

1. The cyclotron frequency ω_c is reduced to some ω_c' because the effective radial binding is reduced by the radial potential hill.
2. A harmonic oscillation is introduced along the magnetic field axis at an angular frequency ω_z.

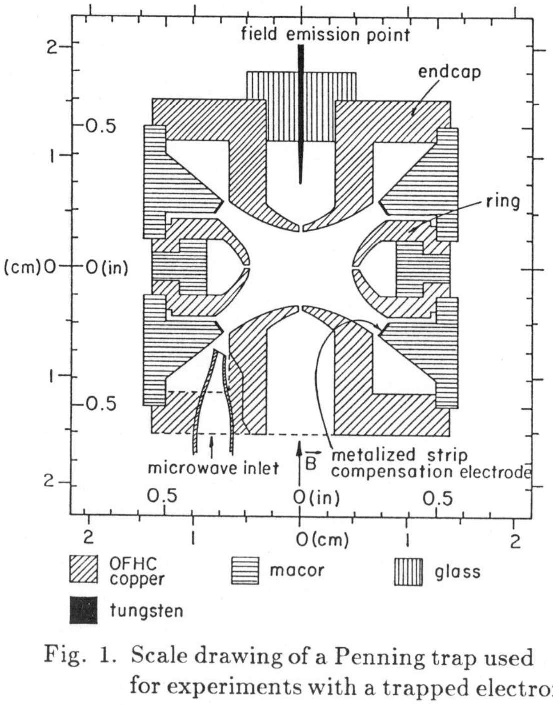

Fig. 1. Scale drawing of a Penning trap used for experiments with a trapped electron.

[1] L. S. Brown and G. Gabrielse, Rev. Mod. Phys. **58**, 233 (1986)

3. The strong magnetic field and the radially outward component of the electric field together comprise a velocity filter. Particles traveling in a circular orbit at a particular velocity thus have zero net force from these two sources, independent of the charge to mass ratio. This circular motion takes place at the magnetron frequency ω_m.

For example, in a 6 Tesla magnetic field and with a 100 volts applied across the electrodes is shown in Fig. 1, a trapped antiproton has eigenfrequencies

$$\omega_c'/2\pi = 90\ MHz, \tag{2}$$

$$\omega_z/2\pi = 4\ MHz, \tag{3}$$

$$\omega_m/2\pi = 100\ kHz. \tag{4}$$

The motions can be detected via oscillatory currents induced between trap electrodes by the oscillatory motion of the trapped particles. A resistor connected between the electrodes transforms the oscillating current into an oscillatory voltage. Unavoidable trap capacitance is tuned out with an inductor so that the resistance is actually realized at only one particular resonant frequency. A sensitive amplifier across the resistor detects the oscillatory voltage. A single trapped particle can be detected nondestructively[2] as illustrated in Fig. 2. The vertical scale indicates the current detected as electrons are loaded into a trap and oscillate along the magnetic field. The horizontal axis is time and the electrons are kicked out of the trap at times indicated by arrows, by reversing the sign of the quadrupole potential for several seconds.

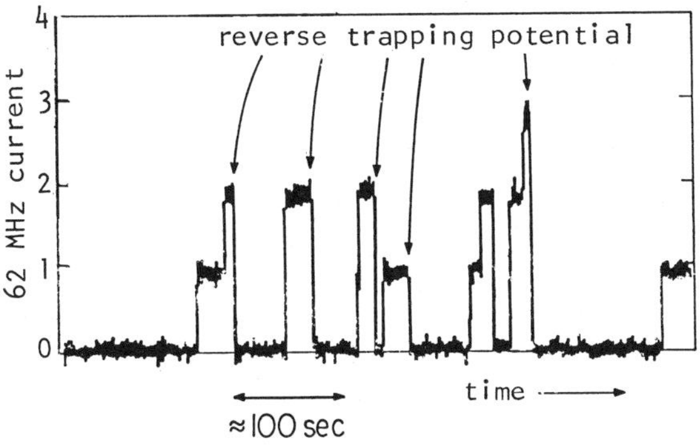

Fig. 2. Measured current induced by the ω_z motion of electrons as they enter the trap. Arrows indicate where electrons are removed by reversing the trapping potential.

[2] D. J. Wineland, P. Ekstrom, H. G. Dehmelt, Phys. Rev. Lett. **31**, 1279 (1973)

B. Comparing Inertial Masses in a Penning Trap

In an ideal Penning trap, the cyclotron frequency ω_c can be simply deduced from measured values of the cyclotron and magnetron frequencies measured in the trap.

$$\omega_c = \omega_c' + \omega_m \tag{5}$$

Masses of 2 different particles can be compared by comparing separately measured cyclotron and magnetron frequencies, ω_c' and ω_m, measured in the same magnetic field. In the example given, Eqs. (2) - (4), the cyclotron frequency in the trap is shifted by 1 part in 10^3. In an imperfect trap, however, various imperfections shift the 3 eigenfrequencies in the trap,

$$\omega_c' \to \bar{\omega}_c' \tag{6}$$

$$\omega_z \to \bar{\omega}_z \tag{7}$$

$$\omega_m \to \bar{\omega}_m \tag{8}.$$

Fortunately, an invariance theorem proved by Lowell Brown and myself,[3]

$$(\omega_c)^2 = (\bar{\omega}_c')^2 + (\bar{\omega}_z)^2 + (\bar{\omega}_m)^2, \tag{9}$$

makes it possible to deduce ω_c in a way which is not sensitive to major trap imperfections, from the three measured eigenfrequencies in an imperfect trap. In particular, the invariance theorem is insensitive to

1. Misalignment of the magnetic field axis and the axis of the electric quadrupole potential,
2. Spatially uniform stray electric fields within the trap, since these merely shift the center of the oscillations within the trap,
3. Harmonic distortions of the electrostatic potential, such as an added potential term which goes as xy.

Since the particles are located near the center of the trap, potential distortions not covered by the invariance theorem are small, of order $(r/d)^3$ or higher, where r is the small distance from the center of the trap, d is a trap dimension and typically $r/d \sim 10^{-2}$. The trap electrodes have axial symmetry about the z axis and reflection symmetry under $z \to -z$, so that many higher order potential distortions are strongly suppressed. In addition, the leading potential distortion allowed by the electrode symmetry is explicitly tuned out[4] in a way which is quantitatively understood.[5]

Surprisingly (to me at least), the invariance theorem is not yet used for most of the mass measurements being made in Penning traps. Also, little advantage is taken of axial symmetry and high order potential distortions are typically not tuned out. I refer here to the rectangular (often cubic) trap configurations used by chemists for less precise mass measurements under the name Ion Cyclotron Resonance (ICR).[6] Complete commercial devices are available and hundreds of papers are in the literature, so it is certainly not possible to do justice to this work. Broadband amplification and

---

[3] L.S. Brown and G. Gabrielse, Phys. Rev. A. **25**, 2423 (1982)

[4] R.J. Van Dyck, D.J. Wineland, P. Ekstrom and H. G. Dehmelt, Appl. Phys. Lett. **28**, 446 (1976)

[5] G. Gabrielse, Phys. Rev. A **27**, 2277 (1983)

[6] D.A. Laude, Jr., C.L. Johlman, R.S. Brown, D.A. Weil and C.L. Wilkins, in Mass Spec. Rev. **5**, 107 (Wiley, N.Y., 1986).

detection is used to detect the response of a cloud of trapped particles to an intense excitation pulse. The response is most often Fourier transformed, hence the name Fourier Transform Mass Spectrometry (FTMS).[7] An often cited "best resolution" achieved is on the order of 10^{-8} for a water sample[8] but an even higher resolution (which I will discuss presently) is claimed for the $^3H - ^3He$ mass difference. The choice of rectangular electrode geometry makes for large departures from an ideal quadrupole potential, and these deviations are especially important when the cloud of particles is highly excited to a large cyclotron radius. The axial and magnetron frequencies are typically not carefully monitored, so that no advantage can be taken of the invariance theorem Eq. (9). It is thus much harder to measure a mass accurately despite high resolution, unless mass doublets are compared. Careful studies of the effect of cloud size and distribution are carried out because these are important systematic effects.

The use of Penning traps for mass spectroscopy is relatively recent. Until recently, the most accurate mass spectroscopy was done in an RF spectrometer built by Smith.[9] A modern version of this machine[10] should compete with the accuracies now being achieved in Penning traps. However, I think that the much smaller Penning trap apparatus will eventually produce the most accurate measurements because it offers a number of advantages for mass spectroscopy. The first is that a good signal-to-noise ratio can be obtained with even a single charged particle. As an example, consider the single electron cyclotron resonance in Fig. 3. The triangular shape of this resonance has been discussed.[11] Here it suffices to observe that this resonance is present and is detected only because of special relativity, even though the electron's kinetic energy is less than 1 eV. This electron also illustrates a second advantage for mass spectroscopy in a Penning trap. Very long confinement times are possible. The electron used to produce Fig. 3 was kept for more than 10 months by itself in a Penning trap.[11] Mass spectroscopy with extremely small samples is clearly possible.

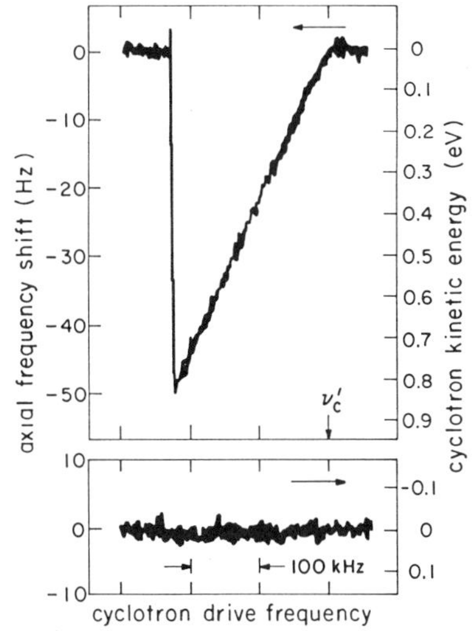

Fig. 3. Relativistic cyclotron resonance.[9]

---

[7] M.L. Gross, D.L. Rempel, Science **226**, 261 (1984).

[8] M. Allemann, H.P. Kellerhals, K.P. Wanczek, Int'l. J. of Mass Spec and Ion Proc. **46**. 139 (1983).

[9] L. Smith, Phys. Rev. C **4**, 22 (1971).

[10] C. Thibault, paper contributed to this school.

[11] G. Gabrielse, H. Dehmelt and W. Kells, Phys. Rev. Lett. **54**, 537 (1985)

A third advantage for mass spectroscopy in a Penning trap is that very different masses can be measured and compared directly. In an RF spectrometer one must measure masses which are nearly equal (i.e., have the same mass number) and from a series of such mass doublets deduce a mass. In a Penning trap one can directly compare very different masses, although if narrow band detection techniques are used this requires a significant change in the tuning of the detection electronics. A good illustration is represented in Fig. 4 where progress in measurements of the ratio of the proton to electron mass M_p/m_e is shown. The upper value is from the 1973 least squares adjustment.[12] It incorporates a number of different indirect measurements of M_p/m_e. All subsequent measurements of this ratio were made using Penning traps. The next two measurements were made at Mainz[13,14] by directly comparing cyclotron frequencies in a Penning trap. Small numbers of electrons and protons were ejected from the trap to a detector and cyclotron excitations were deduced from changes in a time-of-flight spectrum. The three values from Washington[15,16,17] represents the improved accuracy which can be obtained when the particles can be kept in the trap and continuously interrogated with narrow band tuned circuits. The NBS value[18] is derived from a double resonance measurement on trapped Be^+ ions. This measurement uses calculated g values and is thus a much less direct measurement.

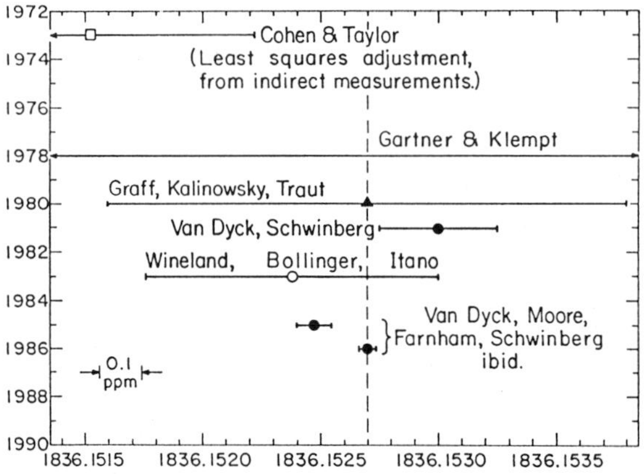

Fig. 4. Measured proton to electron mass ratio.

[12] E.R. Cohen and B.N. Taylor, J. Chem. Ref. Data **2**, 663 (1973)

[13] G. Gartner and E. Klempt, Z. Phys. **287**, 1(1978)

[14] G. Graff, H. Kalinowsky and J. Traut, Z. Phys. **297**, 35,(1980)

[15] R. S. Van Dyck, Jr. and P. B. Schwinberg, Phys. Rev. Lett. **47**, 395 (1981)

[16] R. S. Van Dyck, Jr., F. Moore, D. Farnham and P. B. Schwinberg, Int. J. of Mass Spec. and Ion Proc. **66**, 327 (1985)

[17] R. S. Van Dyck, Jr., F. Moore, D. Farnham and P. B. Schwinberg, Bull. Am. Phys. Soc. **31** (1986)

[18] D. J. Wineland, J.J. Bollinger, W.M. Itano, Phys. Rev. Lett. **50**, 628 (1983)

The convenience of a trap for direct measurements of ion masses is also being used now to measure the masses of unstable nuclei.[19] The object is to check the nuclear mass formula with the masses of unstable nuclei and the trap offers a way to make such measurements without the need to measure a long series of doublets. The range of masses which are of interest is indicated in Fig. 5, where the accuracy desired is plotted horizontally vs. the number of nucleons, vertically. The three dashed vertical lines indicate typical nuclear energies of interest. The binding energy is of order 8 MeV/nucleon, shell adjustments contribute variations of approximately 1 MeV and nuclear pairing effects contribute of order 100 keV. For nuclei of interest, with up to several hundred nucleons (shaded region in figure), this means that a fractional mass resolution of 10^{-7} or slightly better is desired.

The fourth and final advantage I will discuss for mass spectroscopy in Penning traps is that an extremely narrow cyclotron line width can be realized, which is much narrower than that offered by other techniques. Smith's RF spectrometer[9], for example, had a linewidth of 5×10^{-5} (i.e. a resolution of 2×10^{4}) with 5×10^{-6} proposed for an improved version under development.[10] By contrast, linewidths much better than 10^{-8} are routinely observed with trapped electrons, positrons and protons, with resolutions as high as 10^{-10} actually observed in some cases.[20] The basic reason for the narrow line in a trap is that the trap particles occupy such a small region of space that it is relatively easy to obtain a homogeneous magnetic field over this small region. Linewidth is not the only consideration for an accurate measurement, of course. Systematic efforts and statistics will determine how much the line can be split and what accuracy can be obtained. On the other hand, those making precision measurements rarely complain about a line being too narrow.

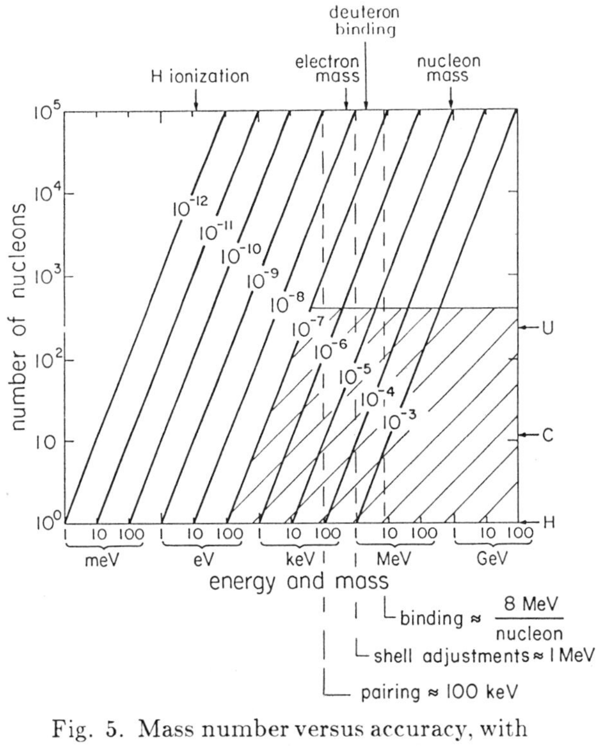

Fig. 5. Mass number versus accuracy, with shaded region of interest for nuclear physics.

[19] Schnatz, et.al., Nucl. Inst. and Meth. (in press).

[20] R. S. Van Dyck, et al. (unpublished)

Fig. 6 illustrates the impact that mass spectroscopy in traps is already making upon the measurement of the mass difference of ^3H and ^3He. This quantity, of course, is of great interest as a part of the effort to measure the neutrino mass. The values plotted are the "adopted values" from the review by Audi, Graham and Geiger,[21] which in some cases have been adjusted from originally published values. Notice in particular the RF spectrometer values from Smith, which are uncertain at the level 5×10^{-9}. The most recent measurements by Lippman et. al.[22] are done in an ICR trap. While I personally am impressed and a bit dubious that an uncertainty of 2×10^{-9} can be achieved given the limitations of the ICR technique discussed earlier, it is clear that mass spectroscopy in traps has already contributed significantly in this case. Moreover, a group at Mainz[23] and at Washington[20] are now in the process of measuring the mass difference in Penning traps so it is very likely that greatly improved values will be presented within a year.

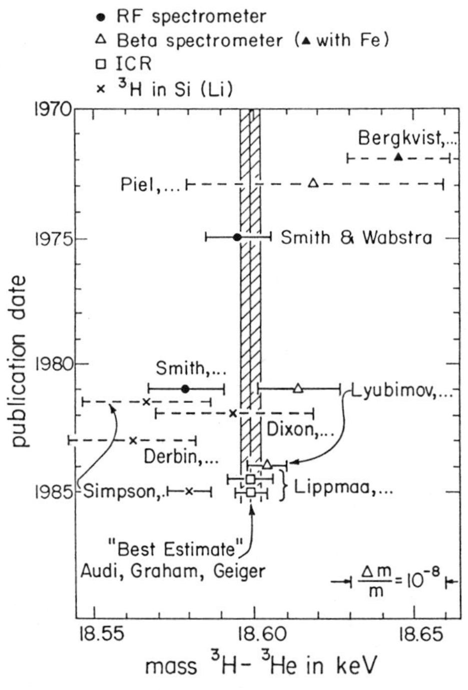

Fig. 6. "Adopted" values and uncertainties in measurements of the mass difference of ^{3}H and ^{3}He, from Ref. 19.

---

[21] G. Audi, R.L. Graham, J.S. Geiger, Z. Phys. **A 321**, 533 (1985).
[22] E. Lippman, et.al., Phys. Rev. Lett. **54** 285 (1985).
[23] G. Werth, et. al. (ununpublished).

C. The Antiproton Mass

Measurements of the antiproton mass[24,25,26,27] are represented in Fig. 7. All of these are deduced from measurements of the energy of x-rays radiated from highly excited exotic atoms. For example, if an antiproton is captured in a *Pb* atom, it can make radiative transitions from its $n = 20$ to $n = 19$ state. The antiproton is still well outside the nucleus in this case, so that nuclear effects can be neglected. The measured transition energy is essentially proportional to the reduced mass of the nucleus and hence the antiproton mass can be deduced by comparing the measured values with theoretical values, corrected for QED effects. The most accurate quoted uncertainty is 5×10^{-5} and is consistent with the much more accurately know proton mass, indicated by the dashed line. It looks like it would be difficult to extend the accuracy realized with the exotic atom method. It might be possible, however, that proton and antiproton masses could be compared directly in a storage ring, from the spatial separation of counter propagating beams of protons and antiprotons at comparable or somewhat improved accuracies.[28]

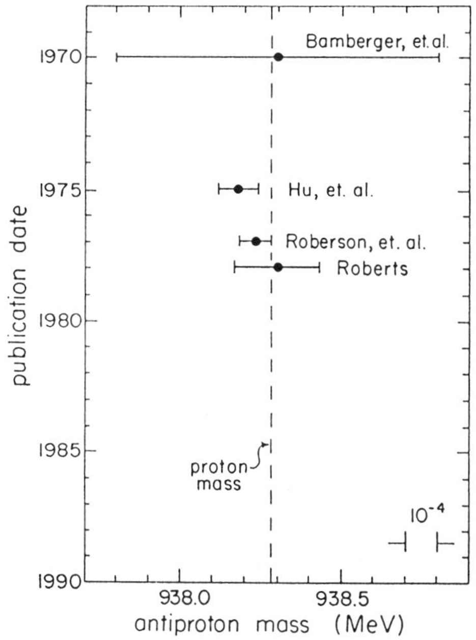

Fig. 7. Antiproton mass measurements.

---

[24] A. Bamberger, et. al., Phys. Lett. **33B**, 233 (1970).
[25] Hu, et.al., Nucl. Phys. **A 254**, 403 (1975).
[26] P.L. Roberson, et. al., Phys. Rev. C **16**, 1945 (1977).
[27] B.L. Roberts, Phys. Rev. D. **17**, 358 (1978).
[28] S. van der Meer, private communication.

Based upon precisions obtained with trapped electrons, positrons and protons, it seems very likely that the measurement uncertainty in the ratio of antiproton to proton masses could be reduced by more than 4 orders of magnitude, to order 10^{-9} or better. A major question, however, is whether or not one should bother. The widely accepted assumption of CPT invariance would insure that antiproton and proton masses are equal. Fig. 8 shows the current status of experimental tests of CPT invariance, taken from the Particle Data Group compilation[29] with several updates. Since CPT invariance implies that a particle and antiparticle have the same magnetic moment (with opposite sign), the same inertial mass and the same mean life, the tests are so grouped. The fractional accuracy is plotted, and baryons, mesons and leptons are distinguished. The neutral kaon system provides a test of CPT invariance of striking precision. Equally striking, however, is that only 3 other tests exceed 1 part per million in accuracy, and these involve leptons only. In fact, there is not even a single precision test of CPT invariance with baryons. The widespread faith in CPT invariance is clearly based upon the success of field theories in general and not upon a dearth of precision measurements.

We note here that it is even conceivable that proton and antiproton masses could be different without a violation of CPT invariance. Precisely stated, CPT invariance relates the mass of a proton in a matter universe to an antiproton in an antimatter universe. A long range coupling to baryon number would not affect the kaon system but could shift differently the proton and antiproton masses, given the preponderance of baryons in our apparatus and universe.

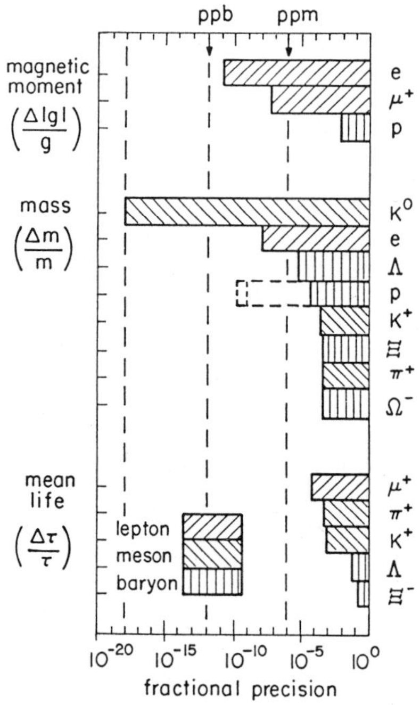

Fig. 8. Fractional accuracy in experimental tests of CPT invariance.

---

[29] Particle Data Group, Rev. Mod. Phys. **56**, S1 (1984).

The scarcity of precise tests of CPT invariance makes the case for a precise comparison of proton and antiproton masses seem to be very strong to me, especially since no precise test at all involves baryons. Such a measurement also satisfies several additional criteria.

1. A big improvement in accuracy is involved, somewhere between four and five orders of magnitude.
2. A simple, basic system is involved.
3. The technique used will be convincing if the masses are found to differ.
4. The measurement will involve a reasonable effort.
5. It will be fun.

The last two criteria are more subjective than the others, but important nonetheless.

Since Vernon Hughes is one of my fellow lecturers at this school, I end this discussion by quoting him and his collaborators, from their published account of one of the early measurements of the antiproton mass by the exotic atom method.[25]

> Formalistically, the equivalence of mass and lifetime of particle and antiparticle is a consequence of the CPT theorem, which also predicts that the magnetic moments of particle and antiparticle are equal in magnitude but opposite in sign. It is clearly of interest to make precise measurements of such basic quantities for the antiproton, not only for the intrinsic value, but also for the knowledge it may yield concerning the basic symmetries underlying the matrix of physical thought.

D. First Slowing and Capture of Antiprotons in an Ion Trap

As you well know, antiprotons are created at energies of several GeV. Precision experiments in Penning traps take place at millielectron volts (meV). An experimental difficulty, then, is to reduce the antiprotons kinetic energy by approximately 12 orders of magnitude, as illustrated in Fig. 9. The first slowing, from GeV energies down to MeV energies takes place within LEAR. The unique capabilities of this machine are well known here, so I will not discuss them further.

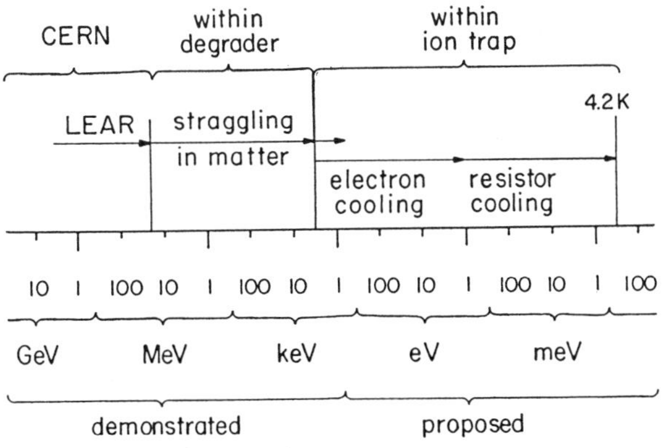

Fig. 9. Antiprotons produced at several GeV energies must be slowed to below a meV for a high precision mass measurement.

I am delighted to report that in the last several months the TRAP Collaboration (PS196) has taken 21.3 MeV antiprotons from LEAR (200 MeV/c) and slowed them down to below 3 keV. At this energy they were caught in the small volume of an ion trap and help up to ten minutes. Members of the TRAP Collaboration are listed in Fig. 10. I should point out that this effort succeeded despite incredible time pressure. The capture of antiprotons, for example, occured during a single 24 hour period.

TRapped **A**nti**P**roton Collaboration

(TRAP Collaboration)

University of Washington

G. Gabrielse

X. Fei

K. Helmerson

S.L. Rolston

R. Tjoelker

T.A. Trainor

University of Mainz

H. Kalinowsky

J. Haas

Fermi National Accelerator Laboratory

W. Kells

Fig. 10. TRAP collaboration.

The experiments went in two stages. In May, we used a simple time-of-flight apparatus to measure the energy distribution of antiprotons emerging from a thick degrader. Since we have not yet finished our analysis, I present in Fig. 11 only a preliminary result taken on line during the May run. The upper graph shows transmitted antiproton intensity versus thickness of the degrader. As degrader thickness is increased, the number of antiprotons drops as more of them are stopped in the degrader. The degrader thickness at the half intensity point is very close to the proton range which is compiled in standard tables. All energies of transmitted antiprotons are included and most of these antiprotons have energies above 3 keV which is the highest energy we could trap.

The lower curve in Fig. 11 is more crucial. Here the number of antiprotons which emerge from the degrader with low kinetic energies (along the beam axis) between 2 and 8 keV is plotted versus degrader thickness. The low energy flux is clearly peaked at the half intensity point of the upper curve. Approximately 1 in 10^4 of the incident antiprotons emerges from the degrader with below 3 keV. These are the particles available for trapping.

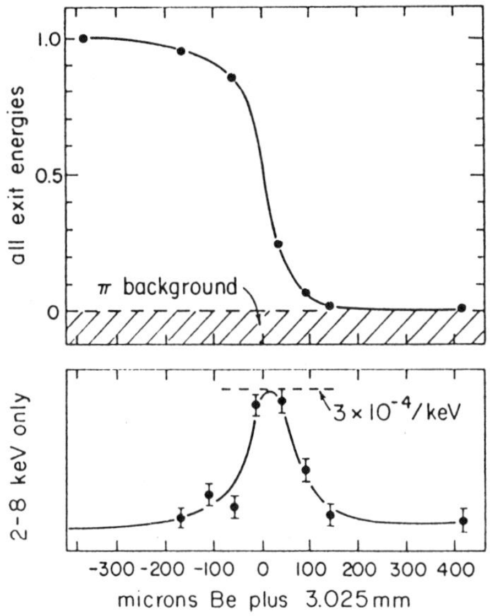

Fig. 11. Transmission of antiprotons through material versus effective thickness of Be.

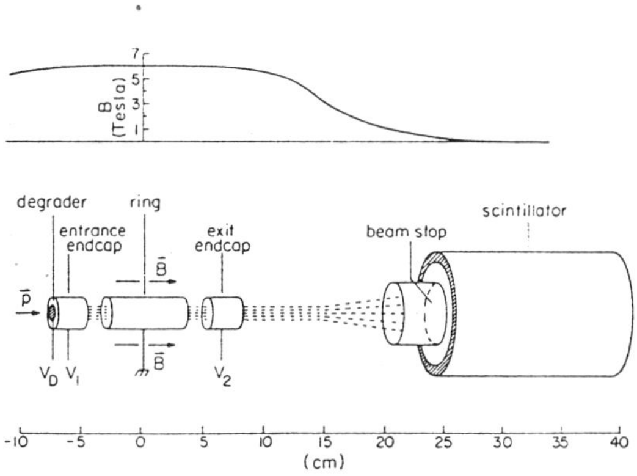

Fig. 12. Ion trap in which first capture of antiprotons took place.

In July we returned to LEAR for a 24 hour attempt to actually catch antiprotons in the small volume of an ion trap. An account of our success is appearing in Physical Review Letters,[30] so I will only briefly summarize. The trap is very simple as is indicated in Fig. 12. The slowest antiprotons leaving the thick degrader are confined in 2 dimensions to field lines of the 6T superconducting magnet (dotted lines in Fig. 12) and are so guided through the series of 3 trap electrodes. As the antiprotons enter the trap, the first ring-shaped trap electrode (the entrance endcap) and the main ring electrode are both grounded. The third cylindrical electrode (exit endcap) is at -3 kV so that negative particles with energy less than 3 keV turn around on their magnetic field lines and head back towards the entrance of the trap. Approximately 300 ns later, before the antiprotons can escape through the entrance, the potential of the entrance endcap is suddenly lowered to -3kV, catching them within the trap. The potential is switched in 15 ns with a kryton circuit developed for this purpose and is applied to the trap electrodes via an unterminated coaxial transmission line.[31] The 3 keV potentials and 15 ns rise times contrast sharply with the several volt potentials and the 100 ns switching times used recently to capture Kr^+ in a few eV well.[19]

After antiprotons are held in the trap between 1 ms and 10 minutes, the potential of the exit endcap is switched from -3 kv to 0 volts in 15 ns, releasing the antiprotons from the trap. The antiprotons leave the trap along respective magnetic field lines and annihilate at a beam stop well beyond the trap. The high energy charged pions which are released are detected in a 1 cm thick cylindrical scintillator outside the vacuum system. A multiscaler started when the potential is switched records the

[30] G. Gabrielse, X. Fei, K. Helmerson, S.L. Rolston, R. Tjoelker, T.A. Trainor, H. Kalinowsky, J. Haas, W. Kells, Phys. Rev. Lett. **57**, 2504 (1986)

[31] X.Fei, R.Davisson and G.Gabrielse, Rev. of Sci. Inst. (in press).

number of detected annihilations over the next $6\mu s$ in time bins of $0.4\mu s$. A second multiscaler records the pion counts over a wider time range with less resolution to monitor backgrounds. This time-of-flight method is similar to but less refined than that used on very low energy electrons and protons ejected from a Penning trap with a 6 volt potential well.[14]

Fig. 13 shows a time-of-flight spectrum for antiprotons kept in the trap for 100s. The spectrum includes 31 distinctly counted annihilations which corresponds to 41 trapped particles when the detector efficiency is included. We carefully checked that these counts are not electronic artifacts. When the high voltage on the exit endcap is switched to release antiprotons from the trap, a single count (occasionally two) is observed in the multichannel scalers. We take this to be time $t = 0$ and always remove a single count from the measured spectra. Otherwise, the background is completely negligible. When the potential of the entrance endcap is switched on just 50 ns before 3 keV antiprotons arrive in the trap, when the magnetic field is off, or when the -3 keV on one of the electrodes is adiabatically turned off and then back on during a 100s trapping time to release trapped antiprotons, no counts are observed.

The potential on the exit endcap is lowered quickly compared to the transit time of particles in the trap in order to maximize the detection efficiency. Even a small number of trapped particles can be observed above possible background rates in the $6\mu s$ window. For trapping times shorter than 100s, however, we actually released so many trapped antiprotons that our detection channel is severely saturated. For a 1 ms trapping time, we conservatively establish that more than 300 antiprotons are trapped out of a burst of 10^8, which corresponds to trapping 3×10^{-6} of the antiprotons incident at 21.3 MeV and 3% of the antiprotons slowed below 3 keV in the degrader. We observe that 5 particles remain in the trap after 10 minutes. This is actually based upon only two trials (since we were reluctant to use up our short time at LEAR holding antiprotons for long times), but both of these trials used a burst of antiprotons from LEAR of comparable intensity to that used for the 41 trapped particles of the 100s spectra in Fig. 2. If a simple exponential decay describes the number of particles trapped between 100 a and 10 minutes, the decay time is 240 seconds. An extrapolation back to the loading time $t = 0$, however, would then indicate that only 62 particles are initially trapped. We clearly observe many more for a trapping time of 1 ms, suggesting that antiprotons are lost more rapidly at earlier times.

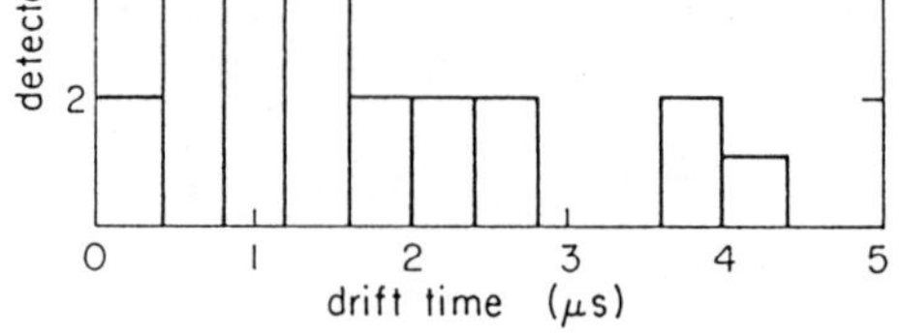

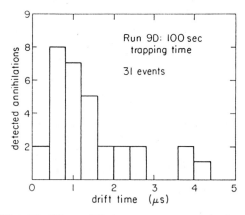

Fig. 13. Time of flight spectra, started when antiprotons were released from the trap.

A key point here is that the rate of cooling and annihilation via collisions with background gas will decrease with decreasing pressure. The background pressure can be made lower by orders of magnitude compared to the present vacuum by cooling a completely sealed vacuum enclosure to 4.2K. We thus expect a very significant increase in achievable trapping times.

E. Cooling Particles Within the Trap

If, as we suspect, we will soon be able to hold antiprotons in a trap for times much longer than 10 minutes, we feel rather confident that we will be able to cool them from approximately 1 keV down to of order meV. When the background pressure is greatly reduced, electron cooling seems to be the most promising method of cooling trapped antiprotons from keV to eV energies. In fact, electrons are probably already confined in the present trap, under the assumption that each antiproton emerging from the degrader liberates several electrons and many of them are trapped. A 1 keV antiproton traveling through a cloud of 1 eV electrons with density of $10^8/cm^3$ loses energy exponentially with a time constant of 1 s or less, which is much shorter than the time antiprotons were held. Although such a calculation of electron cooling rates within a trap was only done recently,[32] and the possibility of spatial separation of trapped electrons and antiprotons must be investigated,[33] such cooling is quite well understood both experimentally and theoretically insofar as cold electron beams have often been used to cool various particle beams traveling along the same axis with the same velocity.[34]

We presently prefer electron cooling as a first step because no resonant frequencies are involved and it thus promises to be the quickest cooling scheme. Once the amplitude of the oscillation along the magnetic field line is sufficiently reduced, the oscillation frequency of this oscillation will become increasingly independent of amplitude. It then should be possible to couple to a resistor such as that used to detect induced currents using a tuned circuit to cancel out the trap capacitance on resonance. The induced current dissipates power in the resistor. This removes energy from the axial oscillation, cooling the axial motion to the temperature of the resistor. Resistor cooling is a well established technique which has been used for many years.

F. Antihydrogen Production

Several lectures at this school are devoted to the production of antihydrogen. The method discussed involves colinear, velocity matched beams of antiprotons and positrons within LEAR. While I would prefer to wait until we have completed a more detailed study which is under way, I think it would be unfortunate if some thoughts about antihydrogen production in traps were not part of this school. If one could make antihydrogen in a particle trap, this would impact LEAR itself very much less than the merged beams approach. Many years ago, we mentioned the possibility of putting positrons and antiprotons into a radio frequency trap at the same time in

---

[32] W. Kells, G. Gabrielse and K. Helmerson, Fermilab-Conf.-84/68 E (1984).

[33] D.J. Larson, J.C. Berquist, J.J. Bollinger, W.M. Itano and D.J. Wineland, Phys. Rev. Lett. **57** 70 (1986)

[34] F.T. Cole and E.E. Mills, Ann. Rev. Nucl. Sci. **31**, 295 (1981)

order to make antihydrogen.[35] At that time, there was no promising proposal for getting antiprotons in a trap, but this of course is no longer the case. More recently, we mentioned the possibility which we have been studying for some time, of using a nested pair of Penning traps to simultaneously capture positrons and antiprotons.[36]

To provide a concrete starting estimate, consider a cloud of positrons of volume density n in thermal equilibrium at 4.2 K. For simplicity, we will assume the positrons are each moving with the average speed $\bar{v} = 10^6 cm/sec$. If we now simultaneously fill the same volume with N antiprotons, with energies below 1 ev, the relative velocity between antiprotons and positrons is the positron velocity $\bar{v}$, since antiprotons are much heavier. The antiproton production rate is thus approximately given by

$$R \approx Nn\sigma\bar{v}. \tag{10}$$

The cross section for radiative recombination to any antihydrogen bound state is well known[37] so that

$$\sigma \approx \frac{1}{10}\sigma_B \approx 10^{-17} cm^2 \tag{11}$$

for $\bar{v} = 10^6 cm/sec$, where σ_B is the Bohr cross section. Suppose we take $N = 10^4$ which is larger than we have already achieved by a factor of 10 but looks to be a realistic expectation. We further assume that $n = 10^8$ positrons/cm^3 can be placed in a trap at in thermal equilibrium at 4.2K. Together this gives a production rate

$$R \approx 10/sec. \tag{12}$$

This rate could possibly be made larger by stimulation with a laser as suggested for the merged beams experiment,[38] perhaps by a factor of 10, depending upon how small the interaction region can be made. Our favorite scheme is to stimulate to $n = 3$ or higher, perhaps even with a diode laser. However, the effective rate is actually lower owing to the duty cycle involved in loading a trap with antiprotons from LEAR.

Such low rates might be sufficient to observe antihydrogen for the first time. When antihydrogen is formed in an ion trap, the neutral atoms will no longer be confined and will thus quickly strike the trap electrodes. Resulting annihilations of the positron and antiprotons could be monitored. However, it is very clear that the very low rate will make further experiments with antihydrogen to be very different than experiments with copious amounts of hydrogen, and much more difficult.

For me, the most attractive way around this difficulty would be to capture the antihydrogen in a neutral particle trap such as has been used for neutrons[39] and neutral atoms[40,41] The objective would be to then study the properties of a small number of

[35] H. Dehmelt, R.S. Van Dyck, Jr., P.B. Schwinberg and G. Gabrielse, Bull. Am. Phys. Soc. **24**, 757 (1979).

[36] G.Gabrielse, K. Helmerson, R. Tjoelker, X. Fei, T. Trainor, W. Kells, H. Kalinowsky, in Proceedings of the First Workshop on Antimatter Physics at Low Energy, edited by B.E. Bonner and L.S. Pinsky, April 1986, Fermilab.

[37] H. Bethe, E. Saltpeter, *Quantum Mechanics of One and Two Electron Atoms*, in Handbuch fur Physik, **35**, 88 (Springer, Springer, 1957).

[38] R. Neumann, H. Poth, A. Winnacker, A. Wolf, Z. Phys.**313**, 253 (1983).

[39] K.J. Kugler, W. Paul and U. Trinks, Phys. Lett. **72B**, 422 (1978).

[40] A.L. Migdall, J.V. Prodan, W.D. Phillips, T.H. Bergeman, H.J. Metcalf, Phys. Rev. Lett. **54**, 2596 (1985).

[41] S. Chu, J.E. Bjorkholm, A. Ashkin and A. Cable, Phys. Rev. Lett. **57**, 314 (1986).

atoms confined in the neutral trap for a long time. To capture an antihydrogen atom directly into a neutral trap would require a neutral trapping well depth of order 4.2 K or 3×10^{-4} eV. This unfortunately is many orders of magnitude deeper than what is being realized. Thus, laser cooling and optical molasses techniques would be required, with Ly α lasers. The technologies which would need to converge in order to permit the study of antihydrogen in a neutral trap is rather imposing, but may be possible. It would also be necessary to minimize the rather strong interactions of the atom and the neutral trap in order cto meaningfully study the properties of antihydrogen.

Since I am speculating in this section, let me make several comments relevant to the discussions about gravity which are part of this school. I find the possibility of measuring the acceleration due to gravity for a antiproton to be very appealing, if it can be done. Another group is presently endeavoring to demonstrate that such a measurement can be done with charged particles.[42] We are investigating a different approach, to see whether such a measurement could possibly be done instead with neutral antihydrogen, in order to reduce the extreme sensitivity to stray charges.

Finally, to avoid the small cross sections involved in radiative capture, it has been proposed to send positronium into a cloud of trapped antiprotons.[43] The cross section is several orders of magnitude larger, but because the positronium can not be confined in the same volume as the antiprotons for a long time, the rate seems to be lower than considered here. There are nonetheless attractive features to this approach and we are studying it further.

G. Acknowledgements

I am grateful for the help of my associates L. Haarsma and S.L. Rolston and to my collaborator W. Kells for useful conversations about the speculations in the last section. My research group is supported by the National Science Foundation, the National Bureau of Standards (Precision Measurements Grant) and by the Air Force Office of Scientific Research.

---

[42] D. Holtkamp, paper contributed to this school.

[43] B.I. Deutch, A.S. Jensen, A. Miranda and G.C. Oades, in Proceedings of the First Workshop on Antimatter Physics at Low Energy, edited by B.E. Bonner and L.S. Pinsky, April 1986, Fermilab.

PRESENT STATUS OF THE RADIOFREQUENCY MASS SPECTROMETER FOR THE COMPARISON OF THE PROTON-ANTIPROTON MASSES (EXPERIMENT PS 189)

Presented by Catherine Thibault

Alain Coc[1], Roger Fergeau[1], Ernst Haebel[2], Heiner Herr[2], Robert Klapisch[2], Gérard Lebée[2], Renaud Le Gac[1], Guido Petrucci[2], Michel de Saint Simon[1], Giorgio Stefanini[2], Catherine Thibault[1] and François Touchard[1]

CSNSM Orsay[1] - CERN[2] Collaboration

INTRODUCTION

The principle and the main parameters of the radiofrequency mass spectrometer for the experiment PS 189 (Fig. 1) have already been discussed and described in detail[1,2,3]. We shall here only recall them very briefly and give the present status of the preparation of the experiment.

The ultimate goal of PS 189 is to compare the masses of p and $\bar{p}$ within an accuracy of 10^{-9} in order to improve by $\sim$ 4 orders of magnitude the stringency of the tests of the CPT conservation for baryons. This will be achieved by comparing the cyclotron frequencies f of H^- and $\bar{p}$ ions rotating in the same magnetic field. They are related to the charge q and mass m of the ions by :

$$f_{H^-} / f_{\bar{p}} = q_{H^-} m_{\bar{p}} / m_{H^-} q_{\bar{p}}$$

In order to measure the ratio of the cyclotron frequencies, the H^- and $\bar{p}$ ions will be alternately injected in a very homogeneous and stable magnetic field. They will then pass twice through a RF modulator so that they can only go across the exit slit of the apparatus when the two modulations have opposite phases. This condition is fulfilled for all ions when the modulation frequency is (n + 1/2) times the cyclotron frequency (n being integer).

(n being integer).

The spectrometer has been designed in order to get a resolving power of 10^6. The required momentum is very low : 20 MeV/c. However, it is now foreseen that LEAR itself will be able to get the $\bar{p}$ beam decelerated down to this momentum. The expected acceptance of the spectrometer is 10^4 times smaller than the emittance presently obtained at LEAR for more than 100 MeV/c. But the emittance which will be achieved at 20 MeV/c will very probably be higher so that the final transmission is not actually known.

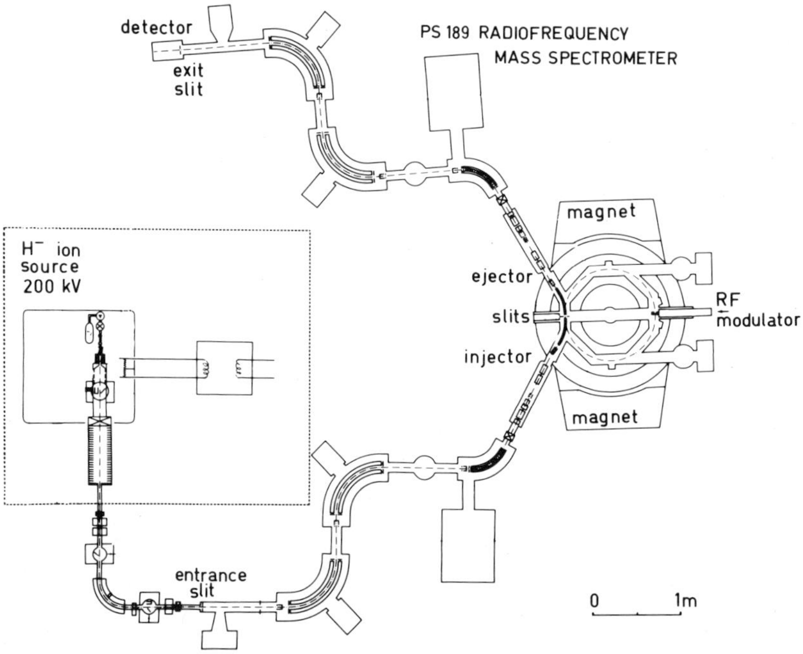

Fig. 1 Scheme of the spectrometer as it is presently for tests with H^- ions in Orsay. The 200 keV (20 MeV/c) H^- are produced by a duoplasmatron source. They are first focussed onto the entrance slit and then guided by an electrostatic line (4 sectors : 2 spherical ,1 cylindrical, and the "injector" which is cylindrical), the purposes of which are to allow a stringent energy selection and to inject the ions inside the magnet without perturbing its homogeneity. The ions are then rotating twice inside the magnet according to an helicoidal path before they are ejected, guided, analysed in energy and refocussed onto the exit slit by an electrostatic line identical to the injecting one. They are then detected when the frequency of the modulator, which acts after the first and third half-turns inside the magnet, is tuned at resonance.

STATUS OF THE MASS SPECTROMETER IN OCTOBER 1986

Magnetic field

The required homogeneity, and the control of gradients in all three directions with an accuracy better than 10^{-6}/cm have been obtained. The stability is better than 10^{-7}/0.3 sec but cannot be further tested without the use of the ions themselves. The isochronism inside the magnet has been studied by a Monte Carlo simulation using the measured field parameters. It has been found to have a FWHM of some 10^{-7} in good agreement with our expectation[3].

Vacuum

A residual pressure of some 10^{-9} mbar has been obtained in the different parts of the apparatus, allowing the compatibility with the LEAR vacuum.

Radiofrequency modulator

The prototype is ready for tests in Orsay. Four new adjustable modulators are in preparation. The radiofrequency synthezizer has been delivered and is presently tested at CERN.

Fig. 2 Overview of the spectrometer in Orsay. The different parts may be identified by comparison with the scheme in Fig. 1.

Beam optics

As can be seen in Fig. 2, all mechanical parts have been assembled in Orsay. The H^- ion source, the injection line and the vacuum chamber inside the magnet are exactly aligned.

The electrostatic optical elements have been adjusted in order to lead the 200 keV H^- ion beam through the injection line. Their setting is very stable and reproducible. The beam is rotating inside the magnet according to the designed trajectory. The $\Delta p/p$ selection agrees with our expectation.

Schedule

We now foresee to complete the beam optics adjustment around Christmas 1986. The whole year 1987 will be devoted to tests using the RF modulation : study of the resolving power, limits of reproducibility and accuracy of the measurements, study of various possible sources of systematic errors (influence of a mismatching of the energy, influence of the harmonic number n,...).

We plan to move the spectrometer from Orsay to CERN at the end of 1987 or very early 1988, and will be able to start its alignment and setting up immediately after its transport.

REFERENCES

1. G. Audi, A. Coc, M. Epherre, R. Fergeau, P. Guimbal, M. de Saint Simon, C. Thibault, F. Touchard, E. Haebel, H. Herr, R. Klapisch, G. Lebée, G. Petrucci, and G. Stefanini, Proposal for high precision mass measurements with a radiofrequency mass spectrometer, CERN/PSCC/81-84 (1981)
2. A. Coc, R. Fergeau, C. Thibault, M. de Saint Simon, F. Touchard, E. Haebel, H. Herr, R. Klapisch, G. Lebée, G. Petrucci, and G. Stefanini, Present status of the radiofrequency mass spectrometer for PS 189, in Physics with antiprotons at LEAR in the ACOL era, U. Gastaldi, R. Klapisch, J. M. Richard, J. Tran Thanh Van ed., Editions Frontières, Gif sur Yvette (1985), p. 675
3. A. Coc, R. Fergeau, C. Thibault, M. de Saint Simon, F. Touchard, E. Haebel, H. Herr, R. Klapisch, G. Lebée, G. Petrucci, and G. Stefanini, Le spectromètre à radiofréquence pour l'expérience PS 189, Rapport LRB 85-01 (CSNSM Orsay)

PROPERTIES OF MATTER VERSUS THOSE OF ANTIMATTER AS A TEST OF CPT INVARIANCE

H. Poth

Kernforschungszentrum Karlsruhe
Institut für Kernphysik
Karlsruhe, Fed. Rep. Germany

The general principles of relativistic field theory require invariance under the combined transformation CPT. The simplest way of testing CPT invariance is to compare the mass, the lifetime, and the magnetic moment of a particle, and of its antiparticle. In this contribution, some prospects for the improvement of the limits[1] derived from various particle–antiparticle systems are presented, updating an earlier paper.[2]

ANTIPROTON MASS

Two experiments[3,4] that are under way at LEAR aim at the determination of the mass of the *free* antiproton with very high precision ($\leq 10^{-9}$), the achievement of which would be a big step forward. Previously the mass of the *bound* antiproton was deduced from X-ray energies of antiprotonic atoms. Using crystal spectrometers or Ge solid-state detectors, the line positions can be measured with an accuracy of better than 1 ppm. It was pointed out by Decker and Pilkuhn that, at this level, novel binding effects (hadronic Lamb shift) should show up.[5] Previously, these effects would have been buried under unknown contributions from residual electrons attached to the antiprotonic atoms. However, it now seems possible to produce 'naked' antiprotonic atoms where electrons are stripped off.[6] A measurement of the mass of the *bound* antiproton with high accuracy would therefore provide new information in addition to CPT tests.

ANTIPROTON LIFETIME

The best limit for the lifetime of the antiproton comes from the measurement at ICE, which gave a value of 10^6 s.[7] As worked out previously,[2] an improvement up to the level of 10^{13} s should be possible.

MAGNETIC MOMENT OF THE ANTIPROTON

The magnetic moment of the antiproton was newly determined from the fine structure splitting of X-ray transitions in $\bar{p}$–Pb.[8] The accuracy was improved by a factor of > 2 and is now at 3×10^{-3}. Again one expects binding effects to alter the magnetic moment of the bound antiproton with respect to a free one.[5] These contributions may reach the level of 10^{-3}. Difficulties in the measurement of the fine structure splitting in antiprotonic atoms arise mainly from the unknown contribution of higher X-ray transitions hidden beneath the X-ray line under consideration. This problem can be overcome by influencing the atomic cascade in such a way that the de-excitation proceeds mainly through circular transitions (for instance by stripping off the electrons[6]). The measurement of the splitting with a solid-state detector should be possible to an ultimate precision of 0.5 eV, which would allow a determination of the magnetic moment to 4×10^{-4}, choosing a heavy target. With a crystal spectrometer one may even approach 5×10^{-5}.

The measurement of the hyperfine structure in antihydrogen has, in principle, an enormous potential for achieving high accuracy in the determination of the antiproton magnetic moment. For hydrogen, this quantity is one of the best measured values in physics. The precision to which it can be determined in antihydrogen depends very much on the production rate of antihydrogen and its velocity.

ANTINEUTRON MASS

The antineutron mass was deduced in a pre-LEAR bubble chamber experiment, with an accuracy of 42 ppm,[9] by analysing the kinematics of the charge-exchange reaction $p\bar{p} \rightarrow n\bar{n}$. This accuracy is based on 20,000 zero-prong reactions producing a sample of 59 events surviving various cuts. As stated by the authors,[9] the error is purely statistical. It is probably justified to assume that this accuracy can be further improved until it is comparable to that of the neutron (2.9 ppm) in a counter experiment at an antiproton storage ring.

ANTINEUTRON LIFETIME

The measurement of this lifetime has not yet been attempted as it requires very slow antineutrons. Low-energy antineutrons are created in the antiproton source of antiproton accumulators, and they can be produced in the charge-exchange reaction $\bar{p}p \rightarrow \bar{n}n$, where the momentum of the incoming antiproton is above 1 GeV/c. It is not inconceivable that antineutrons could be trapped in a magnetic storage device. In contrast to the neutron, decaying antineutrons can probably be more easily detected owing to the outgoing antiproton and its subsequent annihilation. This could allow the antineutron lifetime to be determined directly from the exponential decay. One would thus be free from normalization problems. An accuracy of at least 1% should be achievable.

MAGNETIC MOMENT OF THE ANTINEUTRON

The magnetic moment of the neutron is known with an accuracy of 0.28 ppm.[1] There is no corresponding experimental information for the antineutron. Polarized antineutrons can possibly be produced through polarization transfer in the charge-exchange reaction, as the theory predicts large polarization effects. Given enough analysing power for antineutron-induced reactions, the magnetic moment of the antineutron can be determined by measuring the spin precession of polarized antineutrons in a magnetic field. Here again the achievable precision depends very much on the intensity of the polarized antineutrons.

LEVEL ENERGIES IN ANTIHYDROGEN

The hydrogen and antihydrogen atoms form a conjugate system, but one which is distinguished from other particle–antiparticle systems by the fact that it is composed of elementary particles which are also existing in free space. So any measurement of the properties of antihydrogen is intimately related to the measurement of the properties of antiprotons and positrons, and whilst some of their properties are very well determined (e.g. the positron magnetic moment), others could be much improved. Therefore the measurement of a few level energies in antihydrogen, if done precisely, would provide a great deal of information on the validity of particle–antiparticle symmetries. If there would be any CPT violation at all observed in the electron–positron or the proton–antiproton system, it should also show up in hydrogen–antihydrogen. The spectroscopy of antihydrogen is discussed in detail at this Workshop by R. Neumann.

CONCLUSIONS

The availability of intense antiproton beams provides excellent possibilities for measuring the properties of antiparticle systems very precisely. So far, these possibilities have been exploited only partially, namely for the determination of the magnetic moment and the mass of the *bound* antiproton. In the immediate future we expect to make important progress by measuring the mass of the *free* antiproton. However, it would also be desirable to improve the measurement of the mass and the magnetic moment for *bound* antiprotons, in order to detect binding effects. Apart from this, even moderately precise measurements of the magnetic moment and the lifetime of the antineutron would be of interest as there exist no experimental data. Antihydrogen is a composed particle–antiparticle system which carries all the information on the antiproton and positron. The study of this system is certainly rewarding.

Apart from the experimental possibilities for measuring the properties of antiparticles very precisely, and consequently for testing invariance, it is somewhat disturbing that we have no satisfactory theoretical interconnection between the various limits for the CPT invariance tests.

REFERENCES

1. M. Aguilar-Benitez et al., Phys. Lett. 170B:1 (1986).
2. H. Poth, Proc. 1st Workshop on Physics with Cooled Low-Energy Antiprotons at LEAR, Karlsruhe, 1979, ed. H. Poth (KfK 2836, Karlsruhe, 1979), p. 129.
3. C. Thibault, these Proceedings.
4. G. Gabrielse, these Proceedings.
5. R. Decker and H. Pilkuhn, private communication.
6. L. Simons, these Proceedings.
7. M. Calvetti et al., Phys. Lett. 86B:215 (1979).
8. A. Kreissl et al., Paper contributed to the 8th European Symposium on Nucleon–Antinucleon Interactions, Thessaloniki, 1986.
9. M. Cresti et al., Phys. Lett. 177B:206 (1986).

STOCHASTIC COOLING IN A TRAP

G. Torelli

University of Pisa

I-56100 Pisa, Italy

The aims of my talk is to stress the analogy of the particles motion in a ring and in a trap at least on the point of view of a pick-up and a kicker, and to use this analogy to deduce the cooling rate in a trap on the basis of the similar calculation of the stochastic cooling in a ring.

I assume that the elementary theory of the stochastic cooling is known and I will base my approach on the very clear lectures given by D. Möhl at the CERN Accelerator School 1983.

The stochastic cooling technique is very simple in principle: we measure at the pick-up position the average value of some variable "x" on a sample of N_S particles, we compare this value to the average value of the same variable on all the N particles in the ring, we amplify the difference and we apply this signal to the kicker with a proper delay and phase to correct the measured deviation on the same sample (Fig.1a).

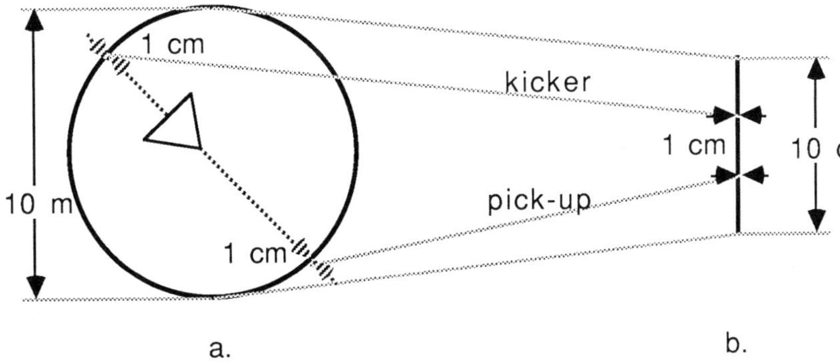

Fig.1

The cooling rate $1/t_x$ can be written:

$$\frac{1}{t_x} = \frac{W}{N} [2g(1- M'^{-2}) - g^2 (M+U)] \qquad (1)$$

where W is the band-width of the pick-up, amplifier and kicker system;
N is the total number of particles in the ring;
g is proportional to the amplifier gain;
U is the noise to signal power ratio;
M'^{-2} takes into account that the correction at the kicker is applied to a sample not identical to the sample that generates the signal at the pick-up, due to the mixing between the samples closed each other;
M is the number of periods needed for a complete rerandomization of the sample after a correction.

The coefficient W/N of formula (1) results from the fact that the effect of a single correction is proportional to $1/N_s$ and that the correction is applied ν_1 times per second. Obviously $N_s=N/n_s$ where n_s is the number of samples, n_s can be written as the inverse ratio between samples time length $T_s=1/2W$ and the revolution period of the particles in the ring $T=1/\nu_o$.

The final result is:

$$\frac{\nu_1}{N} = \frac{\nu_1 n_s}{N} = \frac{2W\nu_1}{N\nu_o} = 2\,\frac{W}{N}\,\frac{\nu_1}{\nu_o} \qquad (2)$$

and the factor 2 will disappear in reducing the cooling rate from quadratic $1/t^2$ to linear $1/t$.

In a ring one has $\nu_1 = \nu_o$ because a proper correction is applied at each revolution, in this case the proportionality coefficient assumes the particular value W/N.

Let me look at the trap now. When I squeeze a ring on a trap (see Fig 1b) two major change on the physical bounds to the particles motion take place: a reduction of a factor 100 or more on the length of the periodic path of the particles and an increase of the same order of the ratio between the transverse dimension and the path length, i.e. the diameter of the ring, let me say 10 m, becomes the length of the trap (L=10 cm) and the pick-up transverse dimension remains substantially the same in a ring and in a trap (d $\approx$ 1 cm).

We can talk of analogies between rings and traps if in the traps d/L>>1. Such a condition can be satisfied in multirings traps (Fig 2).

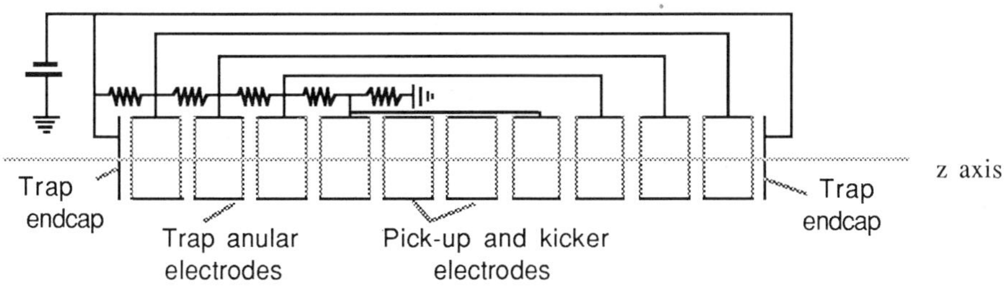

Fig.2.

Due to the changes in the relative dimensions it is impossible in a trap to screen the pick-up from the signal induced by the kicker. This means that pick-up and kicker must work at different time; during a half a period the pick-up measures a signal, the signal is amplified and send to the kicker with a delay of half a period when the pick-up is switched off.

In a more realistic description some half cycles (say n_1) can be required to reset the apparatus after the application of the kick signal. It can be also convenient to measure the pick-up signal for some half cycles (say n_2) to reduce the noise. In this case the correction can be applied only any $n = n_1 + n_2 + 1$ half-cycles and the total number of correction per second ν_1 will be given by:

$$\nu_1 = \frac{2\nu_o}{n} \qquad (3)$$

The third change, the transformation of a continuous unidirectional motion in an oscillation almost harmonic up and down in the same tube, can be compensated using for pick-up and kicker the same positions or symmetric positions with respect to the trap centre where the same particles are present one period or half a period later.

On the basis of this simplified analysis the proportionality coefficient (Form. 2) became:

$$\frac{W}{N}\,\frac{\nu_1}{\nu_o} = \frac{\Omega}{N} \quad ; \quad \frac{2\nu_o}{n\nu_o} = \frac{2W}{nN} \tag{4}$$

and the parameters M and U will obviously appear in the formula with proper coefficients:

$$M \longrightarrow \frac{M}{n} \quad ; \quad U \longrightarrow \frac{U}{n_2^{1/2}} \tag{5}$$

The final result is :

$$\frac{1}{t_x} = \frac{2W}{nN}\,[2g(1- M'^{-2}) - g^2 (\frac{M}{n} + \frac{U}{n_2^{1/2}})] \tag{6}$$

Special consideration needs the band-width W; due to the increase of the d/L ratio the limits in W are essentially due to the number n_s of samples that can be independently measured in a trap; a crude estimation can be obtained comparing the diameter of the pick-up d (of negligible longitudinal dimension) to the full amplitude of the particles oscillation, i.e. comparing the diameter of the trap to its length L:

$$n_s = L/d \quad \text{and} \quad W = n_s \nu_o / 2$$

The parameters M and M' are directly connected to the anharmonicity of the motion in the trap.

A direct derivation of the cooling rate in a Penning trap of standard design is in progress at Genova.

THE CYCLOTRON TRAP AS DECELERATOR AND ION SOURCE

L. M. Simons

SIN, CH-5234 Villigen
Switzerland

INTRODUCTION

It is the aim of this contribution to explain in some detail the underlying physical principles which produce narrow stop distribution of antiprotons in the cyclotron trap. The limiting factors of this method will be considered as well as the requirements for beam quality needed to work under optimum conditions. Moreover the possibility to extract low energy antiprotons from the center of the cyclotron trap will be treated.

DECELERATION IN A CYCLOTRON TRAP

The basic principle of the cyclotron trap is to wind up the paths of decelerating particles in a magnetic cyclotron field. The energy loss due to the deceleration in the target gas together with the focusing properties of the cyclotron field will then yield a concentrated stop distribution. A first insight into the physical principles governing the motion of decelerating particles in a cyclotron field is provided by the quasipotential picture[1]. For a rotational symmetric magnetic field (no electric field, no energy loss) the motion of particles is determined by a quasipotential U(r,z). r,z are cylindrical coordinates. U(r,z) is given by

$$U = 1/2\ m\ (P/r - e/c\ \ 1/r \int_0^r B_z\ (r,z)\ dr\ )^2 .$$

P is a constant of motion, the generalized angular momentum, which depends on initial conditions. The equations of motion are given by

$$m\ddot{r} = -dU(r,z)/dr$$

$$m\ddot{z} = -dU(r,z)/dz.$$

With the help of extremum principles and expanding B_z to first order in the Taylor series, the conditions for focusing in both axial and radial direction requires

$$0 < n < 1$$

where $n = -dB/dr \cdot r/B$ is the field index.

For the field of our cyclotron trap[2] the quasipotential curves in the radial direction in the median plane are shown for decreasing angular momenta. They correspond to decreasing energy of the particles in the field. Below a certain generalized momentum a potential minimum develops for which particles can be trapped. This corresponds to particle momenta of $p < 123$ MeV/c $[p_r = 0,\ r < 0.143\ \mathrm{m}]$.(Fig. 1).

Particles loosing energy at a rate slow compared to the cyclotron frequency will follow the development of the quasipotential. The wall becomes deeper and narrower with decreasing energy and the particles will be guided to the center. When the particles loose energy too rapidly a concentrated stop distribution will not result. In the worst case, eccentric orbits are excited which do not include the geometrical center of the magnetic field. These orbits are characterized by negative angular momenta (Fig. 2). In (Fig. 3) quasipotentials are shown in the axial direction for different equilibrium orbits. Potential minima exist which, contrary to the radial case, become shallower with decreasing equilibrium radius. Nevertheless, a loss of particles is avoided because the axial momenta will also decrease correspondingly due to the overall momentum

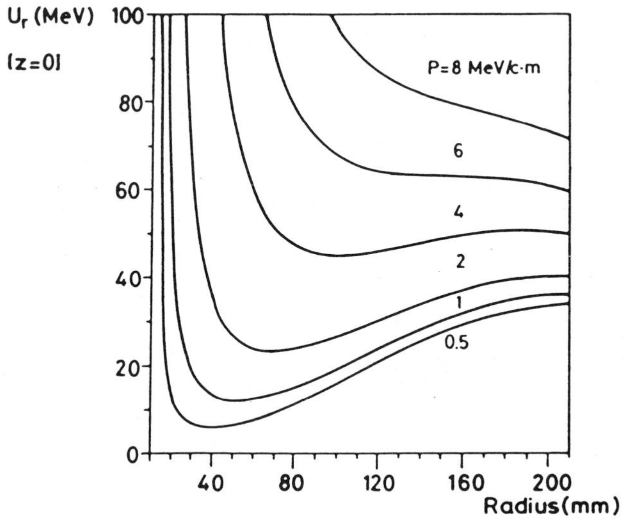

Fig. 1:
Quasipotential in radial direction in the median plane for different positive generalized angular momenta

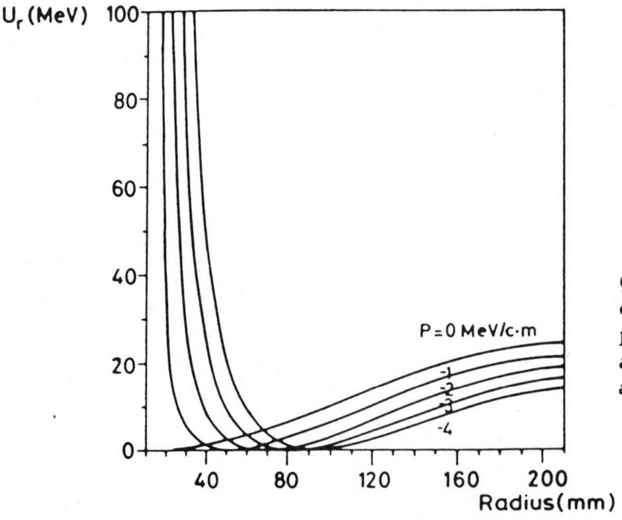

Fig. 2:
Quasipotential in radial direction in the median plane for generalized angular momenta with values at and below zero.

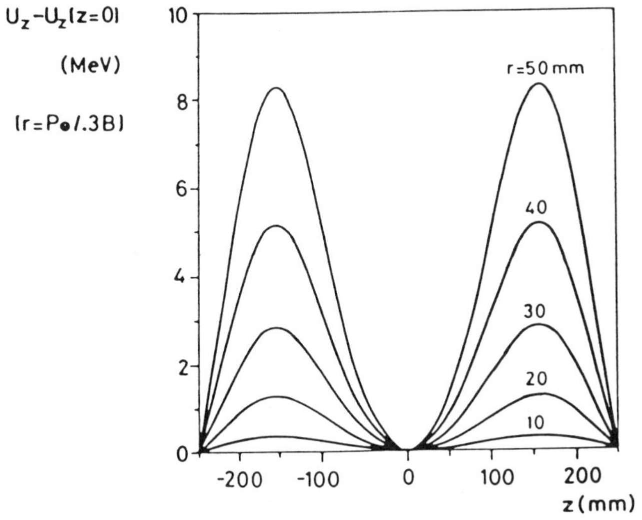

Fig. 3:
Quasipotential in axial direction at the radius r of equilibrium orbits.

loss. As long as the initial amplitudes of oscillation in the z-direction are kept below a particular limit, particles cannot leave the potential well.

Compared to the quasipotential method, the extended Liouville Theorem[3] provides a more quantitative description of the development of radial and axial amplitudes of betatron oscillation. It is used to trace the momentum spread Δp of decelerating particles, which, in turn determines the development of the radial spread Δr_p.

$$\Delta r_p = r \cdot \frac{\Delta p}{p} \frac{1}{1-n}$$

The increase in Δr_p for decelerating particles is counteracted for sufficiently slow deceleration by the shrinking of the amplitudes for betatron oscillations. In this case the conservation of the action integral leads to changes in amplitudes A_r, A_z for radial and axial betatron oscillations:

$$A_r \propto \frac{1}{\sqrt{B_z}} \frac{1}{\sqrt{1-n}}$$

$$A_z \propto \frac{1}{\sqrt{B_z}} \frac{1}{\sqrt{n}}$$

The principle result of this application of the Lioville theorem is that the radial spread of the beam at the beginning of the deceleration process gives the radial extension of the stop distribution. The axial extension of the stop distribution, however, is almost a factor of two of the initial axial width of the beam.

PRACTICAL ASPECTS OF INJECTION AND DECELERATION

For the injection of particles into the trap, it is necessary to decelerate them from the beam momentum to the accepted momentum p_{in}. The deceleration is produced by a moderator sitting in the median plane of the field at a radius $r \approx 110$ mm. Two main problems arise:

1. The transport of the beam through the fringe field to the defining scintillator of a certain radial width Δr should be made without loss of particles.

2. The particles should not hit the moderator-scintillator arrangement in the first few orbits after the injection.

Beam losses would not be a problem if the magnetic field could be shielded up to the moderator. This could be done for beam momenta less than 100 MeV/c. For higher momenta such shielding cannot be done effectively. A defining scintillator of Δr = 10 mm at r = 120 mm requires the accepted particles to lie inside the boundaries of an acceptance band which is shown in Fig. 4. Here the acceptance band of 202 MeV/c particles is plotted for the vertical plane together with the emittance ellipse of the incoming $\bar{p}$-beam at LEAR. Due to scheduling problems in setting up the cyclotron trap in the experimental hall, a better matching could not be done. For perfect matching a radial width of the defining scintillator of only 3 mm is feasible.

In order to minimize the number of particles striking the moderator during the first few orbits, the radial spread for the injected particles should be as small as possible. The deceleration from 202 MeV/c down to 105 MeV/c causes the momentum uncertainty of the LEAR beam to blow from 10^{-3} to $2.5 \cdot 10^{-2}$. Therefore the radial width of the injected beam ($r \sim$ 100 mm, $n \sim 0.8$) is about 13 mm. For a beam momentum of 100 MeV/c which would be decelerated down to about 80 MeV/c the resulting value for Δr would be 2.5 mm. With such a well defined initial radial distribution, excitation of axial betatron oscillations together with radial betatron oscillations makes it possible to avoid hitting the moderator even for very low target gas pressures. Antiprotons in hydrogen or helium could be stopped at pressures well below 1 mbar.

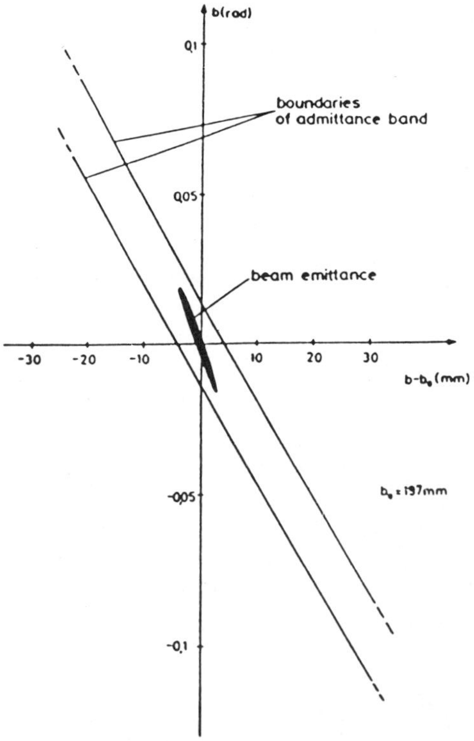

Fig. 4: Acceptance band for 202 MeV/c antiprotons with the emittance ellipse of the LEAR beam. b is the vertical distance from the center of the cyclotron trap (impact parameter).

EXTRACTION

In order to pursue an earlier suggestion[4] to extract antiprotons from the center of the cyclotron trap, we can use the results of a recent SIN experiment which studied the stopping process of negatively charged particles in hydrogen. The experiment provided a measurement of the time T for decelerating muons in hydrogen from velocities of 1/137 c (where c is the velocity of light) to their velocity at atomic capture. The result can be approximated by the formula

$$T\,[\mathrm{ns}] = 200/P\,[\mathrm{mbar}]. \qquad (P\text{: pressure})$$

For the deceleration in He, the time may be a factor of 2 larger[5]. The measured deceleration times support recent theories describing the moderation and captures process of negatively charged elementary particles in hydrogen[6]. These theories predict that the time T of antiprotons in 0.1 mbar H_2 is 2 μs. The mean energy loss during that time would be 9 eV/cm. The radial and axial distribution for these antiprotons is $\Delta r < 10$ mm, $\Delta z < 15$ mm. For axial extraction, a pulsed electrical field with a spill length of 50 ns can be used. The field could be produced by a grid structure. For illustration an electrical field of 500 V/cm and a time between the pulses of 100 ns is assumed. The magnetic field of the cyclotron trap forms a magnetic mirror with B = 43 kG in the median plane (z = 0) at r = 0 and maximum field of 57.5 kG at r = 0, z = 160 mm. Therefore only particles with

$$p_z/\sqrt{p^2-p_z^2} > 0.5$$

can leave the trap. p_z, p: axial, total momentum. Without electrical field p_z is much smaller than p. The field strength can therefore be choosen in a way that only antiprotons below a certain momentum can be extracted. For 500 V/cm only momenta below 1 MeV/c (≙ 0.6 keV) are affected. They will be accelerated to 4.3 MeV/c (10 keV) and can then be transported through a thin carbon foil to a high vaccum region. Then a guiding field of 1 kG is sufficient to guide them further. A momentum spread of 30 % for 4.3 MeV/c is expected. This corresponds to an energy spread of 1.4 keV at 10 keV. The radial spread Δr would be about 10 mm and $\Delta r'$ about 23 mrad.

Hence an efficient conversion of a 5 MeV d.c. beam into a 10 keV pulsed beam could be made. Such a beam would be a natural source for the phase space compressing device as proposed by D. Taqqu[7]. It would also permit construction of cheap beam transport systems in order e.g. to fulfill the needs of several experiments requiring low energy antiprotons.

ACKNOWLEDGMENT

Fruitful discussions with F. Kottmann and K. Elsener are gratefully acknowledged. I thank J. Missimer for carefully reading the manuscript.

REFERENCES

1. A. A. Kolomensky, A. N. Lebedev, Theory of cyclic accelerators North-Holland, Amsterdam 1966.
2. R. Bacher et al., The cyclotron trap: A device to produce high stop densities, in preparation.
3. A. J. Lichtenberg, Phase Space Dynamics of Particles, J. Wiley, New York, 1969, p. 30 ff.

4. P. Blüm, L. M. Simons, Physics at LEAR with Low-Energy Cooled Anti-protons, 1982, eds. U. Gastaldi and R. Klapisch (Plenum Press, New York and London) p. 819.
5. H. Anderhub et al., P.L. 101B, (1981) 151.
F. Kottmann, private communication.
6. J. S. Cohen, P.R. A27 (1983) 167.
7. D. Taqqu, Proc. 3rd LEAR Workshop (1985).
eds. U. Gastald, R. Klapisch, J. M. Richard and J. Tran Thanh Vam (Editions Frontières, Gif sur Yvette, France, 1985).

POSSIBLE EXPERIMENTS WITH ANTIHYDROGEN

Reinhard Neumann

Physikalisches Institut der Universität Heidelberg
Philosophenweg 12, D-6900 Heidelberg, Federal Republic of Germany

1. INTRODUCTION

1.1 Brief Chronology of the Idea

Antihydrogen is the antimatter counterpart of hydrogen and therefore consists of an antiproton and a positron. Till now, no antihydrogen has been produced and identified at any facility in the world. Thus, formation and detection of one antihydrogen atom would be the first artificial realization of atomic neutral antimatter.

During the last twenty years many authors dealt with antihydrogen. A first calculation on hydrogen-antihydrogen interaction was performed in 1968 (G. Steigman, Ph.D. dissertation, Columbia University, unpublished). Extensive $H\bar{H}$ interaction calculations by various authors followed (Morgan and Hughes, 1970 and 1973, Junker and Bardsley, 1972; Kolos et al., 1975; Câmpeanu and Beu, 1983). Antihydrogen formation in an antiproton storage ring by spontaneous radiative antiproton-positron recombination was discussed (Budker and Skrinskii, 1978) at the time when experimental data on spontaneous radiative proton-electron recombination were obtained in Novosibirsk at a small proton storage ring equipped with an electron cooling device. Other considerations dealt with antihydrogen creation in a radio frequency antiproton trap (Torelli, 1980). A first concept for $\bar{H}$ production at LEAR of CERN was worked out (Herr et al., 1984), and the idea of laser-enhanced $\bar{H}$ formation was introduced (Neumann et al., 1983). A CERN-Heidelberg-Karlsruhe Collaboration submitted a proposal on a "Feasibility study for antihydrogen production at LEAR" to the CERN PSCC (Berger et al., 1985). Further recent suggestions concern antiproton polarization via optical pumping of antihydrogen atoms with laser light (Imai, 1985), through a Lamb shift spin filter (Poth, 1986 b, and at this workshop; Wolf, 1986), $\bar{H}$ production in a $\bar{p}$ trap by e^+ Auger capture from positronium (Deutch et al., 1986), and $\bar{H}$ formation in a combination of nested Penning traps (Gabrielse et al., 1986a).

1.2 Reasons to Welcome or Reject Attempts to Form Antihydrogen

To my experience physicists confronted with the idea of antihydrogen formation react very differently: the scale of meanings and judgements varies from enthusiasm to sharp rejection. The reaction of enthusiastic physicists can be formulated like this: "Antihydrogen is a completely new system in the realm of atomic physics. One should by all means produce it, if one can. Every system should be tested if possible, even if one assumes not to find anything new". The judgement of those physicists who reject the idea to produce antihydrogen and study its features - in fact mainly physicists working in the field of elementary particle physics - is based on the postulate of absolute symmetry of particles and antiparticles: "From all what we know about matter - antimatter symmetries hydrogen and antihydrogen must behave absolutely identically. Therefore, it is not worthwhile to produce and study antihydrogen".

I frankly confess that I completely agree with the assumption of identical behaviour. But nevertheless my approach to the problem and my conclusions are different. Specific characteristics of various antiparticles have been and are still being tested though these characteristics are known very well for the respective particles, and should be identical for antimatter: thus, e.g. the g-factor and the mass of the positron, the mass of the antiproton, and the masses of other antiparticles were measured. One looks for antigravity though the structure of the gravitational tensor implies that matter and antimatter behave identically in the gravitational field. Even the CPT theorem is the subject of experimental investigations though it must be considered as the most fundamental basis on which the building of modern physics rests.

Why should a similar interest not be claimed in the case of antihydrogen? Physicists who are more cautious with their judgement ask e.g.: which features of antihydrogen could be measured, in order to test fundamental symmetries? Which accuracy could be reached? The hyperfine structure splitting frequency of the hydrogen ground state has been measured with an accuracy of better than 10^{-12}. The measurement of this splitting in antihydrogen should be of principal interest, the more since in hydrogen a fraction of $3{\cdot}10^{-5}$ is caused by the proton structure. There are less fundamental and less controversal but nevertheless interesting aspects of antihydrogen. It should be possible to produce polarized antiprotons via optical pumping of the antihydrogen ground state.This technique uses the coupling between electron spin and nuclear spin. Electron spin polarization produced by irradiation of circularly polarized light is transferred to the nuclear spin through hyperfine interaction. A rate of polarized antiprotons sufficient for high-energy scattering experiments should be of interest concerning polarization-dependent phenomena.

A further subject is antihydrogen - matter interactions. Several theoretical publications dealt with the hydrogen - antihydrogen interaction potential at very low relative kinetic energy and calculated the cross section for rearrangement annihilation. The cross section for stripping off the positron in thin foils or gas targets should be significantly different from that of hydrogen. A stripped-off positron cannot be replaced while hydrogen atoms exhibit an equilibrium of repeated ionization and electron capture. How is the scattering behaviour of antihydrogen atoms at surfaces as a function of kinetic energy, angle of incidence, kind of material, and surface structure?

Last not least the aim of antihydrogen production would certainly accelerate various technological developments like creation of intense positron beams.

2. PRODUCTION OF ANTIHYDROGEN ATOMS AND BEAM CHARACTERISTICS

2.1 Hydrogen Formation by Spontaneous Radiative Recombination

A specific effect known from electron cooling of stored protons gives detailed information how an experimental configuration for the production of antihydrogen atoms could look like: It is known from experiments at Novosibirsk, CERN, and Fermilab that protons stored in a ring can be cooled by electrons in order to improve the proton beam quality. For this purpose a merging electron beam overlaps the proton beam in a straight section. Protons and electrons propagate in the same direction with identical average velocity $\bar{v}$. In a frame travelling with velocity $\bar{v}$ with respect to the laboratory rest frame, protons and electrons have velocity spreads Δv_p and Δv_e, respectively. Their kinetic energy spreads $(m_p/2)\Delta v_p^2$ and $(m_e/2)\Delta v_e^2$ can be described by temperatures T_p and T_e defined by the equations

$$(m_p/2)\Delta v_p^2 = kT_p \quad \text{and} \quad (m_e/2)\,\Delta v_e^2 = kT_e.$$

Under ideal conditions protons and electrons exchange energy by repeated collisions until thermal equilibrium is reached, expressed by

$$(m_e/2)\Delta v_p^2 = (m_e/2)\Delta v_e^2\,, \quad \text{or} \quad \Delta v_p = \sqrt{m_e/m_p}\;\Delta v_e$$

The latter equation reflects that the average relative velocity between protons and electrons is dominated by Δv_e, and that the protons can be considered to be at rest in the electron gas. The electrons thus populate an energy band in the continuum right above the ionization limit of the hydrogen atom (see Fig.1).

Hydrogen atoms are formed through spontaneous radiative recombination of protons and electrons according to

$$p + e \rightarrow H + h\nu.$$

A photon of energy $h\nu$ carries away the kinetic energy $mv^2/2$ of the captured electron and the ionization energy E_o/n^2 of the electronic state of the created atom, i.e.

$$h\nu = mv^2/2 + E_o/n^2$$

must be fulfilled. Here E_o is the ionization energy of the hydrogen ground state and n the principal quantum number of the electronic state in which the electron is bound. The atoms leave the storage ring tangentially and can be used for diagnostics of the electron cooling process and the cooled beam properties. The total rate of hydrogen atom formation is (Bell and Bell, 1982)

$$R = n_e \alpha_r N_p \cdot \eta \cdot \gamma^{-2}$$

(n_e = electron density, N_P number of stored protons; α_r = spontaneous recombination coefficient; η = ratio of active electron beam length to ring circumference = 0.02 for LEAR ; $\gamma^2 = 1/(1-\beta^2)$ = relativistic factor). Now let us calculate the rate with proton beam and electron beam characteristics as they are realistic for the Low Energy Antiproton Ring LEAR (Lefèvre et al., 1980) and its electron cooling device (Hütten et al., 1984; Poth, 1985) at CERN:

- $N_P = 5 \cdot 10^{10}$ stored protons
- $n_e = 10^8$ electrons/cm^3
- $\eta = 0.02$
- $\alpha_r = 2.2 \cdot 10^{-12}$ cm^3 sec^{-1}
- $\beta = 0.3$ corresponding to an electron kinetic energy $E_{kin} = 26$ keV;

$$\Rightarrow R_{spon} \approx 2 \cdot 10^5 \text{ H atoms/sec.}$$

The conceivable rate of antihydrogen production will be estimated in comparison to the values available in the case of hydrogen. The production of a satisfactory antihydrogen rate will be a difficult task. In particular, the question arises whether a continuous or a pulsed beam would provide a higher rate. For this reason the concept of laser-induced radiative proton-electron recombination (Neumann et al., 1983; Neumann, 1985) will be described in the following which makes pulsed antihydrogen formation an especially attractive scheme.

2.2 Laser-Induced Radiative Recombination

Let us consider an intense laser beam overlapping the proton and electron beams at LEAR, but propagating in opposite direction (see Fig.2). If the laser photon energy fulfills the above mentioned equation of energy conservation in the proton-electron rest frame, e.g. for the principal quantum number n=2, laser light should stimulate radiative electron capture into the n=2 state. This effect is described by the equation

$$p + e + kh\nu \rightarrow H + (k+1)h\nu.$$

k being the number of incoming photons. From the equation for energy conservation immediately follows the relation

$$\Delta v = h\, \Delta\nu / mv.$$

Here, $\Delta\nu$ is the combined spectral width of the inducing light field and the bound final state; Δv represents the electron velocity interval $(v-\Delta v/2\ ,\ v+\Delta v/2)$ out of which the electrons are captured. This interval is a fraction of the total continuum energy band populated by the electrons. For a laser light intensity of 20 MW/cm^2, pulse length of 20 nsec, and laser wavelength resonant with the transition to the n=2 state of hydrogen, calculation gives an induced capture rate $r^{ind}(n=2)$ to n=2, 100 times as large as the total spontaneous capture rate r^{spon} to all states within the laser interaction time, thus providing a gain factor

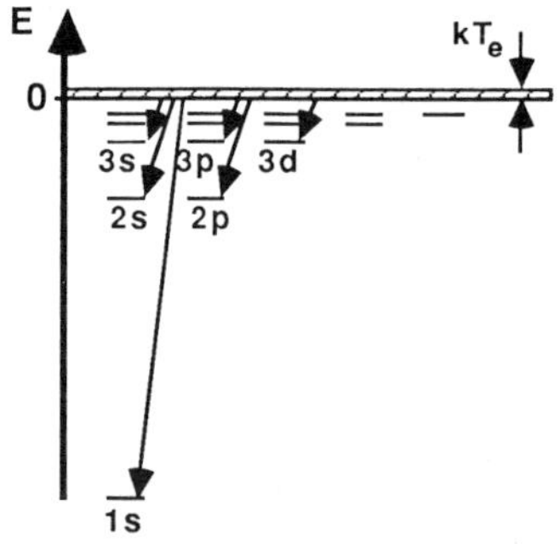

Fig. 1. Hydrogen level diagram with continuum electron band as seen in p-e$^-$ rest frame

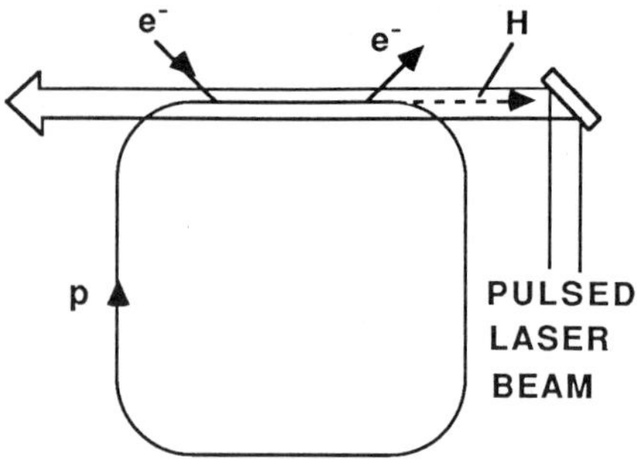

Fig. 2. Scheme for electron cooling of stored protons and laser-enhanced recombination

$$G = r^{ind}(n{=}2) \,/\, r^{spon} = 100\,.$$

The calculation is based on the assumption of an electron energy band width $kT_\perp = 0.2$ eV, $T_\perp$ being the temperature related with the transverse velocity spread of the electrons. A further increase of laser intensity would raise the probability of photo-reionization and thus diminish the gain factor.

Within the laser pulse length of 20 nsec occur on the average $4 \cdot 10^{-3}$ spontaneous recombination processes. A 20 MW/cm^2 laser pulse increases the rate to $4 \cdot 10^{-1}$ per 20 nsec. A laser with a repetition rate of 500 Hz then induces the production of

$$R_{ind} = 200 \text{ H atoms/sec.}$$

This is a small rate in comparison to the spontaneous recombination rate of $2 \cdot 10^5$ atoms/sec. But, one must keep in mind that the latter rate is obtained with a continuous electron beam current of 2.3 A. This indicates that laser-enhanced antiproton-positron recombination could be of interest with pulsed positron beams.

2.3 Production of Continuous Positron Beams

In the following, various schemes of positron beam production (Conti and Rich, 1985) will be considered with respect to the eventual aim of formation, investigation, and use of antihydrogen atoms. Positron beams with high intensities can be obtained either from strong radioactive sources or from e$^+$-e$^-$ pair production by means of a linear accelerator. In the second case accelerated electrons loose their kinetic energy in a target through bremsstrahlung, and photons convert into e$^+$-e$^-$ pairs. For both kinds of sources, the positrons leave the site of creation with broad kinetic energy spreads. However, the $\bar{p}\,e^+$ recombination probability varies drastically with the relative kinetic energy and reaches its maximum, when the relative energy approaches zero. This means that the positrons must be moderated to reach an energy spread of about 1 eV or less.

The best positron moderators are metallic single crystals, e.g. tungsten or copper. Positrons penetrating such a moderator are slowed down, and in most cases eventually annihilate. Only when a positron reaches thermal energy near the surface, it has a good chance to diffuse to the surface and leave the crystal again. This is possible since the work function of many metals is negative and of the order of -1eV for positrons. A fraction of up to $3 \cdot 10^{-3}$ of the incident positrons survives the moderation procedure under optimum conditions and can have an energy spread of 80 meV FWHM. In order to further improve the beam properties, the positrons are not yet immediately accelerated to their final energy but to only a few keV. One

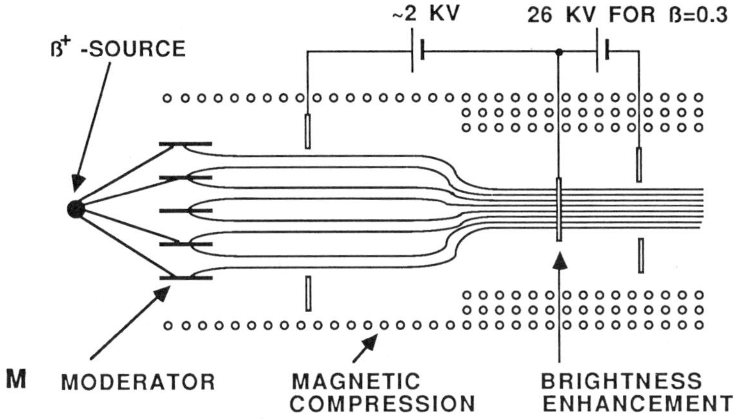

Fig. 3. Arrangement for positron moderation and brightness enhancement (after A. Wolf, private communication)

then focuses them as much as possible on the surface of a second metallic crystal, the remoderator (see Fig.3). The penetration depth of these low-energy positrons is much smaller than that of the unmoderated positrons, their probability of diffusing to the surface and being reejected is typically 25%, thereby decreasing by a factor of up to 20 in beam diameter. The remoderation process is called brightness enhancement (Mills, 1980). It is used to reduce the diameter of a positron beam without increasing its divergence substantially.

Using the present state of the art in moderated continuous positron beams from nuclear beta decay, which characteristics could such a beam have at best? Based on a 10 Ci ^{22}Na positron source, the following parameter values can be obtained (Rich, 1985):

- Rate of unmoderated positrons $3.7 \cdot 10^{11}$ e^+/sec
- moderated rate $3 \cdot 10^8$ e^+/sec
- transverse temperature $T_\perp = 80$ meV FWHM
- longitudinal temperature $T_{||} = 80$ meV FWHM

After acceleration of the moderated positrons to 26 keV one gets with one stage of brightness enhancement a current of $\sim 10^8$ moderated e^+/sec, beam diameter 1 mm, and beam divergence 0.5 mrad. It should be pointed out that the temperature of the moderator and remoderator essentially determines the transverse energy spread of the final positron beam, and that cooling will further reduce it.

2.4 Formation and Detection of Continuous $\bar{H}$ Beam

A calculation of the spontaneous radiative $\bar{H}$ formation rate with a continuous e^+ beam will be based on the following parameter values:

- kinetic energy of e^+ beam 26 keV (equivalent to $v/c = 0.3$)
- energy spread ≈ 0.1 eV
- positron beam current 10^8 e^+/sec after brightness enhancement
- e^+ beam diameter $= 1$ mm
- e^+ density $n_{e^+} \approx 1$ cm^{-3}

Assume $N_{\bar{p}} = 5 \cdot 10^{10}$ antiprotons stored in LEAR and an antiproton-positron overlap region of 1.5 m length equivalent to $\eta = 2\%$ of the ring circumference. With these numbers the antihydrogen formation rate would be

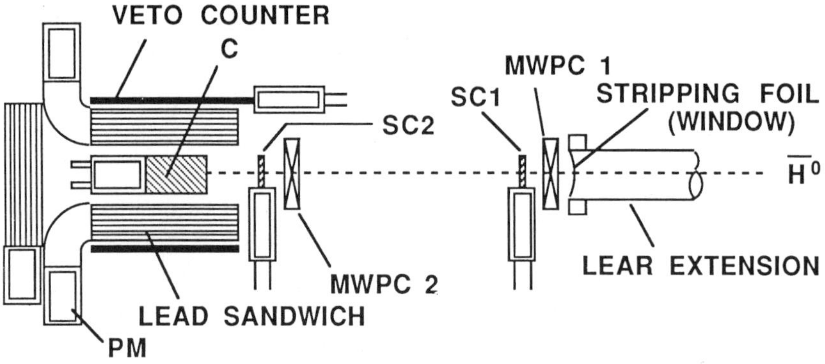

Fig. 4. Detection scheme for antihydrogen atoms (after H. Poth, 1986a)

$$R_{\bar{H}} \approx 7.2/h \approx 170/d\,.$$

Such a rate would be sufficient to demonstrate that antihydrogen formation is feasible with a reasonable investment of effort in the development of a 10 Ci ^{22}Na positron source, an e^+ beam production device, and their installation into a straight section of LEAR. It would certainly not be sufficient to perform laser and microwave precision spectroscopy in order to extract data of energy level splittings and shifts for comparison with hydrogen.

Detection of about 1 antihydrogen atom every 10 minutes with discrimination against spurious events in the detection system should be no major problem. For this purpose the following detector arrangement (see Fig.4) as worked out by H. Poth can be conceived (Poth, 1986 a). An antihydrogen atom created in the $\bar{p}\,e^+$ overlap region will not be deflected by the magnetic field of the next bending magnet but leave the storage ring tangentially. When passing a vacuum window used as stripper foil at the extension of the straight section, the atom looses its positron. The antiproton then traverses a sequence of multiwire proportional counters (MWPC) and scintillation counters (SC) in the order MWPC 1, SC 1, MWPC 2, SC 2. It is decelerated in SC 2, and enters a lead glass calorimeter C surrounded by a series of lead sandwich counters, and by veto counters. An antihydrogen atom will be detected by a characteristic signature, in particular:

- signal coincidence between SC1, SC2 and C
- deposition of right amounts of energy in SC2 (~ 45 MeV) and C (> 1 GeV)
- correct time of flight between SC1 and SC2
- a particle track through MWPC 1 and MWPC 2 corresponding to antihydrogen leaving the ring.

Background caused by cosmic rays, and antiproton-induced reactions in the window material and rest gas of the straight section, must be taken into consideration. An estimate based on $5 \cdot 10^{10}$ stored antiprotons in LEAR and a pressure of $2 \cdot 10^{-12}$ Torr leads to a $\bar{p}$ induced background rate of $8.5 \cdot 10^{-6}$ sec^{-1} which can be neglected. One cosmic ray event per hour with coincident signals in the counters SC1, SC2 and C can be expected. But the SC1 to SC2 time-of-flight condition makes a spurious identification very unlikely.

2.5 Production of Pulsed Positron Beams

As was already indicated by the concept of laser-enhanced antihydrogen formation which is only conceivable in a pulsed mode because of the needed light intensity, pulsed positron beams with a particle flux comparable to continuous beams would be of great advantage. Pulsed signals provide the possibility to eliminate background considerably with gating techniques.

Moderated positrons from a radioactive source have been trapped in an electromagnetic well and then re-expelled with a parabolic potential, thereby forming 8 ns long positron bunches (Mills, 1984). The bunches produced with a repetition rate of 1 kHz had an energy spread of 2 keV, too large for $\bar{H}$ formation. But the experiment demonstrated that a continuous beam can be transformed in a pulsed beam. Its repetition rate could be matched to that of a laser. Combining the bunching device having a conceivable efficiency of 30% with a remoderation stage with an additional loss of a factor of 3, the overall efficiency could be of the order of 10%.Thus, a procedure of positron trapping, re-ejection, and remoderation, starting with $3 \cdot 10^8$ moderated positrons per second from a 10 Ci ^{22}Na source could yield $3 \cdot 10^7$ e^+/sec concentrated e.g. in 500 bunches, each of 10 nsec length with $\beta = 0.3$. Each bunch contains $6 \cdot 10^4$ positrons and fills a 90 cm long cylindrical space volume, 1 mm in diameter, thus having a density $n_{e+} = 8 \cdot 10^4$ e^+/cm^3. This leads to a $\bar{H}$ formation rate

$$R_{\bar{H}} \approx 8 \cdot 10^{-4}/\text{sec}.$$

Overlapping each e^+ bunch with a 20 nsec long laser light pulse of 20 MW/cm^2 intensity could provide the additional gain factor G =100, and thus increase the rate to $R_{\bar{H}} \cdot G = 8 \cdot 10^{-2}$/sec $\approx$ 280/h. As a whole, e^+ beam bunching combined with laser enhancement would multiply the $\bar{H}$ formation rate by a factor of 40 in comparison to continuous formation.

But in order to gain another factor of ~ 100 the sketched scenario can still be complemented by two further devices (Fig. 5):

1) a little positron storage ring for positron pulse recycling (proposed by Herr et al., 1984), superimposing the antiproton ring over a fraction of the antihydrogen formation section;
2) an optical mirror cavity for storage of megawatt laser pulses.

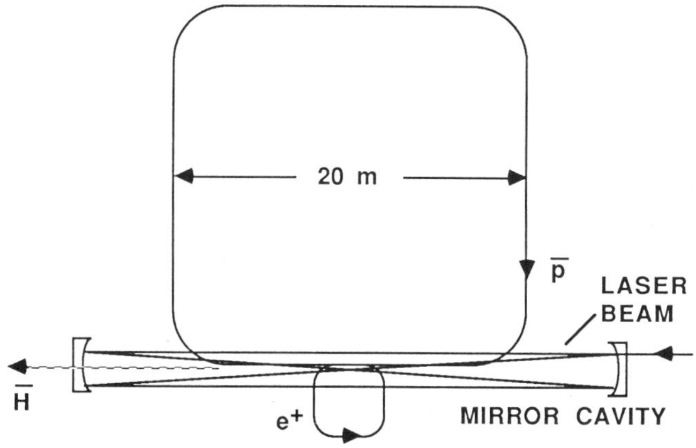

Fig. 5. Arrangement combining antiproton storage, positron pulse recycling, and multiple reflection of megawatt laser pulses

Assuming their installation at LEAR these instruments would act together in the following way: the $\bar{H}$ formation section of LEAR lies within the mirror cavity (Berger et al., 1985, 1986), the path of $\bar{p}$ and e^+ propagation coinciding with its optical axis. A laser pulse fed into the cavity travels back and forth between the highly reflecting mirrors. The light frequency be Doppler-shifted to the frequency needed for induced capture into a bound state, for the light counterpropagating to antiprotons and positrons. If positron pulse and laser pulse have the same recycling time and are synchronized to overlap in the $\bar{H}$ formation zone, e.g. during 100 revolutions without significant intensity losses the formation rate increases to

$$R_{\bar{H}} \approx 8\ \bar{H}/\text{sec}\ .$$

The antiatoms would appear in a pulsed mode. Since LEAR has a lateral length of ~ 20 m, the cavity mirror separation must be ~ 25 m, so a light pulse needs ~ 167 nsec for one cycle. For ß=0.3 the e^+ storage ring must have a circumference of ~ 15 m. With a laser pulse and e^+ pulse repetition rate of 500, antihydrogen atoms will appear in 17 microsecond long time windows every 2 milliseconds. The estimated formation rate of 8 $\bar{H}$/sec should be sufficient for spectroscopic investigations with quite satisfying precision, and by all means large enough for antihydrogen-matter interaction studies like surface scattering, etc..

Additional measures are conceivable that could further increase the $\bar{H}$ formation rate. Referring again to LEAR, such measures are:

- filling the $\bar{p}$ storage ring with the highest possible amount of $5\cdot10^{11}$ antiprotons, a factor of 10 more than used for the estimate above
- enlarging the number of recycling passages of the stored positron and light pulses from 100 to 10^3

This would provide

$$R_{\bar{H}} = 8\cdot10^2\ \bar{H}\ \text{atoms/sec.}$$

It has to be emphasized that the foregoing discussion ignored the production of intense e^+ pulses of high repetition rate by means of linear accelerators. Calculations on the basis of positron density values and beam characteristics as already realized (Howell et al., 1982; Begemann et al., 1982, Gräff et al., 1984) or realizable (Conti and Rich, 1985) with linacs indicate comparable $\bar{H}$ rates as the considerations with radioactive sources, and will not be given here.

2.6 Other Antihydrogen Formation Schemes

Quite different antihydrogen formation schemes, based on trapping of antiprotons have also been proposed. One such concept uses e^+ Auger capture from positronium (Deutch et al., 1986).Antiprotons stored in a Penning trap are confined in a volume of ~ 1 cm^3 with a density of 10^4-10^8 $\bar{p}/cm^3$. The authors assume that at a residual gas pressure of $\leq 10^{-12}$ torr the $\bar{p}$ annihilation background is of the order of 10^{-4} per particle per second. A ^{64}Cu radioactive source placed in the trap wall in combination with a moderator, provides positrons and also a certain amount of positronium. Moderated positronium atoms with a kinetic energy of $E_{kin} \leq 1$ eV, i.e. velocity-matched to 4 keV antiprotons, leave the moderator and enter the $\bar{p}$ volume. A fraction of 3/4 of the emitted e^+e^- atoms is in the triplet state. These atoms travel ~ 2 cm within their lifetime $\tau = 1.4\cdot10^{-7}$ sec. An antihydrogen formation rate is calculated with the following estimates (Deutch et al., 1986):

- e^+ Auger capture cross section $\sigma_A = 10^{-15}$ cm^2 (for $E_{e^+e^-} \approx 1$ eV)
- 50 C ^{64}Cu positron source
- number of positronium atoms per cm^2 and sec passing through the trap $N = 10^8$ e^+e^- cm^{-2} sec^{-1}
- antiproton density in the trap $N_{\bar{p}} = 10^6$ $\bar{p}/cm^3$

These values provide a rate

$$R_{\bar{H}} = NN_p\sigma = 0.1\ \bar{H}\ \text{atoms/sec.}$$

The authors point out that the e^+ Auger capture cross section exceeds the e^+ radiative cross section by about 10^5. They mention a conceivable rate increase of several orders of magnitude by positronium source and $\bar{p}$ trap development. Since the capture cross section should be the same for $\bar{p} + e^+e^-$ and $p + e^+e^-$, the proposed antihydrogen formation scheme could be tested with protons.

It should be remarked that the antihydrogen atoms leave the trap in all directions with $E_{kin} = 4$ keV. They are much too fast to be trapped via their magnetic moment since $E_{kin} < 10^{-4}$ eV would be necessary. Another problem could be the short ^{64}Cu halflife T = 12.8h.

Another group recently captured antiprotons with kinetic energies less than 3 keV in a Penning trap (Gabrielse et al., 1986 b). For this purpose, 21.3 MeV antiprotons from LEAR were decelerated by passing them through 3 mm of material, mostly Be. More than 300 $\bar{p}$ could be trapped out of a 150 nsec burst of 10^8. A fraction of the captured antiprotons was held for 100 sec and longer. The authors emphasize that with this demonstration the principal possibility was opened up to synthesize antihydrogen, by confining both antiprotons and positrons in a radio frequency trap, an idea already stated earlier (Dehmelt et al., 1979), or in a nested pair of Penning traps (Gabrielse et al., 1986a). A cloud of positrons with $E_{kin} \approx 1$ eV, loaded into the central well of the double-trap could cool the antiprotons trapped in the outer well via the same mechanism as electron cooling. The antiprotons see the central positron well as a potential hill and are slowed down when approaching this zone. In order to get an estimate of a conceivable antihydrogen production rate, the authors assume the following parameter values:

- 10^8 trapped positrons in thermal equilibrium at 4.2 K
- 10^4 antiprotons passing per sec through the e^+ cloud with $E_{kin} < 1$ eV,
- radiative recombination cross section (based on the upper two assumptions) $\sigma \approx 0.1\pi\, a_B^2 \approx 9 \cdot 10^{-22}$ cm^2

Evaluation provides a formation rate

$$R_{\bar{H}} \approx 100 \text{ antiatoms/sec.}$$

3. LASER AND MICROWAVE SPECTROSCOPY OF ANTIHYDROGEN

The spectrum of the hydrogen atom (see Fig. 6) offers the possibility to investigate a variety of effects related to fundamental aspects of physics. These effects, causing energy level shifts and splittings were extensively studied and quantitatively calculated by theorists, and measured with high precision by spectroscopists. Insofar, an excellent reference basis exists for similar investigations in antihydrogen. Let us consider the most interesting effects of this kind, concentrating on the 1s ground state and the 2s and 2p excited states.

3.1 The (2s-2p) Lamb Shift, (1s-2s) Splitting Energy, and 2s Hyperfine Structure

Spin-orbit interaction of the electron in the 2p state causes a splitting of ~ 10 GHz into the $2\,^2P_{1/2}$ and $2\,^2P_{3/2}$ fine structure substates. It should also be mentioned that both $2\,^2P_{1/2,3/2}$ fine structure states and the 2s $^2S_{1/2}$ state exhibit small hyperfine structure splittings caused by the proton spin. Within the framework of Dirac theory, $2\,^2S_{1/2}$ and $2\,^2P_{1/2}$ state coincide in energy. But in nature this degeneration is lifted by quantum-electrodynamics, shifting the s state ~ 1050 MHz above the p state. This effect is called Lamb shift. An atom excited to the p state decays within the lifetime $\tau = 1.6$ nsec to the ground state via an electric dipole transition, and emits a Lyman-α photon corresponding to the wavelength λ=1216 Å. The linewidth of Lyman-α light emitted by an unperturbed ensemble of hydrogen atoms in the 2p state, i.e. the natural linewidth, is $\Delta\nu = 1/(2\pi\,\tau) \approx 100$ MHz. This statement ignores broadening mechanisms like Doppler effect, and fine and hyperfine structure splittings. Since 2s and 1s state have the same parity, a (2s-1s) electric dipole transition under emission of one photon is strictly forbidden. The 2s can only decay to the ground state through a two-photon electric dipole transition. But this is a process of higher order, therefore the unperturbed 2s state has a lifetime of ~ 1/7 sec. On the other hand, 2s and 2p have opposite parity, and an electric dipole

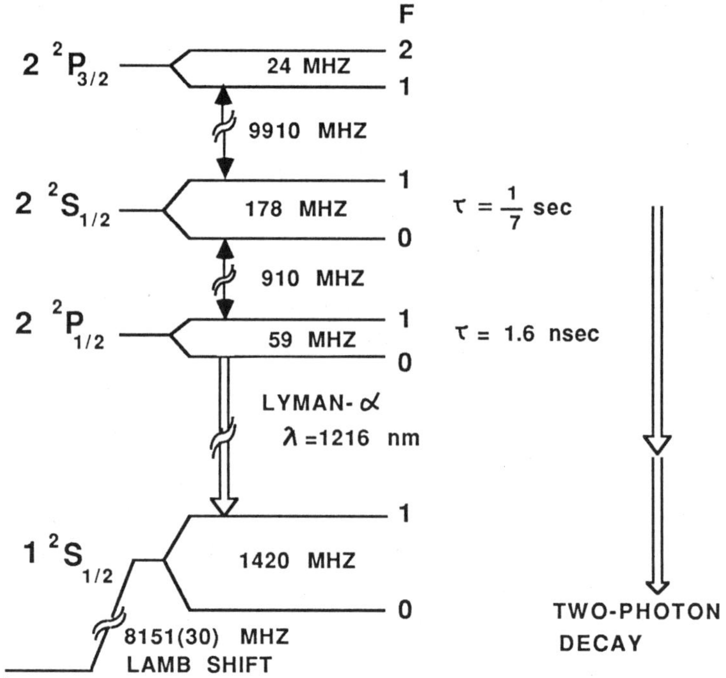

Fig. 6. The n=1 and n=2 energy levels of hydrogen

(2s-2p) transition is therefore allowed. But the 2s lifetime with respect to this decay channel is τ = 163a, since the transition probability is proportional to ν^3, and the (2s-2p) splitting is very small.

Before discussing spectroscopy in antihydrogen, some general questions will be briefly considered for hydrogen. An experimental arrangement for the (2s-2p) Lamb shift measurement looks in principle as follows:
Hydrogen atoms forming a beam are excited to the metastable 2s state by electron impact. Only a very small fraction of ~ 10^{-6} or less can be excited. The atoms then pass a region where they are exposed to a radio frequency (rf) field which induces (2s-2p) transitions. Tuning the frequency into resonance diminishes the ensemble of metastable atoms and causes the emission of Lyman-α photons. The width of the rf induced signal is dominated by the 2p lifetime, if transit time can be ignored, and is therefore ~ 100 MHz. A sophisticated technique with two separated rf field zones (Ramsey, 1956) can further reduce this width to a limited extent. The most accurate experimental values of the splitting frequency amount to 1057.845(9) MHz (Lundeen and Pipkin, 1981) and 1057.8583(22) MHz (Sokolov, 1984), the latter result being based on a method without microwave irradiation. The best theoretical values are 1057.930(10) MHz (Ericson, 1971) and 1057.884(13) MHz (Mohr, 1975).

The 2s state can also be excited from the ground state by absorption of two photons - just as it can decay by two-photon emission - e.g. using the following experimental configuration: An ensemble of hydrogen atoms with a Maxwellian velocity distribution is enclosed in a cell. A monochromatic laser light beam passes the atomic vapour and is reflected onto itself. Tuning the photon energy to about half the (1s-2s) splitting energy gives rise to the absorption of two photons by a fraction of the atoms exposed to the light field, depending upon the transition probability and the light intensity.The atomic ensemble can now be divided in velocity classes. For a given laser light frequency ν, there exists a class of atoms having all the same velocity component $\vec{v}$, e.g. in opposite direction to one of the two laser beams such that they are Doppler-shifted into resonance with this beam, according to $2\nu_o = 2\nu(1 + v/c)$. Here, $h\nu_o = E/2$ is just half the intrinsic (1s-2s) splitting energy. This means that the atoms of this velocity class absorb two photons from the same laser beam. Similarly, the class of atoms with velocity component $-\vec{v}$ in opposite direction to the other laser beam absorbs two photons from that beam. Tuning the laser frequency ν around ν_o produces a two-photon absorption line profile which reflects the velocity profile of the atoms in the vapour. This profile could be monitored e.g. by quenching the 2s state with a small electric field and detection of Lyman-α photons. As soon as ν coincides with ν_o, each atom can absorb one photon from each of the counterpropagating laser rays. The reason is that the atom "sees" the frequency of one beam Doppler-shifted to the red by the same frequency amount as it sees the frequency of the other beam shifted to the blue. Therefore, when scanning the laser frequency over the Doppler-broadened line profile, a narrow Doppler-free absorption peak appears in the center (Vasilenko et al., 1970; Biraben et al., 1974; Levenson and Bloembergen, 1974), whose width is dominated by the laser line width and by transit time broadening.

In the following, the concept of population of the 2s state of hydrogen through two-photon absorption will be applied to a beam of antihydrogen atoms with $\beta = v/c = 0.3$ and a relative longitudinal momentum spread of $\Delta p/p = 10^{-5}$. The (1s-2s) splitting energy $E = 2h\nu_o \approx 10$ eV corresponds to a laser light frequency of about $\nu_o = 1.233 \cdot 10^{15}$ Hz ($\lambda_o = 243$ nm) with the assumption of Doppler-free two-photon absorption. Now the situation is considered that a monochromatic laser beam crosses a monoenergetic $\bar{H}$ beam at a random angle, and that both beams are very well collimated. The formula of special relativity for the Doppler effect is

$$\nu = \nu_o\,(1+\beta \cos \vartheta_o)/\sqrt{1-\beta^2}, \quad \text{with} \quad \cos \vartheta = (\cos \vartheta_o + \beta)/(1+\beta \cos \vartheta_o)$$

where the laboratory coordinate frame is without index, and the rest frame of the atoms is marked with o. In case the light beam crosses the $\bar{H}$ beam at right angles ($\vartheta = 90°$), two-photon absorption takes place at $\nu = \nu_o\sqrt{1-\beta^2}$. For $\beta = 0.3$ follows a frequency shift $\Delta\nu$ of laser frequency ν with respect to ν_o of $\Delta\nu = \nu - \nu_o = \nu_o(\sqrt{1-\beta^2} - 1) = -5.6 \cdot 10^{13}$ Hz, i.e. a red shift. If the laser beam overlaps the $\bar{H}$ beam parallel and the beams travel in opposite directions, the laser frequency in the lab will be in resonance with the atoms at

$$\nu = \nu_o(1-\beta)/\sqrt{1-\beta^2} \approx 9.05 \cdot 10^{14} \text{ Hz}$$

corresponding to a wavelength $\lambda = 331.3$ nm. This means that the two-photon laboratory wavelength is shifted to the red by almost 90 nm with respect to the wavelength in the atom rest frame.

In order to excite a sufficient amount of antihydrogen atoms to the 2s state, one needs a high laser intensity and therefore a pulsed laser. Assuming a pulsed $\bar{H}$ beam the laser pulses must be synchronized with the $\bar{H}$ pulses. It should be mentioned that the laser which excites the atoms to the 2s state cannot be used at the same time for laser-enhanced $\bar{H}$ formation, since the frequencies are very different. In order to maximize the excitation rate and minimize transit time broadening of the absorption profile, it seems favourable to overlap laser and $\bar{H}$ beams parallel. Also, the pulses of the excitation laser can be stored in a mirror cavity and thus fit with an $\bar{H}$ beam produced by positron recycling.

To use the advantage of Doppler shift to longer wavelength, the laser frequency must be tuned thus that the $\bar{H}$ atoms can absorb two photons from the counterpropagating beam as sketched above. But, this means on the other hand that the two-photon transition is not Doppler-free, and the transition rate depends upon the Doppler-broadened profile of the $\bar{H}$ beam caused by the longitudinal momentum spread. The two-photon transition rate per second and atom is given by (Hänsch et al., 1975)

$$\Gamma \approx 7 \cdot 10^{-4}\ I^2/\Delta\omega\ ,$$

I being the light intensity in W/cm^2, and $\Delta\omega = 2\pi\Delta\nu$ the laser bandwidth in MHz. An estimate of the two-photon excitation rate be based on the following parameter values:

- Doppler-tuned laser frequency $\nu \approx 9 \cdot 10^{14}$Hz for $\beta = 0.3$
- longitudinal momentum spread of $\bar{H}$ beam $\Delta p/p = 10^{-5}$, equivalent to a Doppler broadened profile $\Delta\nu_D = \nu\ \beta \cdot \Delta p/p \approx 2.7 \cdot 10^9$Hz = 2.7 GHz
- laser light power 100 kW, beam diameter 3 mm, laser light intensity $I \approx 1$ MW/cm^2

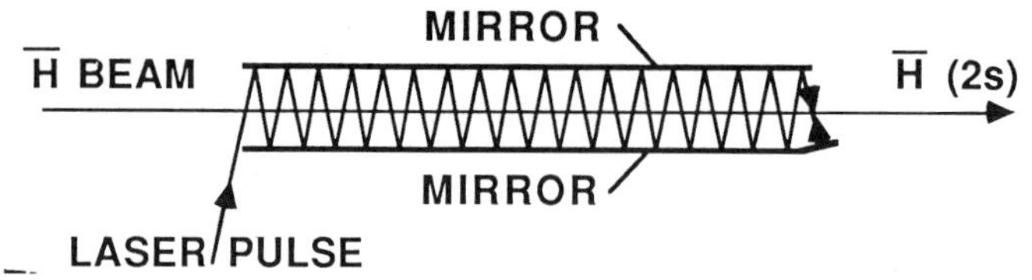

Fig. 7. Mirror configuration for Doppler-free two-photon excitation of antihydrogen

- laser pulse length t = 20 nsec
- Fourier-limited laser bandwidth $\Delta\nu_F \approx 1/t = 50$ MHz

The transition probability per atom during an interaction time of 20 nsec is then

$$\Gamma \cdot t \approx 0.1 \; .$$

The narrowest frequency width a 20 nsec long laser pulse can have, is the Fourier-limited width $\Delta\nu_F \approx$ 50 MHz. Assume an approximately rectangular-shaped longitudinal Doppler profile of the atomic beam. Within the interaction time t = 20 nsec the atoms see a broadened laser bandwidth $\Delta\nu \approx 2/t = 100$ MHz. It means that only a fraction $0.1 \cdot 100$ MHz/2.7 GHz $\approx 4 \cdot 10^{-3}$ of the atoms can be excited if the transition rate is a factor of 5 below saturation condition.

The foregoing consideration indicates that a laser beam and atomic beam crossing configuration for Doppler-free two-photon excitation as shown in Fig.7 may be preferable. The laser pulse travels back and forth between two parallel, plane mirrors of 90 cm length, and is reflected onto itself at the end of the mirrors. The antihydrogen beam with $\beta = 0.3$ passes the mirror gap within T = 10 nsec. The Doppler-free two-photon absorption profile thus experiences a transit time broadening $\Delta\nu_T \approx 1/T = 100$ MHz, in addition to the width caused by the Fourier transform limitation of the laser, and all atoms participate in the absorption process according to the transition probability. However, the laser wavelength must be tuned nearer to $\lambda_0 \approx 243$ nm. Interferometric measurement of the primary laser wavelength in the visible regime could provide the 1s-2s separation energy to a fraction of the two-photon absorption line width. In hydrogen the 1s-2s frequency was measured (Hildum et al., 1986) to

$$\nu(1s\text{-}2s) = 2\,466\,061\,395.6(4.8)\ \text{MHz} \; .$$

From this result, the Rydberg constant was determined to

$$R_\infty = 109\,737.314\,92(21)\ \text{cm}^{-1}.$$

Behind the laser excitation region the $\overline{H}$ beam passes a radio frequency field which induces 2s-2p transitions. The atoms transferred to the 2p state decay rapidly to the ground state since the 2p lifetime is $\tau = 1.6$ nsec. The signal curve as a function of the tuned radio frequency can be monitored behind the rf field via the atoms surviving in the 2s state. If these atoms pass a static electric field (which can be realized as a motional field by means of a magnetic field) the 2s state remains no longer a pure state but partly adopts 2p character, and the emitted Lyman-α photons can be viewed with a photomultiplier. For an rf field transit time significantly longer than the 2p lifetime, the halfwidth of the signal curve is given by

$$\Delta\nu = 1/2\pi\,\tau \approx 100\ \text{MHz} \; .$$

The finite interaction time causes that the atoms observe a radio frequency band of halfwidth (= width of the central maximum) $\Delta\nu_T \approx 1/T$ rather than a narrow δ–function-like frequency similar to the case with laser light, discussed before. Thus, the transit time broadens the rf signal curve. With $v/c = 0.3$ transit time broadening is equal to the natural linewidth for $L \approx 90$ cm. This length corresponds to 5.6 natural lifetimes.

As was mentioned earlier, a very precise measurement of the (2s-2p) Lamb shift has been performed with the separated-oscillatory-field technique (Ramsey, 1956). In such a configuration the atoms are exposed to two rf fields of well-defined relative phase in two separated regions. Through data acquisition at 0° and 180° relative phase an interference signal curve is obtained whose width is mainly determined by the separation of the rf field zones rather than by the natural linewidth. In particular, the signal width can be reduced significantly below the natural width, based on those atoms which are transferred to the 2p state and survive the time of flight T between the two rf field zones in this state (Hughes, 1960). The number of surviving 2p atoms is proportional to $\exp\{-T/\tau\}$. For this reason, the Ramsay technique cannot be exploited in the case of antihydrogen, since the number of atoms is too low, to sacrifice e.g. a factor of 20 in signal rate in order to gain a factor of 2 in line narrowing which is the case for $T = 3\tau$.

Assuming a rate of 100 $\bar{H}$ atoms/sec, a single rf field interaction zone of 1 m length, and a measuring time of ~ 1 day, the 2s-2p Lamb shift frequency of ~ 1000 MHz could be obtained with a precision of 10^{-4}. A measurement of the 2 $^2S_{1/2}$ hyperfine structure (hfs) splitting frequency of ~ 177 MHz would permit a higher precision since both substates are metastable. In this case the separated oscillatory field technique would be useful. A separation of the two rf field zones of 90 m, corresponding to a transit time of 10^{-6} sec for $\beta = 0.3$, reduces the resonance linewidth to ~ 1 MHz. Determination of the centroid of the line profile to ~ 10^{-3}, would provide a precision of ~ $6 \cdot 10^{-6}$. For hydrogen the experimental value (Heberle et al., 1956) is

$$\Delta\nu(2S) = 177\ 556.86(0.05)\ \text{kHz},$$

a fraction of about $3 \cdot 10^{-5}$ being caused by proton structure.

3.2 Laser-Optical Pumping of $\bar{H}$ Ground State

In hydrogen the magnetic interaction between electron spin and proton spin gives rise to a ground state hyperfine structure splitting of $\nu \approx 1.4$ GHz equivalent to the wavelength $\lambda \approx 21$ cm. Spontaneous magnetic dipole transitions from the upper substate with hyperfine quantum number F=1 to the lower F=0 substate cause the famous 21 cm microwave radiation from the hydrogen regions in our milky way and made it possible to map the galaxy structure.The lifetime of an unperturbed F=1 hyperfine state is $1.1 \cdot 10^7$a , due to the very small magnetic-dipole transition matrix element and the proportionality of the spontaneous decay rate to ν^3. There is no way to observe spontaneous transitions under lab conditions.

However, the splitting frequency has been measured through induced transitions by means of a hydrogen maser using a thermal beam of hydrogen atoms. The atoms are distributed in the F=0 and F=1 states according to the statistical weights 2F+1. Under this condition a radio frequency field tuned in resonance with the splitting frequency cannot change the population of the substates through induced transitions. In order to obtain a microwave field induced signal a population inversion must be created. For this purpose the atoms pass an inhomogeneous magnetic field which focuses the F=1 (m_F=+1,0) atoms on the beam axis and defocuses the F=1(m_F=-1) and F=0 atoms. The pure F=1 beam enters a microwave cavity tuned in resonance with the hyperfine frequency. The high-quality resonator induces (F=1 - F=0) transitions without feeding microwave power from outside. These transitions to the lower substate provide a detectable microwave signal. About 10^{13} inverted atoms are necessary to obtain a resonator output power of 10^{-11}W, sufficient to keep the cavity in resonance with the atoms via a feedback circuit.The result of the most accurate measurement (Essen et al., 1971) is

$$\nu = 1\ 420\ 405\ 751.7667(10)\ \text{Hz}.$$

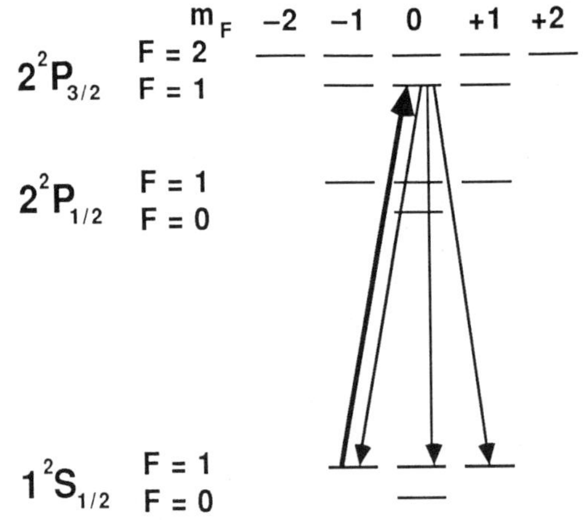

Fig. 8. Optical pumping of hydrogen ground state using light with left-handed helicity

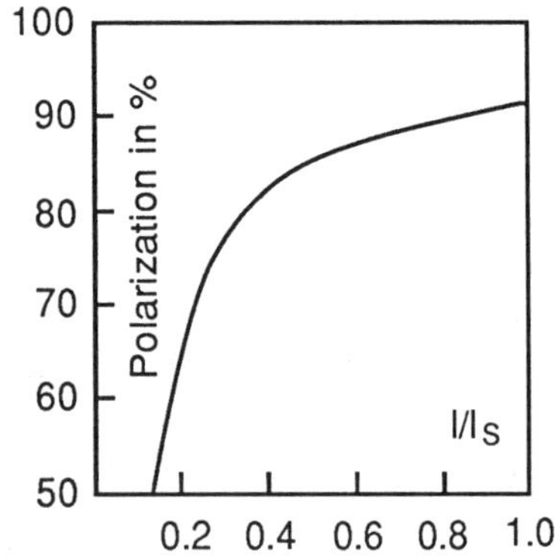

Fig. 9. Polarization versus light intensity I in units of saturation intensity I_S for 10m interaction length

The error of this value is determined by depolarizing collisions of the atoms with the microwave cavity walls rather than by the many orders of magnitude narrower natural line width. The precision of the theoretical calculation including relativistic and radiative corrections, is limited by the contribution of proton structure to the splitting energy. This contribution amounts to 33 ppm and could be calculated with an uncertainty of 3 ppm. Consequently, the theoretical value is almost seven orders of magnitude less precise than the measured value (Gidley and Rich, 1981).

The problem how to tackle a ground state hyperfine structure measurement in antihydrogen must be judged on the basis of the conditions for hydrogen. Obviously an antihydrogen maser is by no means feasible for two reasons:

1) the $\bar{H}$ atoms are much too fast to permit a state selection in a magnet;
2) the flux of atoms is too low by 11 to 12 orders of magnitude to detect hyperfine resonance through the emitted microwave power.

Optical spectroscopy offers a technique to transfer atoms into a selected hyperfine substate: the method of optical pumping, sucessfully applied to alkali atoms. In a static magnetic field a hyperfine structure state with quantum number F splits into 2F+1 magnetic substates characterized by quantum numbers m_F, with $-F \leq m_F \leq F$. But also in zero magnetic field these Zeeman sublevels are physically distinguishable though energetically degenerate. This is illustrated in Fig.8 which displays the 1s and 2p hyperfine structure of hydrogen including Zeeman levels. By absorption of circularly polarized Lyman-α light followed by reemission, the number of atoms populating a selected Zeeman level of the ground state can be increased. Thus, e.g. the absorption of a photon with left-handed helicity (unfortunately called a right-polarized photon since the electric vector rotates clock-wise when observed counter to the direction of light propagation) obeys the $\Delta m = +1$ selection rule while the reemission can follow $\Delta m = 0,\pm 1$. After several absorption and reemission processes an atom is likely to be in the (F=1, m_F=1)-substate. This means that the electron spin orientates itself in a definite direction with respect to the direction of light propagation. Since electron spin and proton spin are coupled through hyperfine interaction, also the proton spin becomes polarized.

Because of the lack of powerful light sources in the range of the very short Lyman-α wavelength, optical pumping of hydrogen atoms has not yet been used to provide ground state inversion or to produce polarized protons. But the polarization scheme was recently proposed

(Zelenskiy et al., 1984) for fast hydrogen atoms with $\beta \geq 0.5$, and briefly later also for antihydrogen (Imai, 1985). For counterpropagating beams and $\beta = 0.6$, the laser wavelength needed for optical pumping of hydrogen is Doppler-shifted to $\lambda = \lambda_0 \sqrt{(1-\beta^2)}/(1-\beta) = 243$ nm. This is just twice the Lyman-α wavelength in the atomic rest frame. The efficiency of optical pumping is estimated based on the following assumptions: Starting with 20 nsec long 1 MW light pulses at the primary wavelength $\lambda = 486$ nm of a dye laser pumped by an excimer laser, pulses at $\lambda = 243$ nm with 20 nsec length, 3mm beam diameter, and 100 kW power can be produced by frequency doubling in a potassium penta borate crystal (i.e. a doubling efficiency of 10%). The laser frequency width $\Delta\nu = 7$ GHz is about equal to the Doppler width of the $\bar{H}$ beam Lyman-α absorption profile for a longitudinal momentum spread $\Delta p/p = 10^{-5}$.

Using the simplifying condition of a (1s-2p) two-level system in hydrogen it can be shown immediately, that the 100 kW laser pulses saturate the Lyman-α transition. The number N_2 of atoms populating the upper state after laser irradiation during a time period t is given (Loudon, 1973) by

$$N_2 = \frac{NBW}{A+2BW} \cdot \{1-\exp(-[A+2BW]t)\},$$

$A = 1/\tau$ and $B = A \cdot \pi^2 c^3/\hbar\omega^3$ being the Einstein coefficients for spontaneous and induced transitions, respectively, $W = n\hbar\omega/V\Delta\omega$ the spectral energy density with n photons in space volume V and frequency interval $\Delta\omega = 2\pi\Delta\nu$, N the initial number of atoms in the lower state. With $BW \approx 10^{13}$ sec^{-1} and $A = 1/\tau = 6.3 \cdot 10^8$ sec^{-1} follows $N_2/N = 1/2$ already after an irradiation time of $t = \tau = 1.6$ nsec. The estimate illustrates that a pulsed dye laser with circularly polarized light can transfer the majority of atoms into the F=1 hyperfine sublevel of the ground state. Zelenskiy and co-authors calculated that 80% proton polarization is reached after an irradiation time period (measured in the laboratory rest frame) of $10\tau/\sqrt{1-\beta^2}$ under saturation condition. The degree of polarization versus light intensity in units of saturation intensity for 10 m interaction length is plotted in Fig.9.

In conclusion, laser optical pumping of antihydrogen atoms at $\beta = 0.6$ could provide population inversion of the ground state hyperfine sublevels, the precondition for a hyperfine structure splitting measurement, as well as an 80-90% polarized, pulsed antiproton beam of e.g. $5 \cdot 10^2$ $\bar{p}$/sec. At this point it should be mentioned that ground state inversion and thus a polarized $\bar{p}$ beam may also be achieved, though to a lower degree, through the recombination with polarized positrons. Positrons from nuclear beta decay are polarized typically with a degree $P \approx v/c \approx 0.7$ but most of the fast and therefore polarized positrons are lost during the moderation procedure because of maximum penetration in the moderator crystal. However, a foil of low-Z material between positron source and moderator decelerates the fast positrons without appreciably decreasing their polarization. A positron beam of typically 50% polarization can be produced in this way but with a beam intensity lowered by a factor of five (Conti and Rich, 1985).

Optical pumping is only the first necessary step of an antihydrogen ground state hfs splitting measurement. Since no realistic concept exists at present for detection of microwave field induced magnetic dipole transitions between the 1s substates, statements on this problem are avoided here.

4. ANTIHYDROGEN - MATTER INTERACTION

4.1 Hydrogen-Antihydrogen Interaction

The hydrogen-antihydrogen system attracted attention already several decades ago as the simplest atom-antiatom pair, and because of conceivable implications of its interactions for cosmological theories (see e.g. Morgan and Hughes, 1973, and references therein). Detailed theoretical treatments of hydrogen-antihydrogen interatomic potential energies and the cross section for rearrangement from $H\bar{H}$ to $p\bar{p}$ and e^+e^- are found in the literature. The interaction potential between hydrogen and antihydrogen was obtained by accurate quantum-mechanical calculations. Using the variational principle, the Born-Oppenheimer approximation, and a

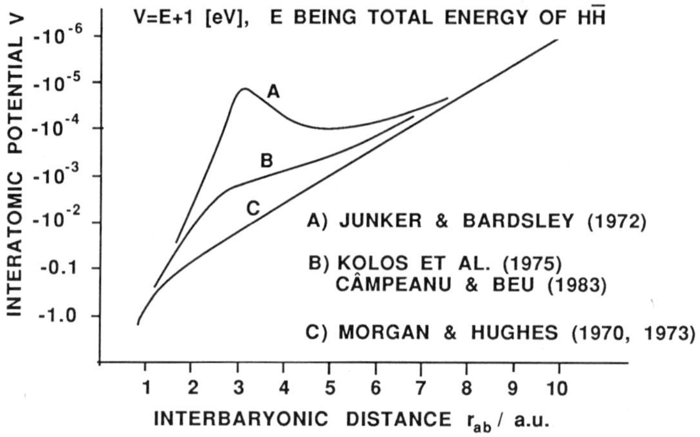

Fig. 10. Calculated interatomic potential of $H\bar{H}$ system

spinless non-relativistic Hamiltonian, Junker and Bardsley calculated a configuration-interaction wave function including 75 configurations (Junker and Bardsley, 1972). They found a small maximum in interatomic interaction energy at a certain p-$\bar{p}$ distance, and speculated that such a maximum may exist for all atom-antiatom pairs. The interatomic potential calculated by Morgan and Hughes, was deduced from second-order perturbation and known limiting forms for large and small interatomic distance r_{ab} (Morgan and Hughes, 1970, 1973). In contrast to the outcome of Junker and Bardsley, the maximum was absent in this result. Also the investigations of Kolos and co-authors (Kolos et al., 1975), based on the Ritz variational principle and Born-Oppenheimer approximation, and of Câmpeanu and Beu (Câmpeanu and Beu, 1983), using another computational technique and a simpler trial function did not reproduce the potential peak. Fig.10 displays the interatomic potential V as a function of the p-$\bar{p}$ distance, as obtained by the various authors. It should be emphasized that the peak in V found by Junker and Bardsley is very small, the difference between maximum at r_{ab} = 3.1 a.u. and minimum at $r_{ab} \approx$ 4.4 a.u. being only 8.8 meV.

A procedure of $H\bar{H}$ formation could follow the concept of co-rotating $\bar{p}$ and H^- beams in LEAR (Gastaldi and Möhl, 1984) worked out for high-resolution spectroscopy of $p\bar{p}$ atoms in flight. The conditions necessary to implement this scheme in LEAR, like a long $p\bar{p}$ beam line as extension of a straight section, an adequate shape of the vacuum chambers at the ring bends, and the possibility of H^- injection have been provided in the LEAR construction. Following the above mentioned authors a short list of characteristics is given, to illustrate a scenario for work with H^- at LEAR:

- LEAR vacuum ~ 10^{-12} torr
- $6{\cdot}10^9$ H^- injected per pulse with 10 mA H^- source
- number of H^- reduced by intrabeam stripping to 10^9 within ~ 1 sec (destruction preferentially of higher velocities)
- after merging $\bar{p}$ and H^- beams, H^- beam lost after 10 sec through beam-beam and intrabeam stripping
- reduction of strip rate, and increase of $p\bar{p}$ formation rate, when switching on electron cooling
- gas stripping main source of H^- losses for less than 10^8 H^-, and lifetime is 40 - 300 sec, depending on β
- $p\bar{p}$ beam accompanied by intense neutral atomic hydrogen beam, variing from ~ 10^9 H^0/sec immediately after H^- injection to ~ 10^9 H^0/sec when only 10^8 H^- left in the ring and 100 sec gas stripping lifetime assumed

Thus, a $H\bar{H}$ formation scheme includes:

- $\bar{p}$ and H^- injection
- deceleration or acceleration of both bunched beams to required energy
- debunching and merging together the two beams
- electron cooling (danger of H^- depletion)
- passage of both beams through e^+ beam region
- antihydrogen formation and H^- ionization by means of laser light in e^+ beam zone
- $\bar{H}$ and H atoms leaving ring together

Many questions remain to be answered, e.g.: will $H\bar{H}$ molecules be formed? How could the annihilation of $p\bar{p}$ and $H\bar{H}$ be distinguished?

4.2 Antihydrogen as Fuel for Rocket Propulsion

The subject of matter-antimatter annihilation as an energy source for propulsion of interplanetary or even interstellar rockets has been dealt with in quite many publications (see e.g. Cassenti, 1985; Forward et al., 1985; Forward, 1985; and references therein). The way to an eventual realization of this extremely ambitious aim is paved with numerous technological problems far beyond present possibilities. Clearly, antiprotons are the antimatter species of choice. Annihilation of a proton with an antiproton converts the masses equivalent to approximately 2 GeV, into other elementary particles and their kinetic energies. A fraction of these particles can be used to heat propellent and thus provide thrust. The physicists produce antiprotons for a rather long time already, and e.g. store them in rings. But the $\bar{p}$ numbers are much too small to yield more than a marginal amount of energy via $p\bar{p}$ annihilation. In addition, these storage facilities are far too large, complicated and energy comsuming to be adaptable for rocket propulsion purposes. Thus, new and finally realistic solutions must be found for efficient production and safe storage of large antiproton amounts. Antihydrogen atoms can probably be stored more easily in larger amounts than charged antiprotons. One thinks of solid antihydrogen kept in a container at very low temperature. Since the frozen antimatter fuel must not contact the enclosure material consisting of matter, it has to be suspended electrically or magnetically. A small antihydrogen vapour pressure will always exist above T=0K, continuously causing annihilation with the wall material. A storage time of more than 10 years was estimated for 1 mg of antihydrogen suspended electrically at 4K (Cassenti, 1985).

Proton and antiproton annihilate into several neutral and charged pions, according to

$$p + \bar{p} \rightarrow m\pi^0 + n\pi^+ + n\pi^- ,$$

m and n being about 1.6. The pions decay as described by

$$\pi^0 \rightarrow \gamma + \gamma, \ \pi^+ \rightarrow \mu^+ + \nu_\mu , \text{ and } \pi^- \rightarrow \mu^- + \bar{\nu}_\mu ,$$

and the muons follow the decay scheme

$$\mu^- \rightarrow e^+ + \nu_\mu + \bar{\nu}_e , \text{ and } \mu^+ \rightarrow e^+ + \bar{\nu}_\mu + \nu_e .$$

Usage of the charged particles to heat propellent is discussed by Cassenti in considerable detail. Suppose the antihydrogen confinement problem is solved, the next aim to be tackled is a combustion chamber for $\bar{H}$ annihilation and heating of a propellent, such as hydrogen. Magnetic fields of the order of 50 kG at the center and 500 kG at the ends of the chamber are necessary to confine the charged annihilation products. Fields of that size are feasible only by means of superconducting coils. A serious material problem to be examined is possible radiation damage of the chamber walls, in particular by gamma rays. The heated propellent leaves the chamber through a Laval nozzle. Cassenti's analysis comes to the result that the transfer of a 10^4 kg spacecraft from low earth orbit of 200 km altitude to geosynchronous orbit and back would require 4 mg of antimatter. Questions of present antiproton production efficiency and necessary steps for its optimization are discussed by R. Forward (Forward, 1985). The author also reviews a variety of schemes maybe applicable to formation, storage, and annihilation of antihydrogen. The review includes laser-enhanced antiproton-positron recombination, slowing and trapping of antiatoms through laser light, and conversion into molecules followed by condensation as a ball of antihydrogen ice. Such a ball with a weight of the order of 20 mg would be trapped due to its diamagnetism in the potential well of a magnetic field. Alternatively the antimatter might be kept electrically as a charged cloud of microcrystals.

The costs for antihydrogen rocket propulsion have been analyzed and compared with those for chemical systems (Forward et al., 1985). The prospects for cost effective antihydrogen propulsion are described as quite favourable.

Acknowledgements. I thank the members of the collaboration working on the proposed 'Feasibility study for Antihydrogen Production at LEAR' (P86) for valuable discussions and advice, in particular H. Poth, A. Winnacker, G. zu Putlitz, H. Pilkuhn, A. Wolf, and P. Blatt, the latter and S. Grafström for preparing the figures, and Mrs. B. Wieland for typing the manuscript.
The financial support of the German Bundesministerium für Forschung und Technologie for the Feasibility Study is also gratefully acknowledged.

REFERENCES

Begemann, M., Gräff, G., Herminghaus, H., Kalinowsky, H., and Ley, R., 1982, Slow positron beam production by a 14 MeV c.w. electron accelerator, Nucl. Instrum. Meth. 201, 287

Bell, M., Bell, J., 1982, Capture of cooling electrons by cool protons, Part. Accel. 12, 49

Berger, J., Blatt, P., Habfast, C., Haseroth, H., Hauck, P., Hill, Ch., Neumann, R., Pilkuhn, H., Poth, H., zu Putlitz, G., Seligmann, B., Winnacker, A., and Wolf, A., 1985, Feasibility study for antihydrogen production at LEAR, Proposal to the CERN PSCC (P86, spokesman H. Poth), CERN/PSCC/85-45, Geneva

Berger, J., Blatt, P., Hauck, P., Neumann, R., 1986, Storage of megawatt laser pulses in a 4.5 m long confocal Fabry-Perot resonator, Optics Commun. 59, 255

Biraben, F., Grynberg, G., and Cagnac, C., 1974, Experimental evidence of two-photon transition without Doppler broadening, Phys. Rev. Lett. 32, 643

Budker, G.I., and Skrinskii, A.N., 1978, Electron cooling and new possibilities in elementary particle physics, Sov. Phys. Usp. 21, 277

Câmpeanu, R.I., and Beu, T., 1983, Hydrogen-antihydrogen interaction potential, Phys. Lett. 93A, 223

Cassenti, B.N., 1985, Antimatter propulsion for OTV applications, J. Propulsion 1, 143

Conti, R.S., and Rich, A., 1985, The status of high intensity, low energy positron sources for antihydrogen production, in: "Proceedings of the Workshop on the Design of a Low-Energy Antimatter Facility in the USA" (Madison, Wisconsin, October 3-5, 1985), in print

Dehmelt, H., Van Dyck, Jr. R.S., Schwinberg, P.B., and Gabrielse, G., 1979, Single elementary particles at rest in space, Bull. of Am. Phys. Soc. 24, 757

Deutch, B.I., Jensen, A.S., Miranda, A., and Oades, G.C., 1986, $\bar{p}$ capture in neutral beams in: "Proceedings of the First Workshop on Antimatter Physics at Low Energy", B.E. Bonner, L.S. Pinsky, eds., Fermi National Accelerator Laboratory, p. 371

Erickson, G.W., 1971, Improved Lamb-shift calculation for all values of Z, Phys. Rev. Lett. 27, 780

Essen, L., Donaldson, R.W., Bangham, M.J., and Hope, E.G., 1971, Frequency of the hydrogen maser, Nature 229, 110

Forward, R.L., Cassenti, B.N., and Miller, D., 1985, Cost comparison of chemical and antihydrogen propulsion systems for high ΔV missions, paper presented at: AIAA/SAE/ASME/ASEE 21st Joint Propulsion Conference, July 8-10, Monterey, California

Forward, R., Antiproton annihilation propulsion, J. Propulsion and Power 1, 370

Gabrielse, G., Helmerson, K., Tjoelker, R., Fei, X., Trainor, T., Kells, W., and Kalinowsky, H., 1986a, Prospects for experiments with trapped antiprotons, in: "Proceedings of the First Workshop on Antimatter Physics at Low Energy", B.E. Bonner, L.S. Pinsky, eds., Fermi National Accelerator Laboratory, p.211

Gabrielse, G., Fei, X., Helmerson, K., Rolston, S.L., Tjoelker, R., Trainor, T.A., Kalinowsky, H., Haas, J., and Kells, W., 1986b, First capture of antiprotons in a penning trap: a keV source, Phys. Rev. Lett. 57, 2504

Gastaldi, U., and Möhl, D., 1984, Co-rotating beams of antiprotons and H^- in LEAR and high resolution spectroscopy of $p\bar{p}$ atoms in flight, in: "Physics with Low-Energy Cooled Antiprotons", U. Gastaldi and R. Klapisch, eds., Plenum Press, New York and London, p.649

Gräff, G., Ley, R., Osipowicz, A., and Werth, G., 1984, Intense source of slow positrons from pulsed electron accelerators, Appl. Phys. A33, 59

Gidley, D.W., and Rich, A., 1981, Tests of quantum electrodynamics using hydrogen, muonium, and positronium, in: "Atomic Physics 7, D. Kleppner and F.M. Pipkin, eds., Plenum Press, New York and London, p.313

Hänsch, T.W., Lee, S.A., Wallenstein, R., and Wieman, C., 1975, Doppler-free two-photon spectroscopy of hydrogen 1S-2S, Phys. Rev. Lett. 34, 307

Heberle, J.W., Reich, H.A., and Kusch, P., 1956, Hyperfine structure of the metastable hydrogen atom, Phys. Rev. 101, 612

Herr, H., Möhl, D., and Winnacker, A., 1984, Production of and experimentation with antihydrogen at LEAR, in : "Physics at LEAR with Low-Energy Cooled Antiprotons", U. Gastaldi and R. Klapisch, eds., Plenum Press, New York and London, p.659

Hildum, E.A., Boesl, U., McIntyre, D.H., Beausoleil, R.G., and Hänsch, T.W., 1986, Measurement of the 1S-2S frequency in atomic hydrogen, Phys. Rev. Lett. 56, 576

Howell, R.H., Alvarez, R.A., and Stanek, M., 1982, Production of slow positron beams with an electron linac, Appl. Phys. Lett. 40, 751

Hütten, L., Poth, H., Wolf, A., Haseroth, H., and Hill, Ch., 1984, The electron cooling device for LEAR, in: "Physics at LEAR with Low-Energy Cooled Antiprotons", U. Gastaldi and R. Klapisch, eds., Plenum Press, New York and London, p. 605

Hughes, V.W., 1960, in: "Quantum Electronics", C.H. Townes, ed., Columbia University Press, New York, p.582

Imai, K., 1985, Polarized antiprotons through antihydrogen formation, in: "Polarized Beams at SSC. Polarized Antiprotons", AIP Conference Proceedings No. 145, A.D. Krisch, A.M.T. Lin, and O. Chamberlain, eds., New York, 1986, p.229

Junker, B.R., and Bardsley, J.N., 1972, Hydrogen-antihydrogen interactions, Phys. Rev. Lett. 28, 1227

Kolos, W., Morgan, Jr., D.L., Schrader, D.M., and Wolniewicz, L., 1975, Hydrogen-antihydrogen interactions, Phys. Rev. A11, 1792

Lefèvre, P., Möhl, D., and Plass, G., 1980, The CERN low energy antiproton ring (LEAR) project, in: "11th Int. Conf. on High-Energy Accelerators", Birkhäuser Verlag, Basel

Levenson, M.D., and Bloembergen, N., 1974, Observation of two-photon absorption without Doppler broadening on the 3S-5S transition in sodium vapour, Phys. Rev. Lett. 32, 645

Loudon, R., 1973, "The Quantum Theory of Light", Clarendon Press, Oxford

Lundeen, S.R., and Pipkin, F.M., 1981, Measurement of the Lamb shift in hydrogen, n=2, Phys. Rev. Lett. 46, 232

Mills, A.P., Jr., 1980, Brightness enhancement of slow positron beams, Appl. Phys. 23, 189

Mills, A.P., Jr., 1984, Techniques for studying systems containing many positrons, in: "Positron Scattering in Gases", J.W. Humbertson and M.R.C. McDowell, eds., Plenum Press, New York and London, p.121

Mohr, P.J., 1975, Lamb shift in a strong Coulomb potential, Phys. Rev. Lett. 34, 1050

Morgan, L.D., and Hughes, V.W., 1970, Atomic processes involved in matter-antimatter annihilation, Phys. Rev. D2, 1389

Morgan, L.D., and Hughes, V.W., 1973, Atom-antiatom interactions, Phys. Rev. A7, 1811

Neumann, R., Poth, H., Winnacker, A., and Wolf, A., 1983, Laser-enhanced electron-ion capture and antihydrogen formation, Z. Phys. A - Atoms and Nuclei 313, 253

Neumann, R., 1985, Laser induced electron capture and related physics, in: "Proceedings of the Workshop on Electron Cooling and Related Applications (ECOOL 1984)", H. Poth, ed., Kernforschungszentrum Karlsruhe Report KfK 3846, July 1985, p. 387

Poth, H., 1985, Electron cooling, Lecture given at the CERN Accelerator School, Oxford, September 1985, printed as report CERN-EP/86-65, 1986

Poth, H., 1986a, Supplements to Proposal P86 (Berger et al., 1985) CERN/PSCC/86-21, CERN/PSCC/86-37

Poth, H., 1986b, Physics with antihydrogen, in: AIP Conference Proceedings (2nd Conference on the Intersections between Particle and Nuclear Physics), D.F. Geesaman,

New York, 1986, p.480
Ramsey, N.F., 1956, "Molecular Beams", Oxford University Press, London
Rich, A., 1985, Private communication
Sokolov, Yu.L., Atomic interferometer method measurement of the Lamb shift in hydrogen (n=2), 1984, in: "Precision Measurement and Fundamental Constants II", B.N. Taylor and W.D. Phillips, eds., Natl. Bur. Stand. (U.S.), Spec. Publ. 617, p.135
Torelli, G., 1980, A device for trapping and cooling to low temperature antiprotons, in: "Proc. 5th Antiproton Symposium, Bressanone, 1980 (CLEUP, Padua, 1980), p.43
Vasilenko, L.S., Chebotaev, V.P., and Shishaev, A.V., 1970, Line shape of two-photon absorption in a standing-wave field in a gas, JETP Letters 12, 113
Wolf, A., Haseroth, H., Hill, C.E., Vallet, J.-L., Habfast, C., Poth, H., Seligmann, B., Blatt, P., Neumann, R., Winnacker, A., and zu Putlitz, G., 1985, Electron cooling of low-energy antiprotons and production of fast antihydrogen atoms, paper presented at Workshop on the Design of a Low-Energy Antimatter Facility in the USA, 3-5 October 1985, Madison, Wisconsin, in press
Wolf, A., 1986, Antihydrogen, Paper presented at the 8th European Symposium on Proton-Antiproton Interactions, Thessaloniki, Greece, 1-5 September 1986, Preprint CERN-EP/86-179
Zelenskiy, A.N., Kokhanovskiy, S.A. Lobashev, V.M., Sobolevskiy, N.M., and Volferts, E.A., 1984, A method of polarizing relativistic proton beams by laser radiation, Nucl. Instrum. Meth. 227, 429

ON THE PRODUCTION OF HIGHLY IONIZED ANTIPROTONIC

NOBLE GAS ATOMS AT REST

R. Bacher, P. Blüm, K. Elsener**, D. Gotta,
K. Heitlinger, M. Schneider, and L. M. Simons*

Kernforschungszentrum Karlsruhe GmbH
Institut für Kernphysik und
Institut für Experimentelle Kernphysik der Universität
P.O.B. 3640, 7500 Karlsruhe 1, Federal Republic of Germany
SIN, Villigen, Switzerland*
CERN, Geneva, Switzerland**

An atomic system consisting of a nucleus, an antiproton and no or one electron is a simple quantum mechanical system. The production of antiprotonic atoms in this ideal state would allow the study of numerous phenomena: for example, the polarizability of the antiproton due to the very strong electric field of the nucleus[1], QED-corrections to the electron binding energy in an one-electron-system[2,3], or the selective refilling of electrons from neighbouring atoms.

In this article, we report the measurement of low-energy-X-rays of antiprotonic Neon, Argon, Krypton and Xenon at pressures less than 50 mbar. The experiment was performed with the 202 MeV/c beam at the LEAR M1-area using the cyclotron trap[4]. We observed the complete ionization of the light noble gases and possibly of Xenon by the electromagnetic cascade, as we supposed previously for muonic atoms[5]. We deduce that antiprotons are captured in very high levels, and thus confirm the earlier analyses of Schneuwly et al.[6] and von Egidy et al.[7] with muonic atoms.

The cyclotron trap provides a high stopping density up to 10^8 $\bar{p}$/sec · g. Low pressures are crucial to suppress X-ray selfabsorption in the target gas at low energies (< 20 keV) and the electron refilling due to collisions with neighbouring atoms. The X-rays were recorded by a thin-window silicon detector flanged directly to the target chamber.

IONIZATION

The cascade[8] of antiprotons proceeds mainly via radiative and non radiative (Auger) dipole transitions. The antiprotons are enriched in the circular orbits ($l = n-1$; n denotes the principal quantum number, l the orbital quantum number), because the antiprotons having reached these levels will remain there through the whole cascade (circular cascade). If electrons are present and if permitted by energy conservation, the Auger process is the dominant mode for cascade steps with $n \geq 7, 9, 13$ in Neon,

Argon and Krypton, respectively. This is also true for the low-energy Auger transitions in Xenon involving the outer electrons (M-, N-, O-shell). But in Xenon the probabilities of the higher-energy Auger transitions ejecting the L- or K-electrons are already of the same order or smaller as the corresponding radiative transitions[9]. Therefore, the observation of radiative transitions from circular orbits ($|\delta n| = 1$) in the Auger dominated part of the cascade is a clear signal for ionization.

Fig. 1 shows a part of the radiative cascade in Neon proceeding almost completely via circular levels. Corrected for the detector efficiency, all lines have the same yield, on an average of 0.55 ± 0.03. The observation of these lines and the identical yields are direct proof that the Neon atom is totally ionized.

The ejection of the last two electrons in Argon can be seen in fig. 2. The X-ray transitions $17 \rightarrow 16$ and $16 \rightarrow 15$ are suppressed. These transitions are induced by the Auger effect. Their energies correspond to the binding energies of the two K-electrons in an Argon atom stripped of all other electrons. The reason for the small, but non vanishing X-ray yield is that the K-shell is already partially depleted with few percent probability by inner $|\delta n| > 1$ transitions.

Fig. 3 demonstrates the ionization of the L-shell in Krypton. The transitions $28 \rightarrow 27$ to $25 \rightarrow 24$ proceed via the Auger effect. Then a radiative cascade emerges, because all L-, M- and N-electrons are ejected, and the energies of the lines are not sufficient to deplete the remaining K-shell. Depletion of the K-shell is possible via the Auger transitions $16 \rightarrow 15$ and $15 \rightarrow 14$. The corresponding X-ray intensities are suppressed. Similar to the K-electrons in Argon the K- and L-shell in Krypton are also partially ionized by higher transitions before steps with $|\delta n| = 1$ are energetically allowed.

For Xenon (fig. 4) the situation is more involved. The suppression of the lines between 6 keV and 10 keV indicates a high degree of ionization in the L-shell. However, the appearence of the radiative transitions is not a strict argument for complete ionization, because the radiative transition rates begin to dominate the Auger rates, independent of the number of electrons present.

To summarize, the spectra show the complete ionization of antiprotonic Neon, Argon, Krypton and probably of Xenon. The ionization is due to the cascade of the antiproton. The depletion proceeds smoothly by peeling off the electron shell from the outer to the inner electrons.

ATOMIC CAPTURE

The atomic capture of antiprotons proceeds via the Auger effect. Following the arguments of the Borde[10], the antiproton is thought to be captured preferably in orbits with maximum overlap of antiproton and K-electron wave function due to the ejection of a K-electron. But the dependence of the capture cross section on the nuclear charge[6] and the X-ray intensities[11] in muonic atoms show the dominant role of the tightly bound outer electrons. Therefore, the measurement of energies and yields of electronic X-rays give some indications of the atomic capture.

Fig. 5 shows the K_α-energy-distribution in Krypton, extended from 11.9 keV to 13.2 keV. This fact can be explained by two effects[12]:

1) The increasing degree of ionization during the cascade increases the electrons binding and consequently the X-ray energies. But the resulting

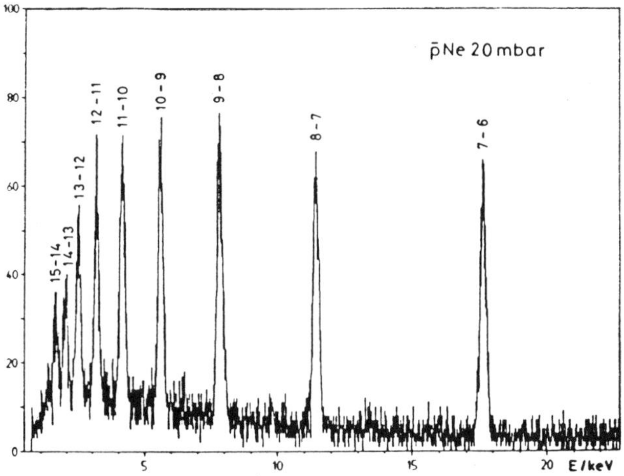

Figure 1

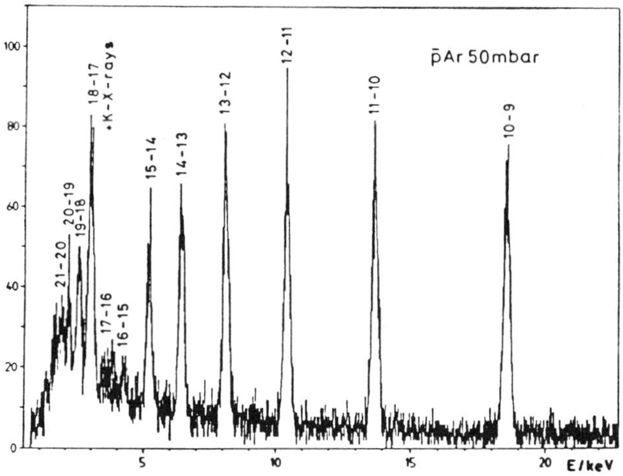

Figure 2

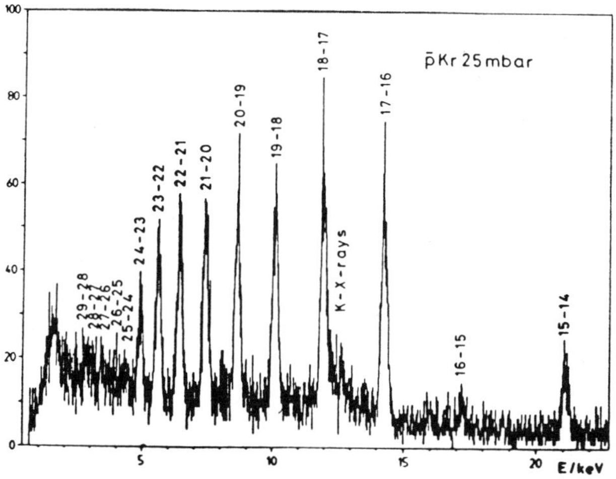

Figure 3

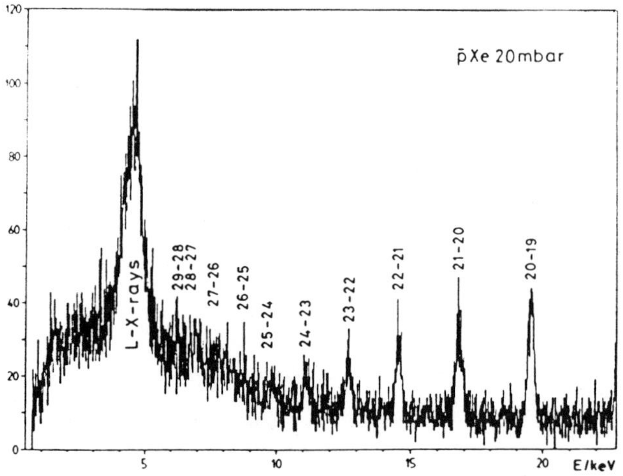

Figure 4

energies lie between 11.9 keV and 12.3 keV[13].

2) The dominant effect is the difference in secreening of the nuclear charge by the antiproton. An electron feels the effective Coulomb field of the nuclear charge shielded by the other electrons and the antiproton. Because of the mass ratio $m_{\bar{p}}/m_{e^-} \approx 2000$, the spacial distribution of an antiproton is a δ-function compared to that of a K-electron. Thus, the lower antiprotonic levels ($n \leq 20$) are well localized within the K-shell. At $n = 43$ the maxima of the antiproton and K-electron distributions coincide. The higher orbits with $n = 70 - 80$ are localized between the K- and L-shell. The screening effect of the antiproton is maximal, if its orbit lies inside the electron shell. Then it shields one elementary charge and the electrons are more weakly bound. For higher levels, however, screening is more and more incomplete. Therefore, if a K-vacancy is created, the energy of the resulting K-X-rays depends on the atomic orbit from which the antiproton starts its Auger transition. The lower antiprotonic levels are related to the lower X-ray energies and the higher levels to the higher energies.

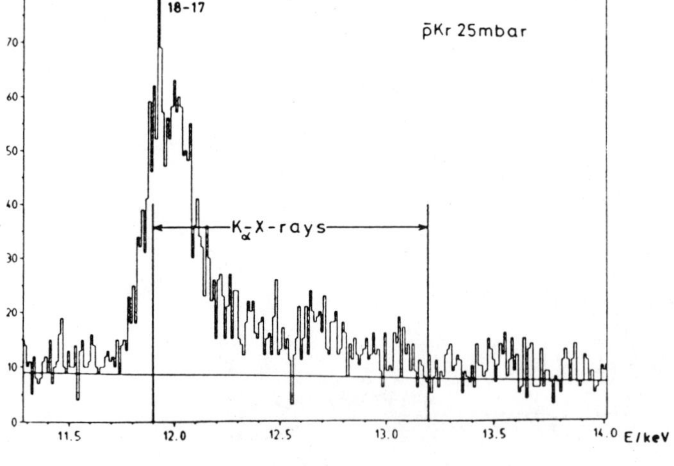

Figure 5

The total K-X-ray yield is 0.294 ± 0.040. This has to be compared with the tabled fluorescene yield $\omega_K = 0.646$[14] of a K-vacancy generated in an atom with complete outer shells. Hence the low yield does not corroborate the assumption of the Borde[10] that the antiproton is preferably captured in levels with n = 43 due to the ejection of a K-electron.

Therefore we conclude from the observed K-X-ray energy distribution and yield, that the antiproton is mostly captured in very high levels (n >> 43) due to the Auger effect involving the outer, weakly bound electrons. This seems to be a general feature of exotic atoms agreeing with the conclusions given in reference 6 and 7.

ACKNOWLEDGEMENT

The authors wish to thank Dr. J. Missimer (SIN theory division) for helpful discussions and careful reading of the manuscript.

REFERENCES

1. T. E. O. Ericson and J. Hüfner, NP B47, (1972) 205.
2. W. R. Johnson and G. Soff, At. Nucl. Data Tables 33, (1985) 405.
3. P. J. Mohr, NIM B9, (1985) 459.
4. L. M. Simons et al., The cyclotron trap: A device to produce high stop densities, in preparation.
5. R. Bacher, D. Gotta, J. Missimer, N. C. Mukhopadhyay and L.M. Simons, PRL 54, (1985) 2087.
6. H. Schneuwly, V. I. Pokrovsky and L. I. Ponomarev, NP A312, (1978) 419.
7. T. von Egidy, D. H. Jakubassa-Amundsen and F. J. Hartmann, PR A29, (1984) 455.
8. Y. Eisenberg and D. Kessler, Nuovo Cimento 19, (1961) 1195.
9. R. A. Ferrell, PRL 4, (1960) 425.
10. A. H. de Borde, Proc. Phys. Soc., London, 67, (1954) 57.
11. F. J. Hartmann, R. Bergmann, H. Daniel, H.-J. Pfeiffer, T. von Egidy and W. Wilhelm, Z. Phys. A305, (1982) 189.
12. P. Vogel, PL 58B, (1975) 52.
13. G. Zschornack, Sov. J. Part. Nucl. 14, No 4, (1983) 349, Akademie der Wissenschaften der DDR, Preprint ZfK-574 (1986).
14. C. M. Lederer and V. S. Shirley, Tables of Isotopes (7th Edition).

ATOMIC COLLISION PHYSICS AT LEAR

PS194 Collaboration, presented by K. Elsener

CERN EP-Division
CH-1211 Geneve 23
Switzerland

INTRODUCTION

Collision processes of charged particles penetrating matter are governed by the Coulomb interaction. At first glance, cross sections are therefore expected to be proportional to the projectile charge squared.

Corrections to this Z_1^2 law are known from models describing precise measurements of ionization, energy loss, inner shell excitation etc.. However, higher order Z_1 effects are generally found to be small in fast collisions, and therefore difficult to investigate.

The low energy antiprotons at LEAR provide a unique way to improve the situation: it is now possible to compare experimental results of heavy negative and positive projectiles, antiprotons and protons, in the MeV region, thus probing directly Z_1^3 effects in atomic collisions. The first experiment of this kind, ionization of Helium by antiproton impact, has been performed and a surprising effect has been found. In the present paper, this new result and some difficulties with its interpretation are briefly discussed. Moreover, antiproton studies of the Barkas-effect and of K-shell excitation are discussed as a possible extension of the ionization work to other areas of atomic collision physics.

IONIZATION OF HELIUM

Ionization cross sections have been measured with 0.5 to 4.5 MeV antiprotons on Helium[1]. In these experiments, a degraded 105 MeV/c LEAR beam passes a low pressure gas cell. The Helium ions are extracted and accelerated towards a channeltron detector. The antiprotons are detected downstreams in a scintillation counter. The time of flight measurement using the ion (start) and antiproton (stop) signals yields clean spectra showing well separated He^{++} and He^{+} peaks[1]. Relative cross sections can be extracted with little uncertainty from these spectra.

The comparison of the antiproton results with earlier proton data shows a dramatic increase in the double ionization (σ^{++}) for antiprotons on Helium, whereas the single ionization (σ^{+}) is the same for negative and positive projectiles. This surprising effect is illustrated in Fig. 1, where the measured ratio $R = \sigma^{++}/\sigma^{+}$ is plotted.

The calculation of the energy dependence of σ^{++} is difficult, and only qualitative results have been obtained so far. Several mechanisms are being discussed (Fig. 2), but none of them contains a large Z_1^3 effect, as demanded by the experimental results. The shake-off process is expected to

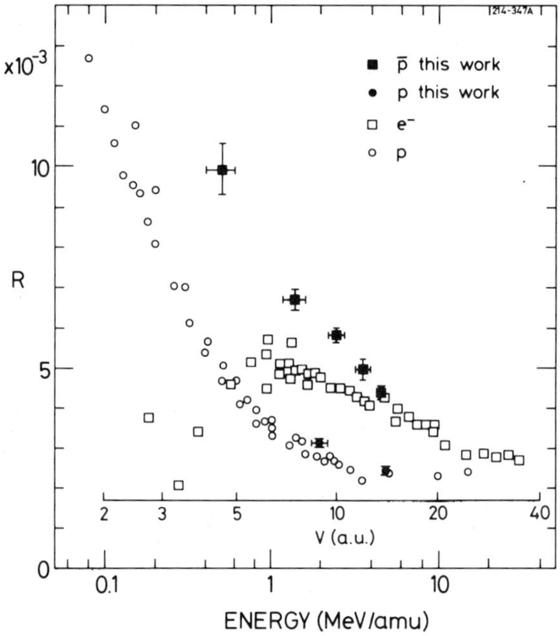

Fig. 1: Measured ratios $R=\sigma^{++}/\sigma^{+}$ of double to single ionzation cross sections of charged particle impact on Helium. Antiproton data are shown together with earlier proton and electron data. (For references to previous measurements see ref. 1).

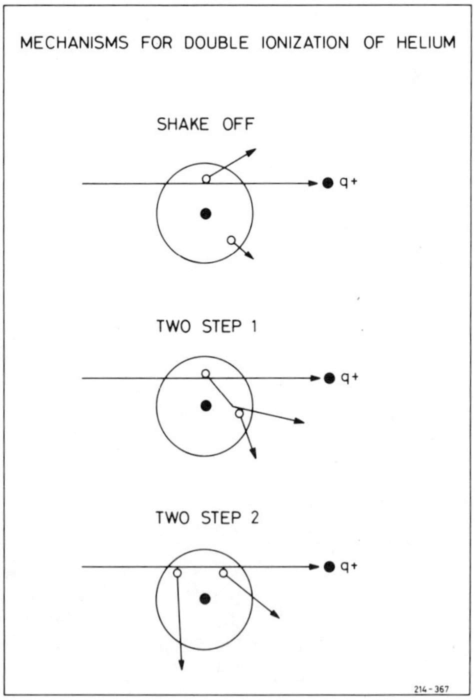

Fig. 2: Schematic representation of double ionization of Helium atoms by fast charged particles.

govern double ionization at high velocities[1], yielding a ratio R = const.. Interferences between the different mechanisms of Fig. 2 are being considered[1,2], which could provide an odd power Z_1 dependence of σ^{++}. The magnitude of such interferences depends, however, crucially on the dynamic electron correlations in the atom: assuming the dipole approximation for fast projectiles ($\Delta L=1$ in a collision), the shake-off process leads to a "collision"-electron in a p-wave final state, and a "shaken-off" electron in an s-wave, if both e^- stem from (uncorrelated) s-wave ground states in the atom. On the other hand, under equal assumptions the two-step-2 process yields both electrons in p-wave final states. Interference between these two processes would therefore be forbidden if the electrons were totally uncorrelated in the Helium atom during the collision process.

The new antiproton results resolve the longstanding problem between electron and proton data at equal velocities (see Fig.1). There is now experimental evidence that the e^-/p difference in the ratio R was due to opposite charge rather than different projectile masses. Further antiproton measurements above 5 MeV (towards the shake-off limit) and below 0.5 MeV are needed to enable a detailed comparison with forthcoming calculations of R. This will provide an insight to the fundamental problem of electron-electron correlations in Helium atoms.

BARKAS-EFFECT IN STOPPING POWER

The surprising antiproton/proton difference in the double ionization cross section once again suggests the importance of accurately measuring the Z_1^3 effects in stopping power. Evidence for such effects has been found already in 1963, when Barkas et al.[3] discovered that the ranges of Σ^- are slightly greater than of Σ^+. It has been difficult, however, to investigate accurately the underlying atomic physics. Only indirect methods were available to test charge effects in energy loss dE/dx. For example, a precise comparison of Z_1=1,2 and 3 stopping power[4] (p, α, Li^{3+} beams) has been critizised for experimental uncertainty (charge pick-up by the projectiles) and because of the interpretation of the results[5] (also Z_1^4 contributions enter the analysis). Theoretical calculations of the Z_1^3 contributions to energy loss have been performed by Jackson and McCarthy[6] and Lindhard[7], but the results differ by close to a factor of two.

Antiproton/proton experiments could help to clarify the situation. Our group's test measurement of energy loss in a 7μm Si surface barrier detector indicates some disagreement with earlier measurements of the Barkas effect. Interpretation suffers from the fact that a degraded $\bar{p}$ beam had to be used at LEAR and dE/dx uncertainties in the degrader obscure the Z_1^3 effect investigated in the Si detector. However, measurements below 1 MeV seem feasible now, as only very low $\bar{p}$ intensities are required. An experiment using a degraded LEAR beam and time of flight techniques, currently being proposed in an addendum to PS194, should provide accurate, unambiguous information on Z_1^3 effects in stopping power.

INNER-SHELL EXCITATION

Coulomb ionization and excitation of atomic inner shells by fast charged projectiles have been studied extensively with positive particles[8]. The theory describing the cross sections, usually called the ECPSSR model, contains a perturbated stationary-state (PSS) approximation with energy loss (E), Coulomb deflection (C) and relativistic (R) corrections. Reliable predictions of e.g. the K-shell excitation cross section σ_K for a variety of projectiles over a wide energy range can be obtained by this model. However, some discrepancies still remain and an empirical correction function has been introduced by the authors of ref.8 to remove them.

The PSS theory, through the consideration of binding and polarization effects, as well as some of the correction factors mentioned above, contain strong Z_1^3 contributions. For example, at low projectile velocity, the target electrons may adjust to the combined Coulomb field of target and projectile. Protons then give rise to an increase in binding energy hence reducing the ionization probability. Antiprotons should show an "antibinding" effect with a correspondingly larger ionization probability. As a consequence of such charge effects, the theory predicts large differences when comparing K-shell excitation by antiprotons and protons. A factor two in σ_K is expected e.g. for 1 MeV $\bar{p}/p$ which penetrate a thin copper foil.

Measurements of K-shell excitation with antiprotons could therefore be a rather simple, but crucial test of the present model, investigating directly the Z_1^3 contributions to the ECPSSR theory. Moreover, it has been suggested[5] that a deeper understanding of the Barkas effect could be gained by studying the polarization of the wave functions through K-shell excitation.

CONCLUSION

The first atomic collision experiments with LEAR antiprotons gave surprising results, a reminder of the fact that even in well established areas of physics a large number of open questions still exist. Many of them can now be attacked in a clean way by experiments using antiprotons. In comparison with the comprehensive data set from proton experiments, LEAR results could therefore resolve some of the most debated problems concerning the interaction of positive and negative particles with gases and solids.

REFERENCES

1. L.H. Andersen et al., Phys. Rev. Lett. 57, 2147 (1986).
2. J.H. McGuire, Phys. Rev. Lett. 49, 1153 (1982).
3. W.H. Barkas et al., Phys. Rev. Lett. 11, 26 (1963)
4. H.H. Andersen, Phys. Scr. 28, 268 (1983)
5. G. Basbas, Nucl. Instr. Meth. B4, 227 (1984)
6. J.D.Jackson and R.L.McCarthy, Phys. Rev. B6, 413 (1972)
7. J. Lindhard, Nucl. Instr. Meth. 132, 1 (1976)
8. see e.g. H. Paul and J. Muhr, Phys. Rep. 135, 47 (1986), and references therein.

MEASUREMENT OF THE ANTIPROTONIC LYMAN- AND BALMER X-RAYS

OF $\bar{p}H$ AND $\bar{p}D$ ATOMS AT VERY LOW TARGET PRESSURES

R. Bacher, P. Blüm, J. Egger*, K. Elsener**,
D. Gotta, K. Heitlinger, W. Kunold, D. Rohmann,
M. Schneider and L.M. Simons*

Kernforschungszentrum Karlsruhe GmbH
Institut für Kernphysik and
Institut für Experimentelle Kernphysik der Universität
P.O.B. 3640, 7500 Karlsruhe 1, Federal Republic of Germany
SIN, Villigen, Switzerland*
CERN, Geneva, Switzerland**

INTRODUCTION

The experiment described here (PS 175) aims at the measurement of the 2p-1s transition in antiprotonic hydrogen under optimal conditions[1]. Theoretical calculations based on potential models[2,3,4,5,6,7] predict hadronic shifts and widths of the ground state of about 1 keV compared to a 2p-1s electromagnetic transition energy of 9.35 keV. Because the energy difference between 2p and 3p levels is 1.735 keV, the 2p-1s transition is separated from other np-1s transitions by more than the width of the ground state. The 2p-1s transition is the only transition satisfying this criterion.

In order to determine the optimal conditions, we first undertook an extensive study of the pressure dependence of the Balmer series over a pressure range from 16 mbar to 300 mbar. From this the population of the 2p level as a function of pressure should be obtained. Moreover in comparison with theoretical predictions about the process of deexcitation of the $\bar{p}p$-system[8] (cascade process), a deeper knowledge of this process could be obtained.

EXPERIMENTAL APPROACH

In order to fulfill the program mentioned above we used a device - the cylotron trap - which concentrates particles in a gas target in a small volume at low pressures[9]. The principle of this device is do wind up the range curve of particles in a magnetic cyclotron field. Particles loosing energy in the target gas spiral to the center. In the usual linear set-up the volume of the stop region for a particle beam is approximately given by $[\delta R]^3$ with straggling widths $\delta R_L = \delta R$, $\delta R_V \approx 2\ \delta R$ in direction longitudinal and transversal to the beam direction. For example the Range R of a $\bar{p}$-beam of 100 MeV/c stopping in H_2 of 10 mbar pressure is R ~ 180 m and the stopping volume is several m^3. The large stopping volume makes the use of high resolution detectors impossible. In the cyclotron trap, however, the volume of the stop region would be ideally of order of magnitude $[R_{in} \cdot \delta R/R]^3$, where the beam is almost tangentially injected at a certain radius R_{in}. Because of betatron oscillations, $\delta R/R$ has to be multiplied with a correction factor f with $1 < f < 10$ depending on beam properties and adjustments. The expected stopping volume for stopped antiprotons in hydrogen gas at 10 mbar is between 1 and 100 cm^3.

A superconducting split-coil magnet provides the cyclotron field. It accepts particles with a momentum up to 123 MeV/c inside a field diameter of 290 mm. The radius of injection is about 120 mm. A target chamber is mounted in the free gap and the bore holes of the cryostat (Fig. 1). The 202 MeV/c antiprotons from the beam pipe enter the target chamber through a 50 µm thick Mylar window and are moderated down to about 105 MeV/c in a scintillator-moderator arrangement. An additional deceleration in a 10 µm thick polyethylene foil placed almost opposite to the main moderator prevents the antiprotons from hitting the moderator again. The x-ray detectors are axially mounted in the bore holes, and are directly flanged to the gas volume. In addition there are 24 scintillation counters placed radially between the target chamber and the cryostat walls. They cover 30 % of 4π solid angle and serve to detect annihilation products. This proved to be useful in optimizing the moderator thickness and to determine the stop efficiency (Fig. 2). At a pressure of 36 mbar He, 30 % of the incoming antiprotons could be stopped in the He gas.

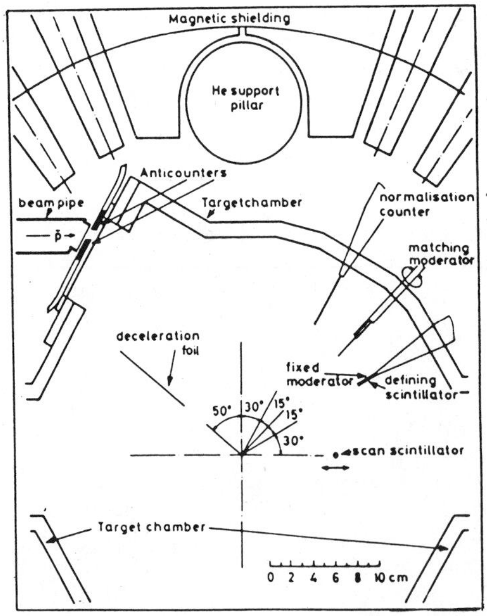

Fig. 1. Set-up for injection at 202 MeV/c beam momentum.

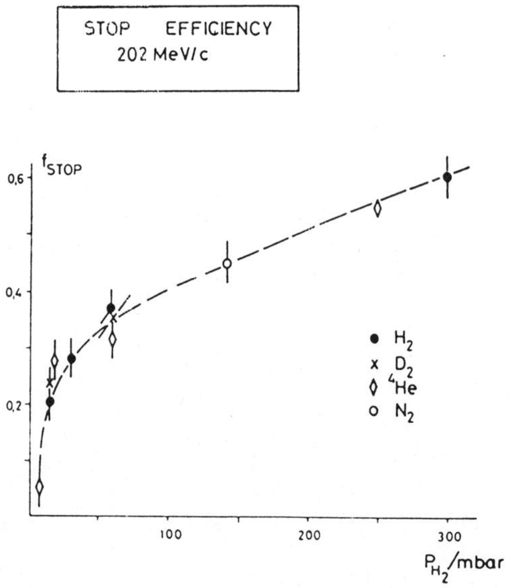

Fig. 2. Stop efficiency for 202 MeV/c antiprotons P = mean equivalent pressure H_2.

MEASUREMENT AND RESULTS

The x-rays have been measured with Si(Li) and Ge solid state detectors. The efficiency and resolution of the detectors for energies above 5 keV could be determined with radioactive sources. In the region below 5 keV a novel method was applied to obtain the response function of the detectors. In antiprotonic N_2 for example the x-ray intensities for transitions $n \rightarrow n-1$ between high principal quantum numbers are constant on the percent level, because the cascade process goes mainly through circular levels with $n, l = n-1 \rightarrow n-1, l = n-2$. The response function of the detectors can therefore be taken directly from the measured $\bar{p}N$ spectra. An example of such a spectrum is shown in Fig. 3. The slope of the efficiency curve is taken from the measured x-ray intensities calibrated to the absolute efficiency measured at 6.4 and 14.4 keV with a ^{57}Co source. Thus absolute yields for the measured L x-rays (Fig. 4) could be determined both for $\bar{p}H$ and $\bar{p}D$ (Fig. 5).

For antiprotonic hydrogen the measured yields are well reproduced by the cascade code of Leon and Borie assuming a kinetic energy of the $\bar{p}p$-system T = 1 eV, and the Stark mixing parameter, k = 2, a consistent picture of the cascade process could be obtained over a pressure range of 2 orders of magnitude.

The intensities of the $\bar{p}D$ L x-rays compared to $\bar{p}H$ x-rays are lower by 30 %. This can only be explained assuming that the strong interaction broadening of the 3d-level is bigger by a factor of 5 than anticipated theoretically[10].

The measured yields of the L x-rays indicate that a measurement of the 2p-1s transition should be performed at pressures below 10 mbar: here the 2p-population is almost maximum. The measurement of the K x-rays was performed at a pressure of 30 mbar because the minimum beam momentum which could be obtained was 202 MeV/c. The x-ray spectrum was measured with a 300 mm^2 Si(Li)-detector covering 10^{-3} of a 4π solid angle. In the spectrum obtained (Fig. 6) from 2.8×10^9 incoming antiprotons (11 spills), the two narrow peaks stem from the $\bar{p}O(8-7)$ and $\bar{p}O(7-6)$ transitions of the water

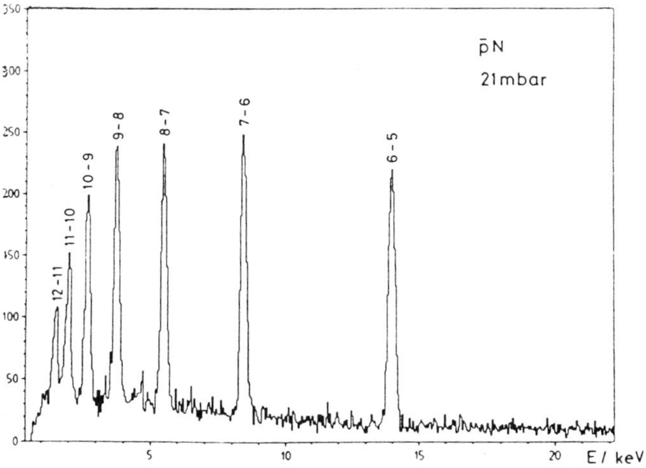

Fig. 3.

Antiprotonic nitrogen spectrum measured at 21 mbar.

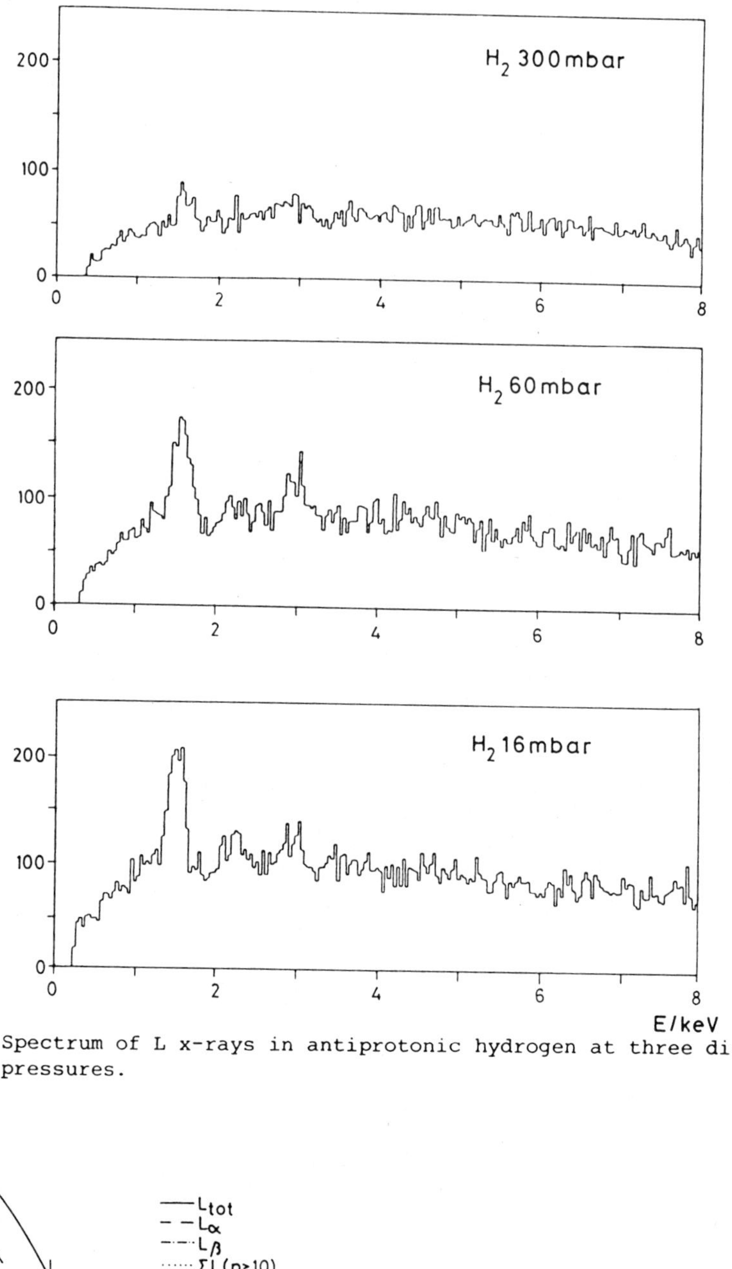

Fig. 4. Spectrum of L x-rays in antiprotonic hydrogen at three different pressures.

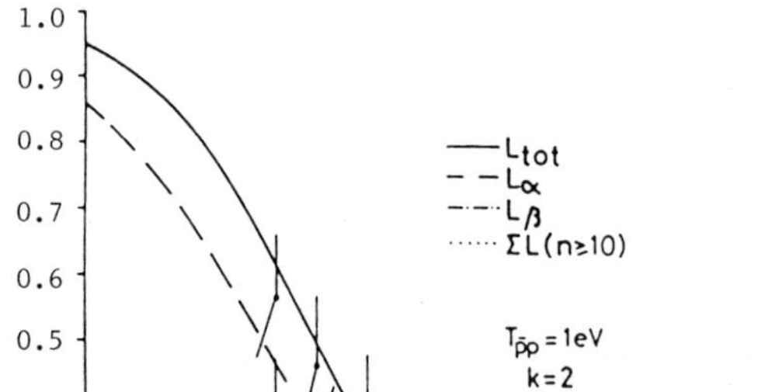

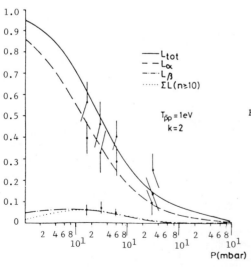

Fig. 5. Measured L-yields in antiprotonic hydrogen compared with the cascade code of Leon and Borie.

evaporating from the walls of the target chamber. The intensities observed correspond to a partial pressure of 10^{-2} mbar H_2O within 30 mbar H_2, which could be confirmed by a mass-spectroscopy measurement. Inside the detector the only electronic fluorescence X-ray produced stems from copper, as confirmed by an irradition measurement. The intensity ratio of the $\bar{p}O$-lines is known from the cascade of a bare antiprotonic oxygen atom. The intensity ratio of $\bar{p}O$ to e^- Cu is known from the uncorrelated spectrum. From this "free-running" spectrum in which no correlation between stopped antiprotons and X-rays is required the slope of the background was determined to be flat. This smooth background originates from the low energy neutral component of the electromagnetic shower produced by annhilation products everywhere in the neighbourhood of the detector crystal.

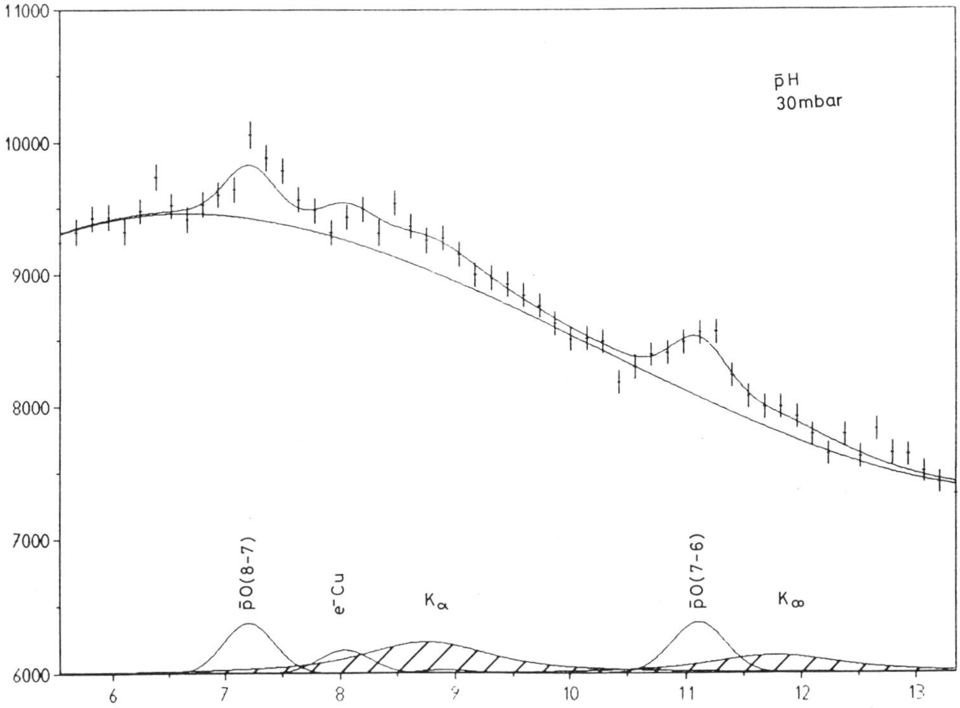

Fig. 6. Antiprotonic hydrogen spectrum at 30 mbar with a fit to the K-transitions and background

At pressures below one atmosphere, the total intensity of the K transition is shared by the 2p-1s and the np-1s transitions with $n \geq 10$. The sharing ratio is determined completely by non hadronic parameters of the cascade which are known from the measurement of the L x-ray intensities. For the computer fit the intensity ratio 2p-1s to np-1s ($n \geq 10$) was fixed to 1.73. The distance of the two K-lines was fixed by the theoretical electromagnetic energy difference. The Lorentz width was assumed to be equal. The background was fitted with a 4th order polynomial. The fit yielded (5200 ± 870) events in the K-structure. The measured hadronic shifts and widths are:

$$\Delta E = (8.75 \pm 0.13)\ \text{keV} - 9.41\ \text{keV}$$
$$= -(0.66 \pm 0.13)\ \text{keV}$$
$$\Gamma = (1.13 \pm 0.23)\ \text{keV}$$

The yields are:

$$Y_{2-1} = (4{,}9 \pm 1.6) \times 10^{-3}$$
$$Y_{tot} = (7.8 \pm 1.9) \times 10^{-3}$$

Using the total L-yield of (48 ± 8) % at 36 mbar yields then a hadronic 2p-width of (37 ± 13) meV.

To summarize the atomic cascade process of $\bar{p}$H is well reproduced by recent calculations. The smaller yields of the L-series in $\bar{p}$D can be explained by a hadronic 3d-width of 30 μeV. The other measured shifts and widths agree with theoretical predictions from potential models[2-7]. However a real comparison to the theory requires data which are an order of magnitude more precise. In $\bar{p}$H an increase of statistics by an order of magnitude is necessary to resolve the hyperfine structure of the ground state. A crystal-spectrometer combined with a small stop distribution at low pressure could measure the shifts and the widths of the 2p-levels of the hydrogen- and helium-isotops. A measurement of these quantities would permit an analysis of the spin and isospin dependence of $\bar{p}$-nucleon forces.

ACKNOWLEDGEMENT

We would like to thank Dr. J. Missimer for claryfying discussions and for reading the manuscript.

REFERENCES

1. P. Blüm, D. Gotta, R. Guigas, H. Koch, W. Kunold, M. Schneider and L.M. Simons, Proposal PS 175, CERN/PSCC/S27 (1980).
   L. M. Simons, Proc. Workshop on Physics at LEAR with low-energy cooled antiprotons (Erice, 1982), eds. U. Gastaldi and R. Klapisch (Plenum Press, New York, 1984) p. 155.
   D. Gotta, ibidem, p. 165.
2. R. A. Bryan and R. J. N. Phillips, Nucl. Phys. B5 (1968) 201.
3. C. B. Dover and J. M. Richard, Phys. Rev. C21, (1981) 1466.
4. J. Cote et al., Phys. Rev. Lett. 48, (1982) 1319.
5. J. M. Richard and M. E. Sainio, PL110B, (1982) 349.
6. B. Moussallam, Orsay preprint 86-44.
7. W. B. Kaufmann, Proc. Kaon factory workshop, T. R. I. 79-1.
8. E. Borie and M. Leon, P.R. A21, (1980) 1460.
9. P. Blüm, D. Gotta, W. Kunold, M. Schneider and L.M. Simons, The cyclotron trap: A device to produce high stop densities, in preparation.
10. S. Wycech, A. M. Green and J. A. Niskanen, PL152B, (1985) 308.

THE MAGNETIC LEVITATION ELECTROMETER: THE SEARCHES FOR QUARKS AND THE ELECTRON-PROTON CHARGE DIFFERENCE

G. Morpurgo

Istituto di Fisica dell'Universita' and INFN-Genova

The original subject of these lectures had to be the electron-proton charge difference;but because such a measurement has been ,infact, a byproduct of a much longer experiment - the search of quarks in matter - and because both experiments use the same instrument and imply similar problems, I thought that it might be advisable to start describing the experiment of search for quarks.

1. INTRODUCTION

1.1-My motivation for the search of quarks in matter was the success of the non relativistic quark model [Morpurgo,1965].Although it is out of the theme,I display in the figures 1 and 2 the predictions of the model on the magnetic moments of the baryons and on the semileptonic decays of the hyperons,because precisely these predictions [Morpurgo,1965, 1968] were among my motivations in starting and continuing our search for free quarks.The agreement between predictions and data (10% usually,20% in the worst case) are still,to my mind,very striking and a sort of mistery.

1.2-One reason why we started the search of quarks in matter in the way I am going to describe was that the different (indirect) methods suggested or used at that time had the defect,in my opinion,of introducing too many assumptions;if free quarks were found,these methods would have been o.k., of course, but if they were not, as it was the case, one did not really know if this was so because free quarks were not present or because some of the assumptions introduced were not correct.

Our idea was to measure simply the "residual charge" of a grain of matter as heavy as one could.By "residual charge" we mean the minimum value of the electric charge of the grain reached after the grain is neutralized as much as possible extracting from it or adding to it a convenient number of electrons.

If a grain of matter does not contain a quark its charge is:

$$Q = ne \qquad n=0,\pm1,\pm2.... \qquad (1)$$

and its residual charge Q_R=0;on the other hand if a grain contains in its interior one or two quarks, its charge is:

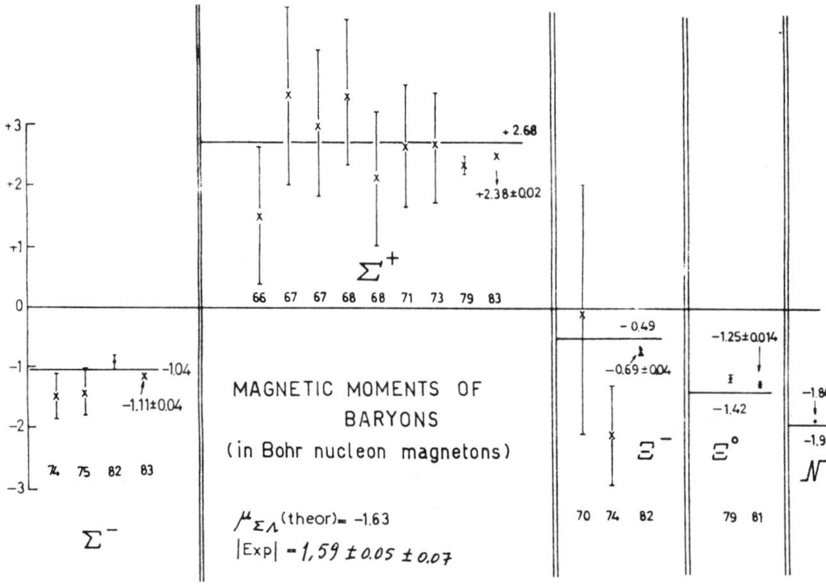

Fig.1- The values of the magnetic moments of the octet baryons calculated with the non relativistic quark model (Morpurgo 1965) using as inputs only the magnetic moments of the u quark and of the s quark. The calculated values are compared with the time evolution of the experimental values. For more details compare [Morpurgo, 1983b].

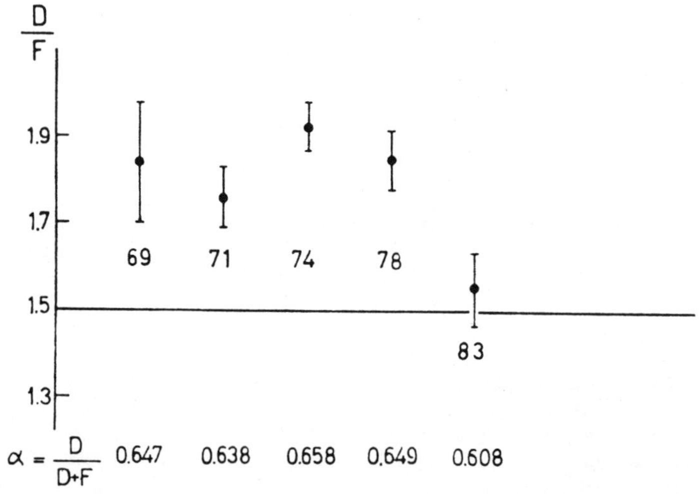

Fig.2- The experimental values of the D/F ratio in the semileptonic decays of the hyperons optained from various experiments compared with the non relativistic quark model prediction of 3/2 (same ref. as that of fig.1).

$$Q = ne \pm \frac{1}{3}e \qquad (2)$$

and the residual charge is $Q_R = \pm(1/3)e$. Thus, if no quarks are contained in the object, we reach -extracting or adding electrons to the object (kept isolated)- a state with $Q_R=0$; if the object contains 1 or 2 quarks we reach $Q_R=\pm(1/3)e$.

To keep an object isolated one might follow the original technique of Millikan, having the object suspended electrically between the plates of a capacitor (fig.3). But the electric field necessary to prevent it from falling is too high to be practical if the mass of the object is larger than 10^{-9}g [the droplets used by Millikan were in the range of 10^{-10}g].

Therefore our idea was to use a magnetic field to levitate the object in vacuum and - once the object is there - to measure its charge by an electric field; this is the principle of what we called "the magnetic levitation electrometer".

Consider an object O levitating in vacuum in the middle of two plates (fig.4), and apply to the plates an electric field E (along the x axis). The levitating object can be thought as a pendulum, capable of oscillating with some natural frequency ν_x along the x axis near to the equilibrium position. If the x component of the force on the object is simply:

$$F_x = QE \qquad (3)$$

the displacement (Δ_i) of an object of mass M and charge Q_i from is equilibrium position under the action of the electric field is:

$$\Delta_i = \frac{Q_i E}{4\pi^2 \nu_x^2 M} \qquad (4)$$

For a given object the equation (4) can be simply written:

$$\Delta_i = kQ_i \qquad (5)$$

where k is a constant $(=E/(4\pi^2\nu_x^2 M))$. Clearly, on measuring the sequence of displacements Δ_i that correspond to successive extractions or additions of electrons from or to the levitating object, one obtains the sequence of the corresponding Q_i and thus the residual charge Q_R of the object.

As a matter of fact it is convenient to use instead of a static electric field, a square wave alternating one:

$$E_x(t) = E\eta(t) \qquad (6)$$

where $\eta(t)=+1$ in the first half of each period and $\eta(t)=-1$ in the second half; the field alternates at the natural frequency ν_x of the magnetic valley.

Then the equation (5) is simply rewritten as:

$$A_i = f\Delta_i \qquad (7)$$

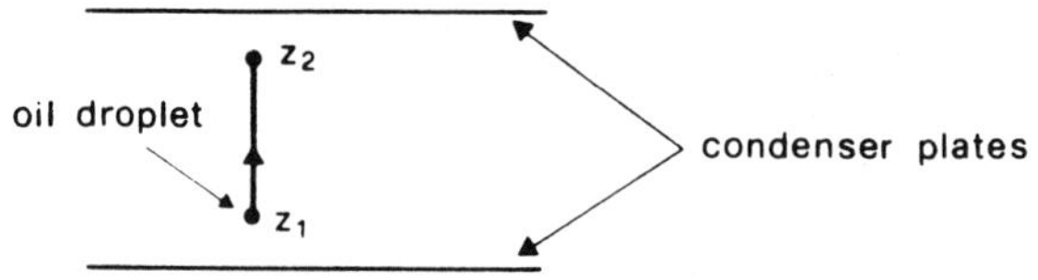

Fig.3-A scheme of the Millikan set up.

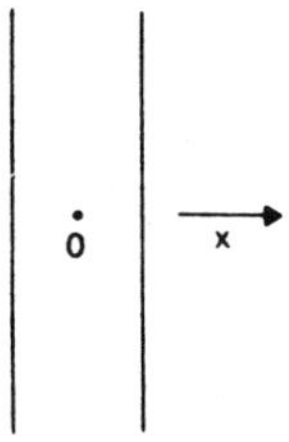

Fig.4-A scheme of the magnetic levitation set up.

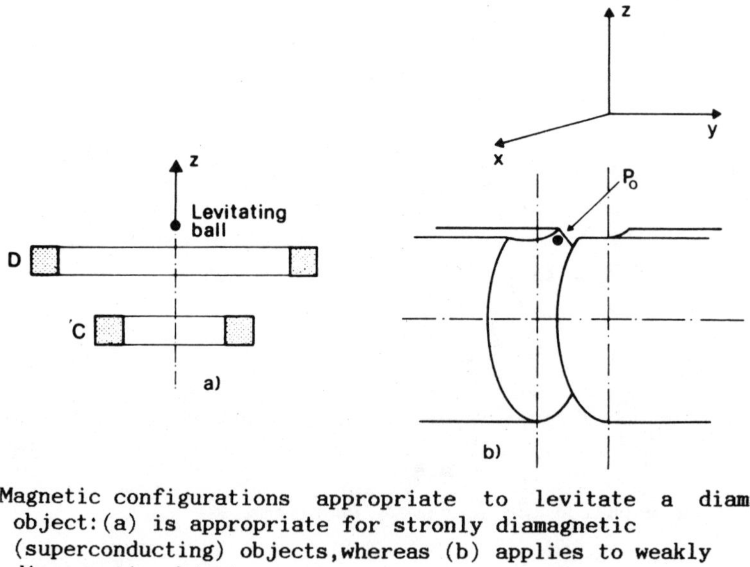

Fig.5-Magnetic configurations appropriate to levitate a diamagnetic object:(a) is appropriate for stronly diamagnetic (superconducting) objects,whereas (b) applies to weakly diamagnetic objects.

where A_i is now the amplitude of oscillation of the object when its charge is Q_i ;f in (7) is the usual amplification Lorentz factor arising from the resonance.The value of f depend on how near is the frequency of the electric field to the natural mechanical frequency ν_x of oscillation.At resonance it is:

$$f = \frac{4N_{1/2}}{\ln 2} \tag{8}$$

where $N_{1/2}$ characterizes the damping of the oscillator;$N_{1/2}$ is the number of free oscillations (in the absence of the driving force) after which the amplitude of oscillation has decayed to a half.

The use of an oscillating electric field increases of course the sensitivity, and thus the mass M of the object that can be measured. Moreover -this is equally important- it eliminates automatically -as we will see- some forces simulating a spurious charge of the object [I will discuss this whole subject of the spurious effects in sect.4].

Let me finally note some typical orders of magnitude.For:

$M=10^{-4}$g , ν=1 Hz , E_x=2 kV/cm , Q=1 electron charge , $N_{1/2}$=20

one gets for the amplitude of oscillation:

$A \approx 1\ \mu$

Of course the important quantity is not the amplitude of oscillation itself but the signal to noise ratio.I will come to this later.

1.3-Before concluding this introductory section I list in the following table I all the magnetic levitation experiments performed so far (I have omitted only some unfinished ones).The first experiment (by Gallinaro and myself,1966) used the room temperature of graphite;it was followed , one year later, by the first ferromagnetic levitation apparatus by Stover et al,the aim of which was infact not the search of quarks but the electron-proton charge difference.Five years later (in 1971) we note the preliminary results of the low temperature diamagnetic levitation experiments by the Stanford group.These were accompanied by other measurements of the same type -by the same group-and by a number of ferromagnetic levitation experiments.The process of making the ferromagnetic levitometer a real instrument was a slow one (we spent 10 years on it) because it was necessary to understand several spurious forces that might simulate an electric force acting on a residual charge.As I have said I will comment on these forces in sect.4;before doing this it is necessary to have a more detailed view of the instrument.

2.DIAMAGNETIC AND FERROMAGNETIC LEVITATION

We have mentioned the diamagnetic and ferromagnetic types of levitation.The difference is as follows.The diamagnetic levitation is a static equilibrium condition;a diamagnetic object is sustained against gravity from the repulsion produced on it by the gradients of an appropriately shaped magnetic field (compare the fig.5).For instance in the fig.5b the edges of the pole caps of an electromagnet produced a gradient of the magnetic field directed upwards that can be of strenght sufficient to counteract the gravity if a material having a sufficient

T A B L E   I

MAGNETIC LEVITATION EXPS

YEAR	METHOD	SUBSTANCE	MASS OF OBJECT g.	TOTAL MASS g.	SIZE OF OBJECT (R. in mm)	AUTHOR
1966	DIAMAG.	GRAPHITE	10^{-9}	5.10^{-9}	10^{-2}	Gallinaro and Morpurgo
1967	FERROMAG.	IRON	4.10^{-6}	10^{-5}	5.10^{-2}	Stover et al.
1970	DIAMAG.	GRAPHITE	3.10^{-7}	$2.5\ 10^{-6}$	7.10^{-2}	Morpurgo et al.
1971	DIAMAG. (Low T)	NIOBIUM	10^{-5}	10^{-5}		Fairbank et al.
1974	FERROMAG.	IRON	3.10^{-5}	$3.6\ 10^{-4}$	10^{-1}	Garris and Ziock
1977	FERROMAG	IRON	3.10^{-5} – 1.10^{-4}	10^{-3}	$1.5\ 10^{-1}$	Morpurgo et al.
1977	DIAMAG. (Low T)	NIOBIUM				Fairbank et al.
1979	DIAMAG. (Low T)	NIOBIUM				Fairbank et al.
1980	FERROM.	IRON	→ $3\ 10^{-5}$ to			Marinelli and Morpurgo
1981	DIAMAG. (Low T)	NIOBIUM	1.10^{-4}	1.10^{-3}	1 – $1.5\ 10^{-1}$	Fairbank et al.
81-82	FERROM.	IRON		$3.7\ 10^{-3}$	1 – $1.5\ 10^{-1}$	Marinelli and Morpurgo
1983	FERROM.	IRON		$0.7\ 10^{-3}$	1.10^{-1}	Ziock et al.
1985 1986	FERROM.	IRON + NIOBIUM		**$6.5\ 10^{-3}$** Iron $0.7\ 10^{-3}$	1.2 – $1.5\ 10^{-1}$	Smith et al.
				$\approx 1.3 \cdot 10^{-2}$g Total		

diamagnetic susceptivity can be found.

At room temperature the material with the highest specific diamagnetic susceptivity is pyrolitic graphite and it was indeed the graphite that we used in our initial experiment (compare the table I).

Of course the diamagnetic susceptivity of graphite at room temperature is still very small (a few times 10^{-6}) with respect to that of a superconductor, (e.g.Nb) which,as well known, expels completely the magnetic field from its interior and is therefore a complete diamagnet.Of course to work with superconductors implies to operate at low (liquid He) temperature.

Quite differently the ferromagnetic levitation is not a static equilibrium condition;a general theorem (Earnshaw) states that no stable equilibrium position can exist for a ferromagnetic substance inside a static magnetic field.To levitate a ferromagnetic object we need a feedback process, exemplified in the fig.6.The vertical position of the object is measured by its shadow on an appropriately positioned photodetector, and the current in the coil producing the magnetic field is increased if the object tends to fall,it is decreased if it tends to raise.

At first sight the ferromagnetic levitation appears to be more complicated than the diamagnetic one.However when,around 1972,we decided to turn from the diamagnetic to the ferromagnetic levitation our motivation was the following:

a)It appeared to us important to work at room temperature,in order to guarantee a flexibility to the instrument that would have been difficult to achieve with the need of a cryostat.

b)We argued (compare Gallinaro,Marinelli and Morpurgo,1976 - Marinelli and Morpurgo 1982) that the variety of substances that could be explored for the presence of quarks could be much larger in the ferromagnetic than in the diamagnetic instrument;we suggested infact that one could use alloys of ferro and non ferromagnetic substances,and still have levitation.A similar idea has been adopted recently by the Rutherford group (Smith et al,1985 -compare the table I);using a ferromagnetic levitation electrometer they have measured Nb spheres covered with a ferromagnetic material.

3.THE GENERAL SCHEME OF THE INSTRUMENT

The general scheme of the instrument is apparent from the fig.7.The coils A and B provide the feedback levitation of the object O illuminated by the lamp L. Usually the levitated object is a sphere and in the following we will have in mind the spheres that we used,namely steel spheres with a diameter of 2/10 and 3/10 mm and a corresponding mass -respectively- of $3 \cdot 10^{-5}$g and $1.1 \cdot 10^{-4}$g. As one sees from the table I a diameter of 3/10 mm is essentially the maximum that has been used in all the experiments so far;there is a reason for this which we shall discuss later.

The "vertical" photodiode V.D. detecting the shadow of O (through a system of lenses and the semitransparent mirror M) controls the feedback levitation current.[It is the photodetector that has already appeared in the scheme of fig.6].The "horizontal photodetector" H.D. measures the x oscillation of the object and transfers the signal to the lock-in amplifier;H.D. governs also the coils D that create the damping.Also visible in the fig.7 are the plates P_o and P_1 between which the alternating square wave high voltage is applied. The square wave is generated in the periodic interchange of the high-voltage supplies;this interchange is effected by a system of high voltage reeds piloted by a clock;the same clock governs the reference channel of the lock-in. Of course the plates are inside a vacuum tight chamber (the levitation

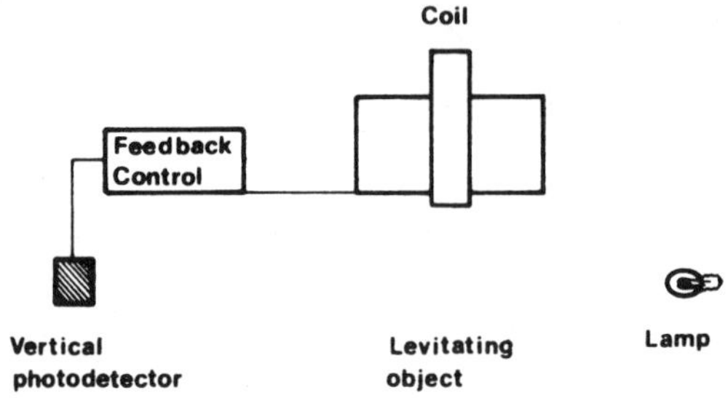

Fig.6-Principle of the feedback levitation of ferromagnetic objects.

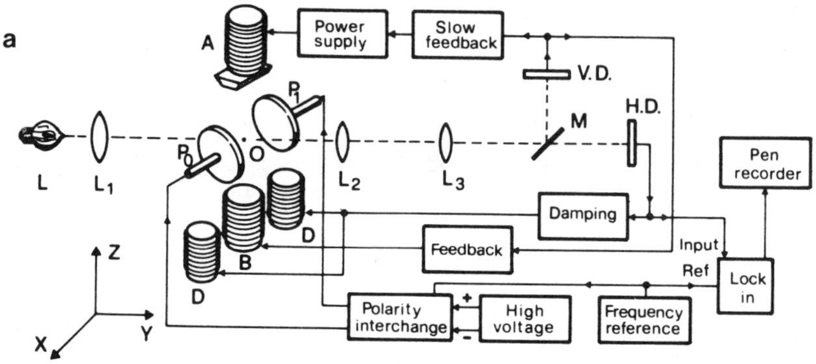

A schematic view of the apparatus. L = lamp; L_1, L_2, L_3 = lenses; M = half transparent mirror; H.D. and V.D. = horizontal and vertical photodiode system; O = levitating object; A = main coil; B = feed-back coil; D damping and auxilary coils; P_0 and P_1 electric plates. The boxes showing the electronics are self-explanatory. We have not shown the vacuum tight box containing the plates, the ultraviolet lamp, the TV transfer system and several other minor details. The figure is not in scale. The distance between the lower part of A and the upper part of B is 15 cm. The distances between the optical components are – approximately – as follows: L_1–O = 30 cm, O–L_2 = 11 cm, L_2–L_3 = 53 cm, L_3–M = 6 cm, M–V.D. = 16 cm, M–H.D. = 20 cm (the distances from M refer to a 3/10 mm sphere and are taken from the center of M).

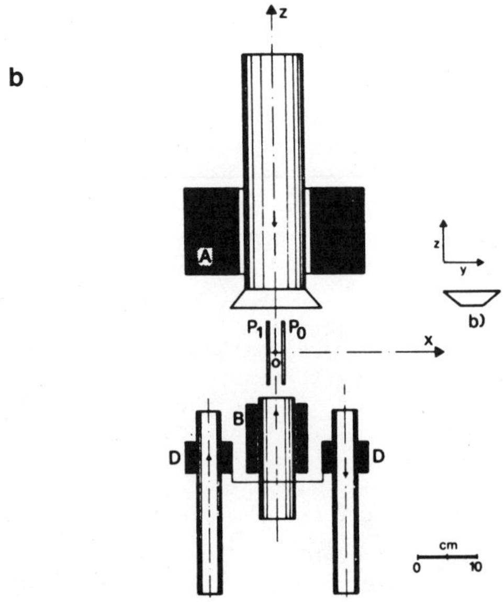

Fig.7-Scheme of the magnetic levitation electrometer,with a view (fig.7b) of the positioning of the main magnets.

chamber ,fig.8 and 9).Many other components , in particular the four coils used for spinning the levitating object [we will speak of this later] are not shown in fig.7 for simplicity.

Of course I cannot deal here with many aspects of the instrument, both because of time and because this would become too technical.An extensive presentation of the instrument,of its operation and of the measurements is given in the paper already mentioned:Marinelli and Morpurgo,1982.

Here I will confine to three points:

a)the noise rejection.

b)the discharge of the levitating object and the stability in charge.

c)the stability of the magnetic valley and of the damping system.

<u>The noise rejection</u> is performed using a lock-in amplifier having as reference signal the clock piloting the electric field.It is the large integration time of the lock-in (τ=300 sec) that mainly provides the noise rejection;with respect to such a filter,the mechanical filter,constituted by the oscillating object with a damping $N_{1/2}$ of the order from 10 to 20,is very wide.

Typical graphs giving the signal to noise ratio are shown in the fig.10.The plots there give,both in conditions of low seismic noise and in conditions of higher one,the amplitude of oscillation (the lock-in signal) corresponding to two states of charge between which there is a variation of charge by one.The upper graph shows that the charge levels are determined to better than $2\cdot 10^{-2}$ electron charges.These small errors are possible in view of the huge charge stability that allows to have levels with a given charge lasting many hours ;no electrons leaves the levitating ball or attaches to it for many hours [once we had no spontaneous change for more than 80 hours].Note, incidentally, that in fig.10 the transition between the two charge levels, which is of course instantaneous,appears to take ≈20 minutes, due to the integration time of the lock-in.

<u>The stability in charge;the discharge of the object.</u>We have just said that an essential point for decreasing the noise error on the determination of a level of charge is the charge stability [moreover this stability was vital in the long period of preparation for understanding the spurious effects].To reach this stability in charge we had simply to insert many red filters between the lamp L illuminating the object and the object itself;sometimes we had also to decrease the electric field on the object;to go much beyond an alternating field of ±2000 V/cm tends to lower the stability in charge.

As to the preparation of the object -how to decrease its charge from the high value that it has immediately after the levitation to the few units of charge necessary for starting the measurement- one can summarize this as follows. Immediately after the levitation the sphere has usually a charge of the order of 10^4 to 10^5 electron charges;things can be arranged so that this charge is negative,by inserting some appropriate auxiliary potential during the levitation act.To reduce this high value to that of one or two electron charges,necessary for starting the measurement,one uses a small UV lamp.With some practice it is not difficult to dose the light so that at the beginning the ionizazion is fast and later the electrons are extracted one by one.Also one can find methods to reverse the direction of change in charge (add electrons) if this proves to be necessary.

<u>The stability of the magnetic valley and the damping.</u>An important requirement is to keep constant the natural mechanical frequency of oscillation ν_x of the object, so that,in the course of time,this frequency continues to be accorded with the frequency of the clock producing the alternating electric field and acting as reference signal

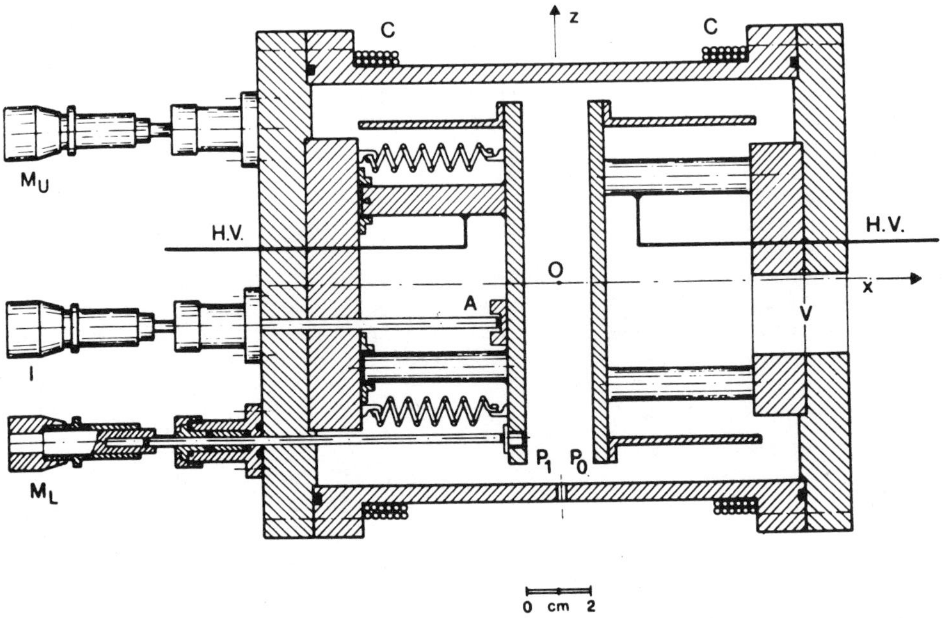

Fig.8-The levitation chamber:a section in the x,z plane.

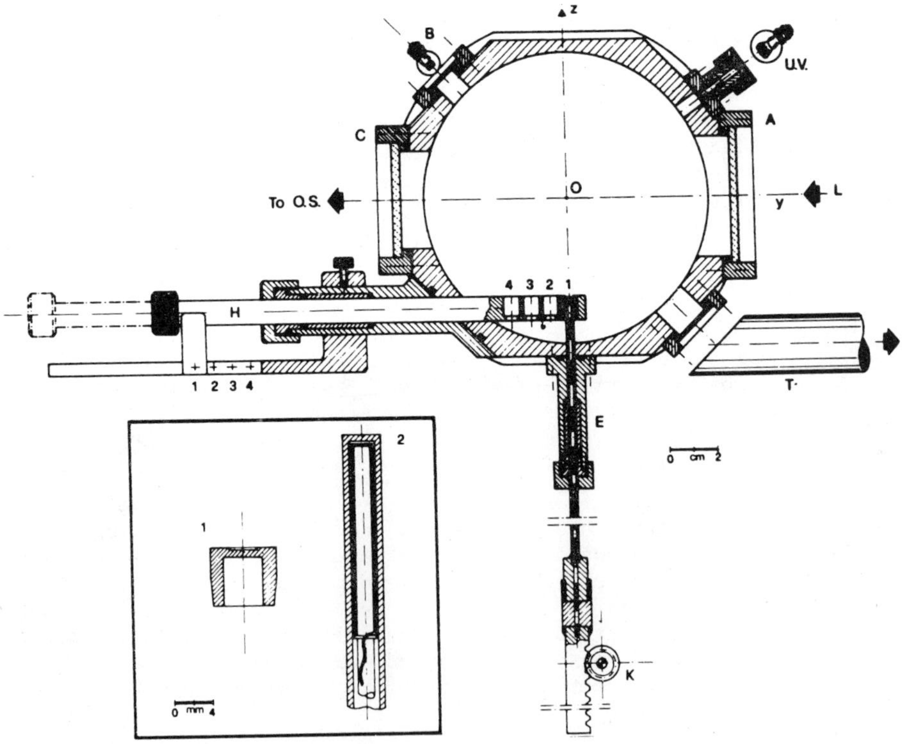

Fig.9-The levitation chamber:a section in the y,z plane.

for the lock-in. We had never really any problem on this, particularly if one operates with a comparatively broad mechanical resonance ($N_{1/2}$ =10÷20); nor we had problems in keeping constant the damping coefficient γ of the object. The equation of motion of the object in the x direction -normal to the plates- has the general form:

$$M\ddot{x} + \gamma\dot{x} + kx = F_x(t) \tag{9}$$

Here the damping term $\gamma\dot{x}$ is introduced as follows: the horizontal photodetector measures the abscissa x(t) of the object and produces a corresponding signal; the derivative of this signal is performed, essentially by an RC circuit, and the $\dot{x}$ signal is fed into the two lateral coils of fig 7 (coils D) so as to create a force on the object proportional to $-\dot{x}$; this introduces the term $\gamma\dot{x}$ in the equation (9). One can show that the damping produced in this way satisfies the following important property: for a given external force acting on the sphere, the signal produced by the x oscillation at resonance is independent from the intensity of illumination of the sphere.

I have inserted this very incomplete list of technical points only to give some limited idea of the type of problems characterizing the experiment. For a complete review there are of course many other problems; but, at this point it is appropriate to move to another aspect of the experiment, namely the spurious effects simulating a residual charge.

4. THE FINDING AND ELIMINATION OF THE SPURIOUS EFFECTS SIMULATING A RESIDUAL CHARGE

If the only force acting on the object were:

$$F_x = QE_x(t) \tag{10}$$

the experiment would have been very easy. But this is not so: there are other forces acting on the object, linear in the applied electric field, that can simulate the presence of a charge on the object. These forces are clearly dangerous because, if they are not recognized, one can reach the conclusion that the object has a non zero residual electric charge when this is not so.

<u>1. The patch or Volta force.</u> A conducting sphere in an external electric field $\underline{E}$ polarizes itself getting an induced electric dipole moment $\alpha\underline{E}$, where:

$$\alpha = 4\pi\epsilon_o r^3 \tag{11}$$

In (11) ϵ_o the dielectric constant of vacuum (8.59 10^{-12}MKS) and r is the radius of the sphere; α is called the polarizability of the sphere. As a consequence of the above induced electric dipole moment, a force arises on the sphere, the x component of which is:

$$F_x^{(pol)} = \alpha\underline{E}\cdot\frac{\partial\underline{E}}{\partial x} \tag{12}$$

Assume, first, that the electric field $\underline{E}$ acting on the sphere is only the

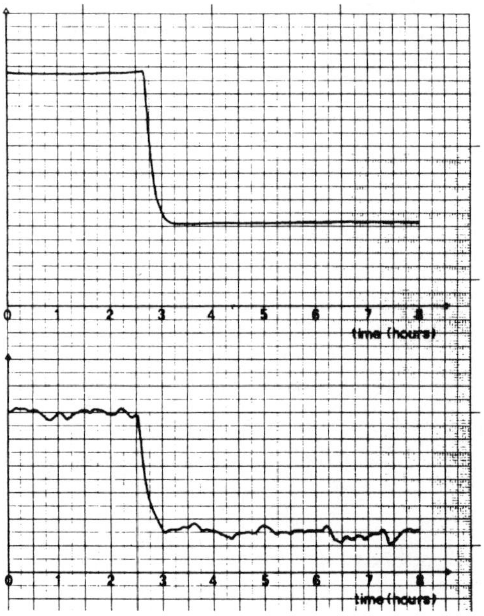

Fig.10-The change of charge by one unit |e| for two different objects, in the case of small seismic noise and in the presence of more noise.

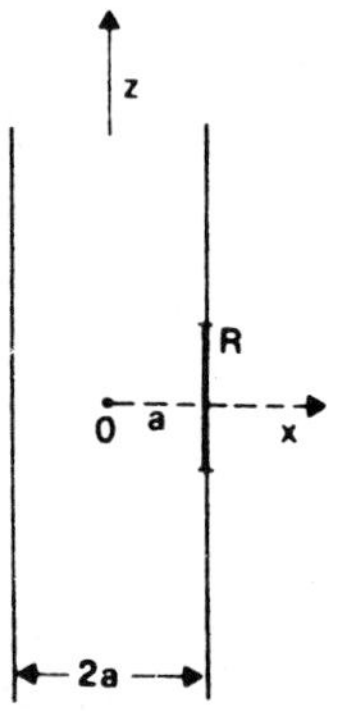

Fig.11-The geometry and notation to calculate the patch effect.

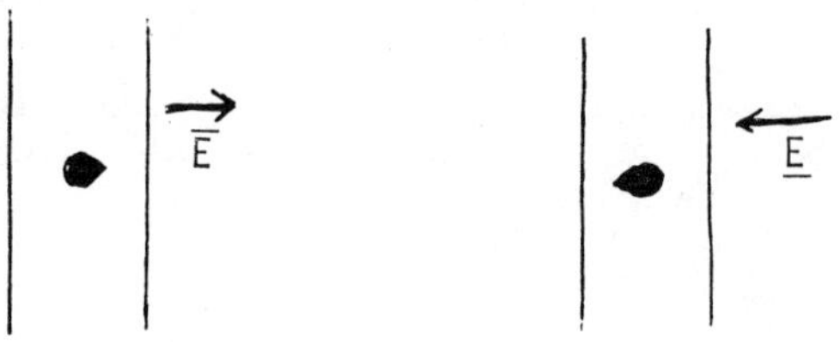

Fig.12-Explaining the reason why it is necessary to spin the sphere around the z axis.

applied alternating electric field; to make clear this point I will call for a moment $\underline{E}^{(a)}$ the applied electric field, the suffix (a) referring to "applied". In this case the equation (12) could be rewritten:

$$F_x^{(pol)} = \alpha\underline{E}^{(a)} \cdot \frac{\partial \underline{E}^{(a)}}{\partial x} \qquad (13)$$

Because the force given by (13) is quadratic in the electric field $\underline{E}^{(a)}(t)$ it produces an oscillation at a frequency $2\nu_x$ and therefore does not contribute to the signal at frequency ν_x. [As a matter of fact there are in addition two other reasons why the force (13) is irrelevant: a) for a pure square wave $E^{(a)2}(t)$ = constant and even the $2\nu_x$ component vanishes; b) in our set up $(\partial\underline{E}^{(a)}/\partial x)$ is very small due to the parallelism of the plates].

However the total electric field $\underline{E}$ acting on the sphere is not only the alternating $\underline{E}^{(a)}(t)$ field that we have just considered. No matter how clean the plates are, we cannot avoid that different regions of their surface have slightly different potentials, corresponding to slightly different work functions. Such differences in potential between different regions of the plates are usually small - of the order of a few times 10 mV [of course if the plates are definitely dirty they may reach larger values]. Still these potential differences are there; they produce what we called a Volta field and is often called a "patch" electric field. Let us indicate this field by $\underline{E}^{(V)}$. This means that the total electric field acting on the object is:

$$\underline{E} = \underline{E}^{(a)}(t) + \underline{E}^{(V)} \qquad (14)$$

and inserting (14) in (13) we find:

$$F_x^{(pol)} = \alpha\,(\underline{E}^{(a)}(t) + \underline{E}^{(V)}) \cdot \frac{\partial}{\partial x}(\underline{E}^{(a)}(t) + \underline{E}^{(V)}) \qquad (15)$$

It is apparent from the equation (15) that we get in addition to the force (13) a term linear in the applied electric field, and precisely:

$$F_x^{(V)}[\text{linear in } \underline{E}^{(a)}] = \alpha \left[\underline{E}^{(a)} \cdot \frac{\partial \underline{E}^{(V)}}{\partial x} + \underline{E}^{(V)} \cdot \frac{\partial \underline{E}^{(a)}}{\partial x}\right] \qquad (16)$$

The second term in (16) is negligible compared to the first because the gradients of the applied field are negligible. One has therefore:

$$F_x^{(V)} = \alpha\,\underline{E}^{(a)} \cdot \frac{\partial \underline{E}^{(V)}}{\partial x} \qquad (17)$$

Clearly the force (17) corresponds to an effective charge of the sphere; let us call this charge $Q^{(V)}$; it is given by:

$$Q^{(V)} = 4\pi\epsilon_o r^3 \,\frac{\partial E_x^{(V)}}{\partial x} \qquad (18)$$

This charge $Q^{(V)}$ behaves in the same way as a residual charge of the object. The first time we discovered its existence was in the graphite experiment (Morpurgo, Gallinaro and Palmieri 1970) when we saw that the apparent residual charge decreased on measuring the same grain at different increasing distances between the plates [the plates were then movable].

As one sees from the eq. (18) $Q^{(V)}$ increases with r^3 (that is with the volume of the sphere) and with $(\partial \underline{E}^{(V)}/\partial x)$, the x gradient of the patch field. How large is this gradient? It can be shown that, calling 2a the distance between the plates and assuming a circular patch centered at the center of one plate with radius R (fig.11), the most "malefic" patch (that is the patch producing the largest possible $(\partial \underline{E}^{(V)}/\partial x)$ at the object, midway between the plates) is obtained, for a given patch potential, when the radius R is related to a by:

$$R = a\sqrt{\frac{2}{3}} \tag{19}$$

If we call $V_o^{(1)}$ the potential difference between the above circular patch and the rest of the plate, the corresponding maximum value of $(\partial E_x^{(V)}/\partial x)$ at the object is:

$$\text{Max}\left.\frac{\partial E_x^{(V)}}{\partial x}\right|_{\text{at } 0} = \frac{0.56\ V_o^{(1)}}{a^2} \tag{20}$$

The corresponding spurious charge is thus:

$$Q_{max}^{(V)} = 4\pi\epsilon_o r^3 \cdot 0.56\ V_o^{(1)}/a^2 \tag{21}$$

where r and a are expressed in meters, $V_o^{(1)}$ in Volt and $Q^{(V)}$ in Coulomb.

What is a typical value of $V_o^{(1)}$? Using a Monroe electrostatic non contact Voltmeter (mod.162) we could establish that the value of $V_o^{(1)}$ for a patch on brass plates cleaned by the standard procedure and exposed to the air is usually not larger than 30 mV, and perhaps somewhat less. [Other materials for the plates do not change sensibly this value]. Inserting this value in (21) and setting $a=10^{-2}$ m [the distance between the plates in our set up is 2 cm] we get:

$$Q_{max}^{(V)}(\text{expected}) = 0.37\ |e| \text{ for spheres with } \phi = 0.3 \text{ mm} \tag{22a}$$
$$Q_{max}^{(V)}(\text{expected}) = 0.11\ |e| \text{ for spheres with } \phi = 0.2 \text{ mm} \tag{22b}$$

In a moment we shall see how these estimates compare with the data.

<u>The tilting effect.</u> Whereas, as I stated, we found and understood very early the patch effect, a second spurious effect, that I will describe now, was much harder to find. It is the so called "tilting effect". I refer for all the details on its story to the report by Marinelli and Morpurgo (1982) and proceed directly to the facts.

2.1-Spinning the sphere. Let me first say that we realized soon that to have reproducible results one had to spin the levitating object around a vertical axis, an axis parallel to the magnetic field. If this is not done the presence of a permanent electric dipole moment of the sphere can produce a rotation of some angle of the sphere around the vertical axis when the electric field has a given orientation and a rotation of the reverse angle after half a period when the orientation of the electric field is reversed (fig.12). If then there is some irregularity in the shape of the sphere -this is usually not so, but a tiny piece of dirt can be attached to the ball- this produces on the horizontal photodetector a signal in phase with the electric field that can be confused with a residual charge.

To spin the sphere we use four coils positioned outside the levitation chamber; they produce a rotating high-frequency magnetic field at the sphere and act infact as the stator of a motor of which the sphere is the rotor. In this way a very stable spinning of the sphere at about 400 Hz is produced.

2.2-The tilting. When the ball spins the average d_x and d_y components of the permanent electric dipole moment of the sphere vanish, but still the vertical d_z component can be non zero (fig.13). Then the alternating electric field gives rise to a torque on d_z tending to tilt the electric dipole moment towards one plate for half a period and towards the other during the next half period. It is a very tiny tilting, because it is counteracted by the fact that the magnetic moment of the sphere (the permanent one) tends to stay aligned with the magnetic field, which is vertical. A tipical tilting angle is 10^{-5} rad. In spite of its smallness the tilting of d_z produces the following effect: the permanent magnetic moment μ also tilts and because the magnetic field at the position of the object is inhomogeneous a magnetic force on the object is produced by the tilting. This force is in phase with the alternating electric field. The x component of this force is:

$$F_x^{(tilting)} = \underline{\mu}\cdot\frac{\partial \underline{H}}{\partial x} \cong \mu_x \frac{\partial H_x}{\partial x} \tag{23}$$

Now:

$$\mu_x = |\underline{\mu}|\theta \equiv \mu\theta \tag{24}$$

where θ is the tilting angle. The tilting angle θ is determined by the condition of equilibrium mentioned above between the magnetic torque and the electric torque:

$$\mu\theta H_z = d_z E_x^{(a)} \tag{25}$$

so that the force F_x is:

$$F_x^{(tilting)} \cong d_z \frac{E_x^{(a)}}{H_z} \frac{\partial H_x}{\partial x} \cong - \frac{d_z}{2} \frac{E_x^{(a)}}{H_z} \frac{\partial H_z}{\partial z} \tag{26}$$

[Note that in a cylindrical geometry it is:

$$(\partial H_x/\partial x) \cong -1/2\ (\partial H_z/\partial z) \qquad (27)]$$

It is clear from the equation (26) that $F_x^{(tilting)}$ is linear in the applied electric field and therefore it simulates a charge:

$$Q^{(tilting)} = \frac{d_z}{H_z}\frac{\partial H_x}{\partial x} \qquad (28)$$

Thus the whole measured residual charge is:

$$Q = q + Q^{(V)} + Q^{(tilting)} \qquad (29)$$

Introducing,as we shall do often in the following,the quantity:

$$q' \equiv q + Q^{(v)} \qquad (30)$$

we can also rewrite (29) as:

$$Q = q' + Q^{(tilting)} \qquad (31)$$

In the equations above q is the true electric charge of the object and $Q^{(V)}$, $Q^{(tilting)}$ are the pseudocharges discussed in this section.

There is an important difference between $Q^{(V)}$ and $Q^{(tilting)}$.The point is that while we can do little to eliminate the patches on the plates,we can on the other hand measure $Q^{(tilting)}$ and thus eliminate it.

For doing this it is sufficient,when a given sphere is levitating and being measured,to change substantially the value of $\partial H_x/\partial x$ at its position.Two coils,not shown in fig.7, have been mounted to do this.In this way one obtains for each sphere a set of charge levels, a set corresponding to one value of $\partial H_x/\partial x$,the other set corresponding to the other value (this of course doubles the time needed for a measurement,but this is all).From a comparison of the two levels one can eliminate $Q^{(tilting)}$ in (29).The details are again given in (Marinelli and Morpurgo,1982).

I would like to insist that the presence of $Q^{(tilting)}$ is a slight complication ,but is not a serious inconvenient.As a matter of fact,after we did realize the presence of $Q^{(tilting)}$,it has bee an effect related to $Q^{(tilting)}$ that provided -unexpectedly- a striking check of our results.I will explain this check in the next section.

5.THE PERMANENT ELECTRIC DIPOLE MOMENT OF THE SPHERE

The equation (26) shows that if one knows $\partial H_x/\partial x$, H_z and $E_x^{(a)}$ and if one measures $F_x^{(tilting)}$ one can determine the z component of the permanent electric dipole moment of the sphere.We shall call the value of d_z obtained in this way d_z^M , where M stays for "magnetic";indeed this determination of the permanent electric dipole moment of the sphere is largely based on magnetic measurements.

On the other hand if we create at the position of the sphere a gradient of the E_x field in the z direction $(\partial E_x^{(a)}/\partial z)$ [we achieve this by producing during the mesurement a small known inclination of one of the plates] then an additional force given by:

$$F_x^{(\text{dipole})} = \underline{d}\cdot\frac{\partial \underline{E}^{(a)}}{\partial x} = d_z \frac{\partial E_x^{(a)}}{\partial z} \qquad (32)$$

is produced on the sphere;measuring this force [due, we repeat, to the inclination of the plate] we obtain a measurement of d_z , which we shall call d_z^E (E stays for "determined electrically").If everything is o.k., the values of d_z^M and of d_z^E should be equal.

The fig.14 provides this comparison for the 25 balls for which these measurements have been performed.[As a curiosity the maximum permanent d_z found is ≈ 1 (electron)x(meter);most d_z's are 20 times smaller].

It is also interesting to consider repeated measurements of d_z^E and of d_z^M made on a given sphere during its levitation at different times;infact it was found quite often that the "permanent" electric dipole moment of a given sphere drifts with time (electrons redistribute) and it is of interest to see (fig.15) that the same variation is obtained with electric or magnetic measurements [Incidentally this drift with time of d_z is responsible for a corresponding drift in the residual charge in some objects,which we could not understand until we discovered $Q^{(\text{tilting})}$].In conclusion it is evident that the measurements reproduced in the figures 14 and 15 provide a strong check of internal consistency of the whole technique.

6.THE RESULTS OF THE MEASUREMENTS

The first thing that one can extract from the measurements is,for each sphere, the value of q_R' ,that is (eq.(30)),the value of the true electric residual charge,plus the Volta charge (the tilting charge can be forgotten after it has been measured and subtracted).I first present the values of q_R' for our experiment and later I will show the results of the experiments performed by other groups (see the table I).

The figures 16 and 17 display the values of q_R' for all the spheres that we measured in our experiment.For the balls with diameter 0.2 mm the figure 16 gives directly the value of q_R' obtained from the measurement.One sees that the maximum residual charge found for them is $0.14|e|$.This value is in agreement with the expectation from the eq. (22b).It is also rewarding that both Ziock et al. and the Rutherford group (Smith et al.) found an almost identical value for the patch spurious charge in their plates (which are built with materials different from ours).

For the balls of diameter 0.3 mm the measured value of q_R' were *divided*, before plotting them in fig.16 by a factor $(1.5)^3$– arising from the ratio between the cubes of the radii - so as to reduce the patch charge to a value comparable to that of the 0.2 mm spheres.It appears

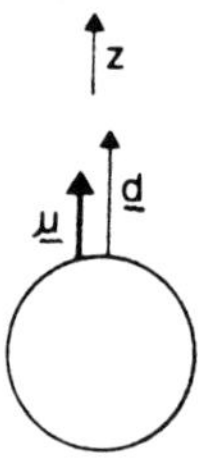

Fig.13–Introducing the tilting phenomenon.

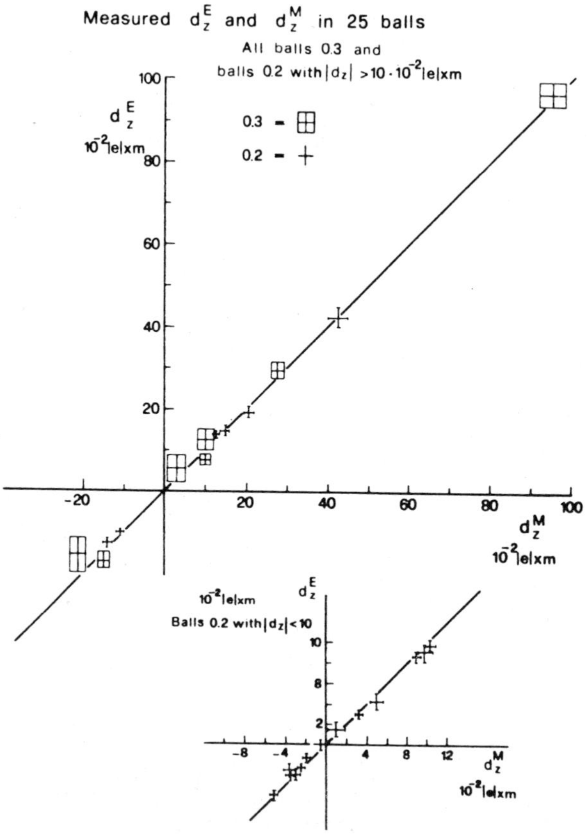

Fig.14–The measurements of d_z^M and of d_z^E.

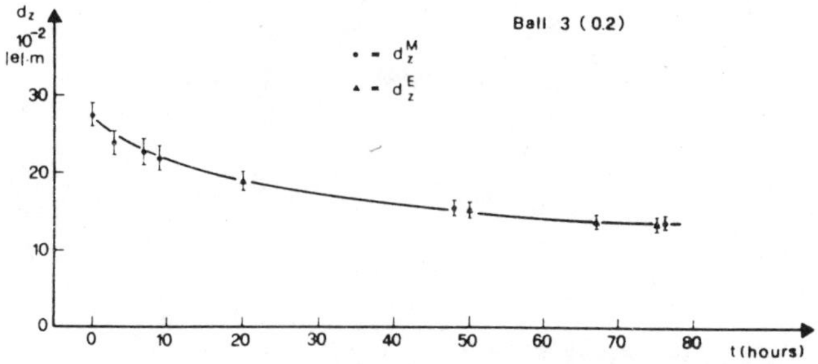

Fig.15–An example of drift with time of d_z^M and d_z^E.

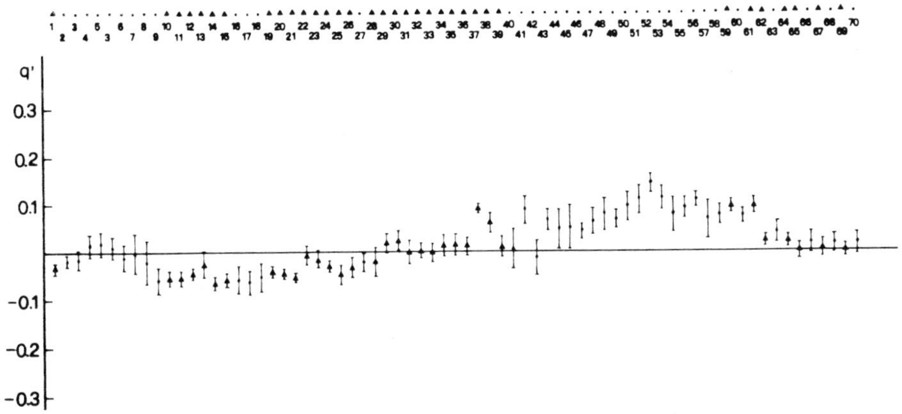

Fig.16–The values of q'_R for all the spheres measured in our experiment. The values corresponding to the balls with ϕ=0.3 mm have been divided by $(1.5)^3$.

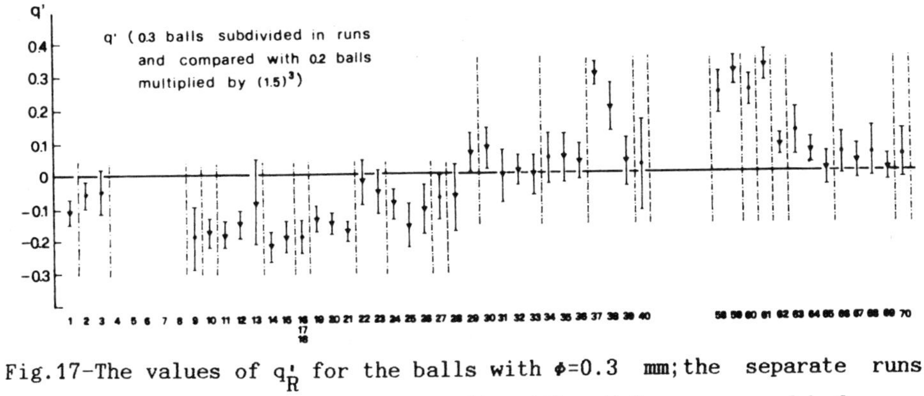

Fig.17–The values of q'_R for the balls with ϕ=0.3 mm; the separate runs have been indicated. The balls with ϕ=0.2 mm measured before or after a run have been also reported, with their q'_R multiplied by $(1.5)^3$.

that the patches on the plates change with time rather smoothly, except when a "cleaning" is performed.

The fig.17 shows the values of q'_R for the 0.3 spheres and the values of q'_R (this time *multiplied* by $(1.5)^3$) for the 0.2 mm balls measured immmediately before or after a 0.3 mm sphere.While for the 0.2 mm balls the values of q'_R are all much less than 1/3 (this is already evidence for the fact that they do not contain quarks), several values of q' for the 0.3 balls are near to 1/3, in agreement,again, with the expectation from the eq.(22a).This was the reason why we costructed our instrument so as to have sets of measurements of different balls performed without opening the levitation chamber [we called "runs" these sets of measurements];inside a run the patches should stay approximately constant and the differences between subsequent values of q'_R should therefore be approximately independent from the patches;they should therefore be equal to the differences between the q_R.This procedure [to perform runs] is unnecessary for the 0.2 mm balls,but is necessary for the 0.3 mm ones.

The differences between successive values of q'_R for all the measured 0.2 balls are reported in the fig.18 [to each error of a 0.2 mm ball we added a "systematic" error of ±0.01e; the statistical errors are then less than those displayed in the fig.18].Clearly no quarks emerge from the fig.18 .For the 0.3 mm balls we display in fig.19 the differences between subsequent measurements inside a run.Again there is no evidence for quarks.We can thus conclude from our measurements that no quarks are present in 3.7 mg of Iron.

Let me now show the analogous results from other groups who also have used the ferromagnetic levitation electrometer, with some modifications that I will mention later.

a)The fig.20 reproduces the result of an experiment by Ziock et al., performed with iron balls of diameter 0.2 mm.The total amount of matter explored was 0.7 mg.No quarks were found.

b)The fig.21 reproduces results from the magnetic levitation electrometer of the Rutherford group (Smith et al.).It shows that no quarks have been found inside 4.7 mg of Nb coated with Iron;in addition the same group found no quarks in 0.7 mg Iron.It is also of interest (fig.22) to show the residual charges due to the patch effect found by the Rutherford group.More recently the same group has increased the negative evidence on quarks exploring 1.8 mg of Nb spheres (coated with iron);also they have coated ferromagnetic balls with several substances (private communication) demonstrating the effective applicability of the method to various substances.

As to the well known observations of the Stanford group,they are displayed in the fig.23;clearly they do not receive confirmation.I recall,in this respect, that we remarked already that the frequent gain and loss of quarks from the Fairbank's balls (compare the fig.23) was apparently in contradiction with our results in Iron;infact if the quarks could tranfer so easily from the Nb sphere to the plastic holder on which the sphere is deposited before or after a levitation,the quarks should be abundant not only in Nb (or in the plastic holder) ,but everywhere, and presumably also in our Iron.This was of course an indirect argument.Now the null results in Nb by the Rutherford group on a quantity of Nb ≈6 times larger than that of the Stanford experiment, are definitely in contrast with Fairbank's findings.It is unfortunate ,in this respect, that no information has been made available during the past 5 years by the Stanford group on the results of the so called "blind tests",a bias free procedure to analyze the data (these tests were announced as forthcoming by W.Fairbank at the San Francisco and Lisbon Conferences in 1981).

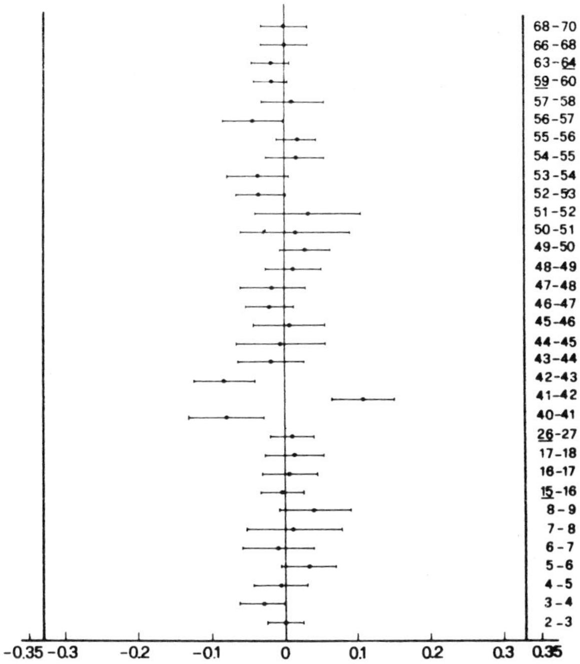

Fig.18-The differences $q'_i - q'_{i+1}$ between successive mesurements of the q' for all the 0.2 mm balls.

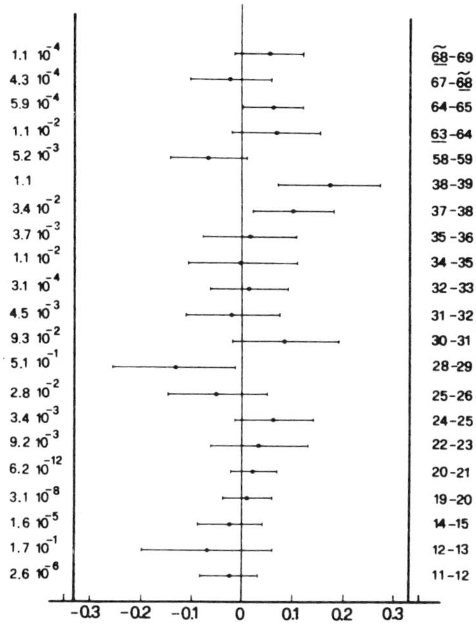

Fig.19-Same as for fig.18 for the 0.3 mm balls inside runs.

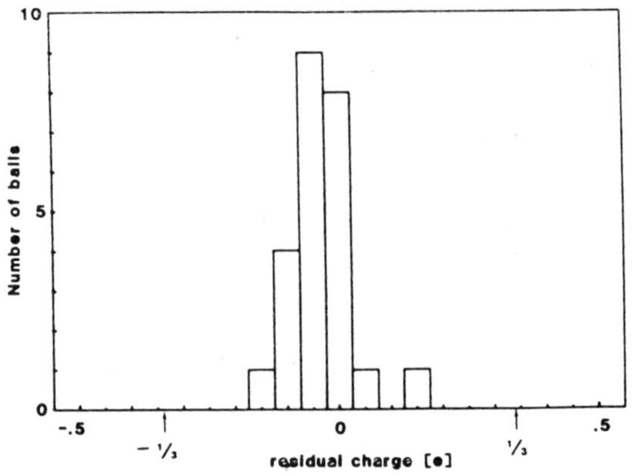

Fig.20–The q_R' distribution in the experiment by Ziock et al.

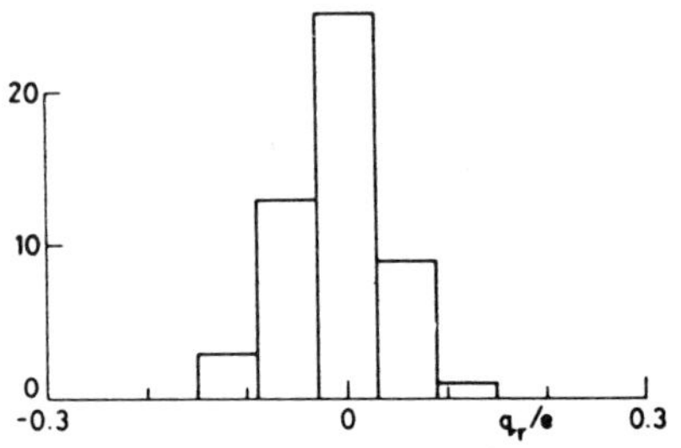

Fig.21–The q_R' distribution (differences between subsequent measurements) in the 1985 data from Smith et al (Rutherford eperiment).

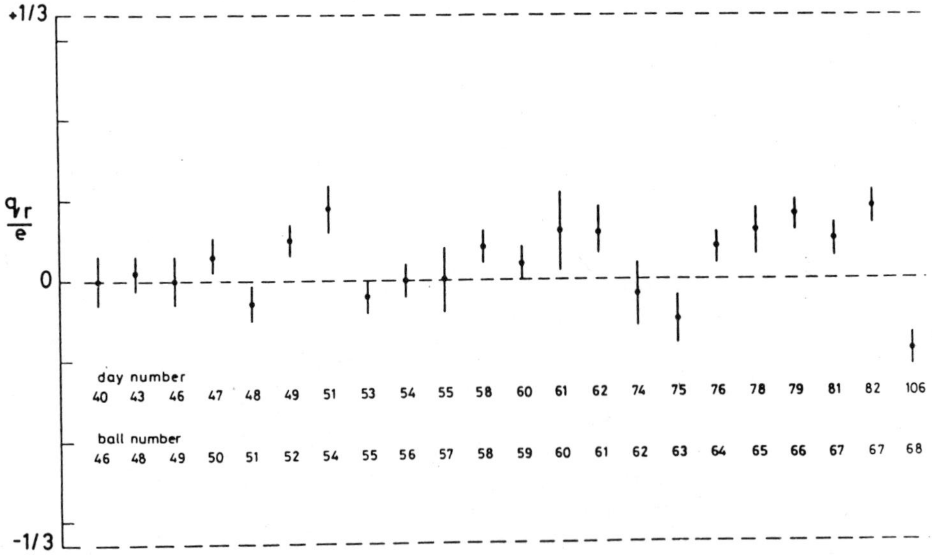

Fig.22–The time distribution of q' in the Rutherford experiment.

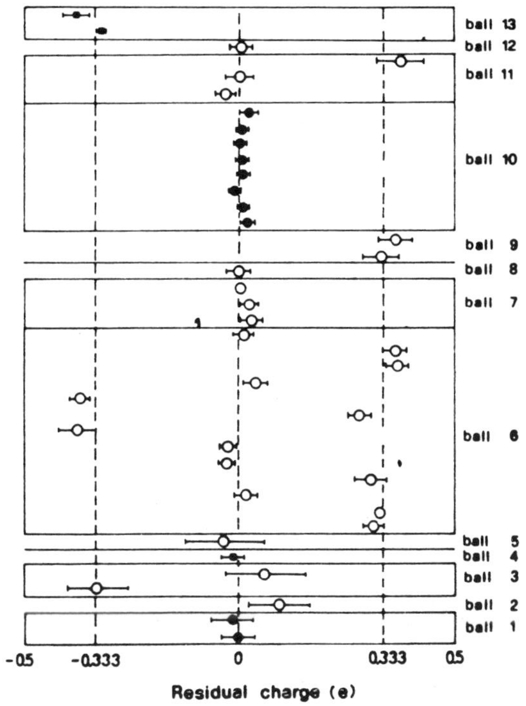

Fig.23-The results of the Stanford experiment.

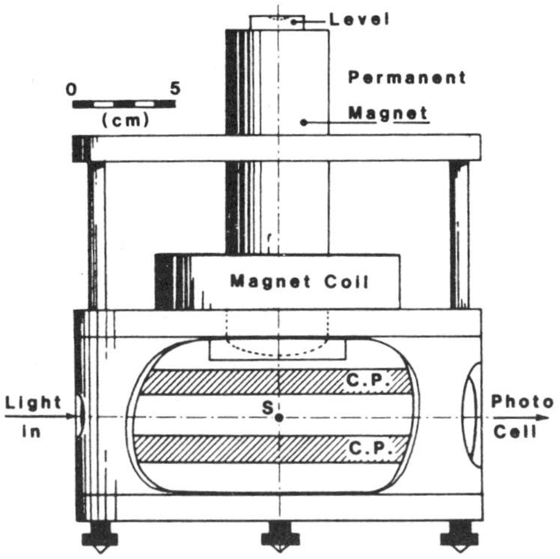

Fig.24-The apparatus of Ziock et al. showing its (vertical) geometry.

I noted above that the ferromagnetic levitation electrometers of Ziock et al. and of Smith et al.do have some differences with respect to ours.The difference is that in both the geometry is such that the magnetic field used to levitate the object is parallel to the applied electric field.In the instrument of Ziock et al. both these fields are vertical, as shown schematically in the fig.24.In the Rutherford instrument they are both horizontal as shown in the fig.25 [in this instrument the distance between the plates is 2.5 cm and the plates are correspondingly wider].It can be shown-compare Marinelli and Morpurgo,1982- that in a geometry in which $\underline{H}$ and $\underline{E}$ are parallel the tilting effect disappears.This is an advantage:the determination of q_R' does not require a subtraction between two measurements, corresponding to the two values of $(\partial H_x/\partial x)$ necessary to determine $Q^{(tilting)}$ and thus needs half of the time.[Note,however, that in the geometry of fig.24, having a bad signal/noise ratio,this advantage is completely cancelled by this bad signal/noise ratio].

It would be interesting to know - it is not clear at the time of writing- if in a geometry with $\underline{H}$ and $\underline{E}$ parallel the spontaneous rate of change in charge is so low as in our set up.Note that in principle in a vertical geometry like that used by Ziock there should not be even the need of spinning the object.Fairbank,who used in his diamagnetic levitometer a vertical geometry similar to that of the ferromagnetic levitometer of Ziock,did not infact spin his spheres.However Ziock has shown that in his case one gets a continuous spectrum of residual charges (and not the spectrum of fig.20 centered around zero) unless the balls spin around the vertical axis [also the Rutherford group has provided spinning].

7.FUTURE PERSPECTIVES:INCREASING THE MASS EXPLORED

As one sees from the table I the total amount of matter explored so far with the magnetic levitation electrometer is of the order (1/100)g.With the existing instruments one can hope with some effort to arrive to ,say,10^{-1}g, but it is hard to go much beyond.Of course one can try to develop other instruments;for instance a technique proposed by Hendricks,Hirsch and Hagstrom [Hendricks et al.1979] is intended to reach a rate of a gram per day.However its development has been apparently more difficult than foreseen,since the instrument was stated to be almost ready in 1981.

One can ask if the magnetic levitation electrometer can be developed so as to be capable to establish the presence of quarks inside objects much heavier than the ones in use today.[Then one could also improve the upper limit to the electron-proton charge difference,as we shall discuss in a moment].

The real problem in increasing the mass of the individual objects measured with the magnetic levitation electrometer is the patch effect.The eq.(18) gives the spurious charge due to the patch effect as:

$$Q^{(V)} = 4\pi\epsilon_o r^3 \frac{\partial E_x^{(V)}}{\partial x}$$

To increase the size of the object without increasing $Q^{(V)}$ at the same time, one has to increase the distance 2a between the plates in order to reduce $(\partial E_x^{(V)}/\partial x)$ at the position of the levitating sphere.This reduction goes as $(1/a^2)$ [eq.(20)].Therefore on doubling the radius r of

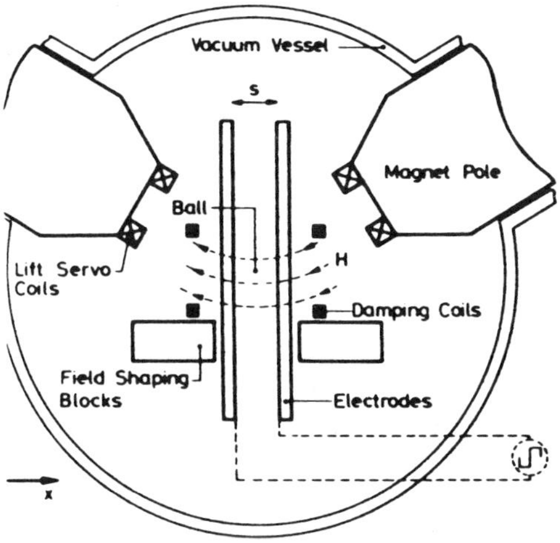

Fig.25–The Rutherford apparatus,showing $\underline{E}$ and $\underline{H}$ parallel and horizontal.

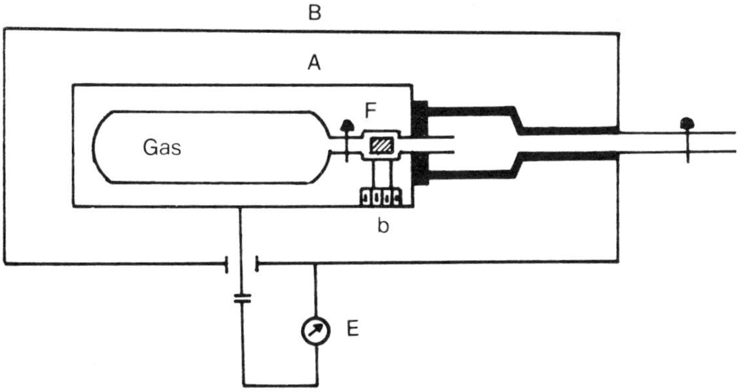

Fig.26–A scheme of the gas efflux method experiment.

the object (thus increasing its mass by 8) we must increase the distance between the plates by $\sqrt{8}$, almost three times. To prevent the occurrence of unwanted edge effects from the plates one must then increase the diameter of the plates in the same ratio, that is from our value of 12 cm to 35 cm. In order to keep the same electric field at the position of the ball (but presumably the electric field should be increased to maintain the same signal/noise ratio) the voltage to each plate must be brought from the present value of +2000,-2000 to +6000,-6000 Volt and must alternate between these two values with an accuracy better than one Volt. These numbers show that at some cost, and with *much* ingenuity it may be possible to extend the magnetic levitation electrometer to balls with a diameter 0.6 mm , double than the present one (the mass of each ball would then be ≈1 mg, if Iron). But further extensions appear unlikely, unless one does not find a way to reduce substantially the patch effect.

As already stated the maximum patch effect observed in our experiments (and also in those of the other groups) corresponds to a value of $(\partial E_x^{(V)}/\partial x)$ of the order:

$$\frac{\partial E_x^{(V)}}{\partial x} = \frac{0.5\cdot(30\ \mathrm{mV})}{a^2}$$

where the "30 mV" represents the patch potential. If the 30 mV above could be reduced to 3 mV this "miracle" would allow to operate with the plates at the same distance with balls having a volume 10 times larger. Of course another possible way could be not to reduce the patch potential, but to be able to measure it with great accuracy during the experiment.

8. THE ELECTRON-PROTON CHARGE DIFFERENCE

We now come to our last point, that is a discussion of how these measurements provide at the same time an upper limit to the electron-proton charge difference. We know of course that the electron charge is very nearly equal to (minus) the proton charge, but so far it remains an experimental problem to know how near they are. At the moment the relative values of the charges of proton and electron could be arbitrary, according to theory; in other words we have no theory that explains why they are equal. Only if some kind of superunification were true and if, therefore, the charges of the particles-say of the leptons and the quarks- were the eigenvalues of a generator of a group, then they would be related; equal charges should have to be really equal. But, as I repeat, at the present stage, for instance in the standard model, this is not so. There one has:

$$q = t_3 + \frac{y}{2} \tag{33}$$

where y can, a priori, be arbitrary for each particle [except that the need for cancellations of the infinities due to "anomalies" requires some relations].

The story of the measurements of the electron-proton charge difference is very long and I will not report it here. I mention only the first experiment on the subject performed in 1925 by Piccard and Kessler [and repeated later by Hillas and Cranshaw and by King]. The idea of this experiment was the so called "gas efflux method". An insulated container A (fig.26) inside a Faraday cage B is filled with a gas; this gas is prefiltered, in order to avoid the presence of electrified dust. The electrometer E measures the static potential difference between A and B (B is earthed) when the gas is inside A.The gas is then allowed to go out from the container A.If each atom or molecule of the gas has a tiny charge (due to a tiny difference in the absolute values of the charges of the electron and proton) the total charge contained in A will change after the efflux and the electrometer E will signal this change.Of course to increase the sensitivity of the method it is necessary that electrified grains of dust possibly still present, or ions, are not allowed to leave the container A because they might produce a large signal not due to the effect that one intends to measure.This is accomplished inserting at the exit of the vessel an electrostatic filter F;an internal battery (bat.in fig.26) creates a potential difference of some 50 Volt between the two electrodes of the filter.

These experiments established that the electron-proton charge difference in a molecule of N_2 is less than $(2\pm2)\cdot10^{-21}|e|$ [for a precise definition of the electron-proton charge difference see below].[Some unpublished experiments by King listed in a paper by Dylla and King,1973,apparently go beyond this limit by a factor 30].

In addition to the technique sketched above,two other techniques have been used;one started by Hughes (1949) based on the electric deflection of atomic or molecular beams,and the other by Dylla and King (1973), based on the electro acoustic effect.I will not discuss them here also because I am not an expert;I sketched above the gas efflux method only because there have been statements,long ago,that the results of the gas efflux experiments had relevance on the experiments of search for quarks;this is not so,because the presence of the filter in the apparatus of fig.26 keeps the quarks inside A and no effect from them can result.

The paper by Dylla and King mentioned above contains a table summarizing all the experiments made until 1972.

Obviously a measurement of the residual charge of a levitated object can set a limit on the electron proton charge difference;as I stated already, Stover et al.built their magnetic levitation electrometer with this aim [the reason why their instrument gave results for the residual charge affected by a spurious effect,is presumably to be found in the tilting effect,that we discovered much later].If, for instance,the sum of the charges of a proton and an electron is, say,$10^{-20}e$ and an object contains 10^{19} protons and exactly the same number of electrons, its total charge is (1/10)e.The above argument assumes that the charge of the neutron is zero;if it is not (at present the best upper limit -Gahler et al,1982- is $(-1.5\pm2.2)10^{-20}e$,the argument continues to be valid provided that we replace the "sum of the charges of an electron and a proton" with the "sum of the charges of the electrons,protons and neutrons in an atom, divided by the number of protons".It is now conventional to divide ,instead,by the number of nucleons;so that for a piece of matter the "electron-proton charge difference Δ" is simply:

$$\Delta = \frac{q}{N} \tag{34}$$

where q is the residual electric charge and N is the number of nucleons;of course it is implied in (2) that all the sources of spurious charge have been eliminated from q (and that no quark is contained in the object) so that q is effectively the residual <u>true</u> electric charge of the object,attributable to the effect under investigation.

The whole problem of determining Δ consists [compare Marinelli and Morpurgo 1984] in determining q.After subtraction of $Q^{(\text{tilting})}$ our experiment gives us not q, but q', the sum of the true residual charge and the patch charge $Q^{(V)}$:

$$q' = q + Q^{(V)} \tag{35}$$

Of course one might be tempted to argue in the following way: let us take a particular measurement in which q' is very small [for instance we have some for which $q'=2\cdot10^{-2}e$] and determine Δ dividing such a q' by the number of nucleons in the sphere. However this would not be correct. If q' for some sphere is measured to have the above value of $2\cdot10^{-2}$ this might happen because q, the quantity in which we are interested, is, say $20\cdot10^{-2}e$, and $Q^{(V)}$ is $-18\cdot10^{-2}e$. Thus the value of q' measured for one sphere cannot be taken as providing a determination of q. Therefore we must have information on $Q^{(V)}$ in order to know q; but of course, while for an individual sphere we may know q' with an error of 0.01e or 0.02e we cannot hope to be able to determine independently $Q^{(V)}$ with an error comparably small.

However there is a way out: the point is that all the 0.2 mm spheres that we used are equal to a part in a thousand, and the same is true for all the 0.3 mm balls. Thus the charge unbalance due to the electron-proton charge difference should be equal (once the results of the 0.3 mm balls have been normalized to those of the 0.2 mm balls) for all the balls to a part per thousand. If, therefore, we sum all our results on 71 different measurements, and divide by 71, the contribution from the electron-proton charge difference under measurement remains, while the contribution from the $Q^{(V)}$s, if they are random, averages to zero.

Thus, on making this assumption of:

$$\langle Q^{(V)}\rangle_{av} = 0 \tag{36}$$

we get:

$$q = \langle q'\rangle_{Av} = (1.6\pm1.6)10^{-21}e \tag{37}$$

How was obtained the error in (36) ?We have 71 measurements, as I said, but they cannot be all considered as independent measurements of the patches; it is visible from the fig.16 that many patches are correlated. On examining the data and taking into account the number of times that we did a "cleaning" operation, the number of independent situations of the plates was around 10 and our error in the result (35) has been based on this number. From (35), taking into account that a sphere of ϕ=0.2 mm contains $1.97\cdot10^{19}$ nucleons we get:

$$\Delta = (0.8\pm0.8)\ 10^{-21}\ e \tag{38}$$

This is not all ,however. There is infact no compulsory reason why the $Q^{(V)}$ should average to zero even in an extended sequence of measurements, as that discussed above, lasting approximately 9 months; a zero average can be likely, but is not compulsory. For instance one cannot exclude that a particularly resistant patch has formed on a plate and that our cleaning procedure produces simply a random variation of this fixed patch around a non zero value. As a consequence the fact that the average q' from the above measurements is almost zero does not strictly allow to exclude a sizeable value of q (and therefore of Δ).

To exclude the possibility of this fixed patch,precisely compensating the effect of the elctron-proton charge difference,we need another series of measurements,similar to the previous ones but performed with a different pair of plates.If indeed a non zero average patch of the amount needed to cancel precisely the effect of the electron proton charge difference was present during the above measurements,there is no reason why it should be present,with exactly the same value,in a different sequence of measurements performed with different plates.

I do not want here to be too long;but both the data we had from the measurements made in the preparatory stage performed using different plates; and the data from the other experiments listed previously produce another example of $\langle q' \rangle_{Av}$ with the same order of magnitude as that in (35);and thus confirm the result (36).This result is only two times better than the best published value.But there is perhaps the possibility,if this should prove of interest, to use the method for improve it,perhaps, by an order of magnitude.

ACKNOWLEDGEMENT

I thank Prof.M.Marinelli for having helped in writing a large fraction of this paper -in particular we wrote together the Secs.4 to 8- and for frequent conversations on this subject also after the completion of our experiments.

REFERENCES

1965,G.Morpurgo,Physics,2,95.
1966,G.Gallinaro and G.Morpurgo-Physics Letters 23,609.
1967,R.W.Stover et al.-Phys.ReV.164,1599.
1968,G.Morpurgo,Rapporteur talk:"The non relativistic quark model",in proc.XIV conf.on high energy physics (Vienna),p.225 [CERN sci.serv.Geneva 1968].
1971,W.Fairbank and A.F.Hebard-Proc 12th Int Conf.Low T Phys (Tokyo).
1973,H.Dylla and J.King-Phys.Rev.A7,1224 (this paper should be consulted for all the necessary references to the electron-proton charge difference before 1973).
1974,E.D.Garris and K.Ziock-Nucl. Instr.and Methods,117,467.
1976,G.Gallinaro,M.Marinelli and G.Morpurgo,report INFN AE/76/1.
1977,G.Gallinaro,M.Marinelli and G.Morpurgo,P.R.L.38,1255.
1977,W.M.Fairbank et al,P.R.L.38,1011.
1979,W.M.Fairbank et al,P.R.L 42,142
1979,C.Hendricks,R.Hagstrom and G.Hirsch,Berkeley Lab. preprint 9350.
1981,W.M.Fairbank et al,P.R.L.46,967.
1981,M.Marinelli and Morpurgo,Proc.of the Lisbon Conf. of the European Phys.Soc.(Dias de Deus,ed.).
1982,M.Marinelli and G.Morpurgo,Physics Reports 85,161.[This paper contains a very detailed description of the instrument,a story of the whole subject and a complete set of references up to 1982;it should be consulted for any further detail].
1982,R.Gahler et al.,Phys.Rev.D25,2887
1983,G.Morpurgo-Invited paper presented at the meeting "The frontiers of Phyics" ,Bologna October 1983 (INFN-GM-83-1)(unpublished).
1984,M.Marinelli and G.Morpurgo,Phys.Lett.137B,439.
1985,P.F.Smith et al.Phys.Lett.153B,188.
1986,P.F.Smith et al.Phys.Lett.171B,129.

CP VIOLATION

Lincoln Wolfenstein
Physics Department
Carnegie – Mellon University
Pittsburgh, PA 15217

1. THEORIES WITH CP VIOLATION

The best known and verified part of particle physics is quantum electrodynamics (QED). This theory is invariant under each of the discrete symmetries

P : Parity (space inversion)

T : Time-reversal

C : Charge conjugation (particle-antiparticle exchange)

It was natural in trying to invent theories of the strong and weak interactions to copy this feature of QED. With the discovery of P violation in beta-decay in 1957 it was necessary to revise these ideas.

It was immediately clear that it was easy to formulate theories which violated C or P or T. However, it was impossible to write a local relativistic field theory that violated the combined symmetry CPT (C times P times T). Therefore, CP violation would imply T violation. The V-A theory that successfully described weak interactions after 1957 violated both C and P in a maximal way but was CP and T invariant. To violate CP or T in a Lagrangian it is necessary to introduce couplings that are complex rather than real. In general this means that in its final form the Lagrangian contains more than one coupling, that these have a relative phase between them, and that this phase cannot be removed by a redefinition of fields.

Theories of elementary particles today are gauge theories. The Lagrangian L of a gauge theory has the general form

$$L = G + H + Y$$

G is the standard gauge theory L modeled after QED; this means it contains the free field L with derivatives replaced by gauge-invariant derivatives. For a product gauge group (like SU(3) x SU(2) x U(1)) there is a coupling constant g_i for each member of the product. One can always choose g_i as real and G is always CP and T invariant. H is the Higgs potential containing terms quadratic or quartic in the Higgs fields. Y is the Yukawa interaction coupling the Higgs fields to the fermions. While the interaction terms in G are totally prescribed by gauge principles the terms H and Y are completely arbitrary. The origin of CP violation in gauge theories can come from

(1) Complex couplings in Y

(2) Complex couplings in H

(3) H and Y may have no complex couplings but the vacuum expectation values (VEV) of the Higgs fields obtained by minimizing H may be relatively complex. This is referred to as spontaneous CP violation.

In the minimal version of the standard SU(2) x U(1) electroweak theory there is only one Higgs field. As a result hermiticity assures that the terms in H $(\mu^2\phi^+\phi, \lambda(\phi^+\phi)^2)$ satisfy CP invariance and the vacuum expectation value can be chosen real. Thus the only possible source of CP violation is Y. Considering only the neutral Higgs field Φ_o the Yukawa interaction Y can be written

$$Y = \frac{1}{2}\sum_{ij} f_{ij}\{\bar{D}_{Li}D_{Rj}\Phi_o\} + g_{ij}\{\bar{U}_{Li}U_{Rj}\Phi_o^+\} + h.c.$$

Here $U_i(= u, c, t)$ and $D_i(= d, s, b)$ are quark fields and L and R denote chirality states. The original Lagrangian has no mass terms so that the basic fermion fields are chiral fields, that is pure L or pure R; the L fields, of course, are doublets under SU(2) while the R fields are singlets. Mass for the fermions (as well as for the W and Z bosons) arises because of the vacuum expectation value (VEV)

$$< \Phi_o >= v$$

We obtain the physical Higgs field ϕ by the replacement $\Phi_o = v + \phi$ so that Y becomes

$$M_d + M_u + Y(\phi)$$

$$M_d = f_{ij} v \bar{D}_i D_j$$

$$M_u = g_{ij} v \bar{U}_i U_j$$

and $Y(\phi)$ are the terms linear in ϕ that give the fermion interaction with the physical Higgs. (We have made the standard choice that v is real, which has no physical significance.)

To obtain the quark mass eigenstates we diagonalize M_d and M_u. Without loss of generality we have chosen f_{ij} and g_{ij} as hermitean matrices. (This is because we can redefine the right-handed quarks. For a nice discussion see P. H. Frampton and C. Jarlskog, Phys. Lett 154B, 421 (1985)). Then

$$V_d^+ M_d V_d = diagonal$$

$$V_u^+ M_d V_u = diagonal$$

The mass eigenstates are related to the original flavor eigenstates by the unitary matrices V; for example

$$D_{Lj} = (V_d)_{ji} d_i$$

$$U_{Lj} = (V_u)_{ji} u_i$$

Since the eigenvalues of the mass matrices are real the only places CP violation (arising from f_{ij} and g_{ij} being complex) shows up is in V_u and V_d. Notice that $Y(\phi)$ couples ϕ to quarks with a coupling proportional to the mass and so is CP invariant. However the gauge coupling

$$\sum_i gW^\lambda \bar{U}_{Li}\gamma_\lambda D_{Li} + h.c.$$

which is CP-invariant becomes when expressed in terms of mass eigenstates

$$\sum_{ij} gW^\lambda \bar{u}_{Lj}\gamma_\lambda U_{ji} d_{Li} \tag{1.1}$$

where

$$U = V_u^+ V_d \tag{1.2}$$

The unitary matrix U, called the Kobayashi-Maskawa (KM) matrix, is the only place CP violation can show up in the standard model. Although originally the CP violation appears in Y, after spontaneous symmetry breaking it shows up only in the W interaction.

We now look at the unitary matrix U. For the case of two generations an arbitrary unitary matrix is described by one mixing angle and three phases. However, there is still the possibility of redefining some of the quark fields u_j or d_i by changing their phase. This is because all the other terms in the final Lagrangian except for the W interaction (1) conserve flavor. Thus we can set

$$s \to se^{i\alpha}$$

$$c \to ce^{i\beta} \tag{1.3}$$

The conservation of electric charge also allows us to change the phase of all u fields relative to d fields. By these three phase transformations, which are simply redefinitions of fields, we can transform U into a real matrix. Thus in the two generation case there can be no CP violation in the standard model. For this case CP conservation can be considered an accidental symmetry.

With three generations the same argument shows there exists one phase which cannot be removed. A convenient parameterization of U due to Maiani is

$$U = \begin{pmatrix} U_{ud} & U_{us} & U_{ub} \\ U_{cd} & U_{cs} & U_{cb} \\ U_{td} & U_{ts} & U_{tb} \end{pmatrix} =$$

$$\begin{pmatrix} C_\beta C_\theta & C_\beta S_\theta & S_\beta e^{-i\delta} \\ -C_\gamma S_\theta - C_\theta S_\beta S_\gamma e^{i\delta} & C_\gamma C_\theta - S_\theta S_\beta S_\gamma e^{i\delta} & C_\beta S_\gamma \\ S_\theta S_\gamma - C_\theta C_\gamma S_\beta e^{i\delta} & -C_\theta S_\gamma - C_\gamma S_\theta S_\beta e^{i\delta} & C_\gamma C_\beta \end{pmatrix}$$

where $C_\theta = cos\theta$ and $S_\theta = sin\theta$. All CP violation depends on the phase δ. Experimental data on strange particle and B decay rates determine the values of U_{us} and U_{cb} and set a limit on U_{ub}. Given these values I have made the empirical observation that the mixing angles have a hierarchical structure such that we can expand in powers of $\lambda = sin\theta = .22$ with

$$sin\gamma = A\lambda^2$$

$$sin\beta e^{-i\delta} = A\lambda^3(\rho - i\eta)$$

The exact value of U_{cb} depends on the theoretical calculation of the decays $B \to X_c + e + \nu$ where $X_c = D, D^*$ and other states containing the c quark. A recent analysis by Isgur, Grinstein and Wise yields the value $A = .9 \pm .2$, where the uncertain error compounds theory and experiment, both of which need to be improved. Analysis of the decays $B \to X_u + e + \nu$ gives an uncertain upper limit on U_{ub}, yielding the estimate $(\rho^2 + \eta^2) < 0.5$. Keeping only powers up to λ^3, U takes the simple form

$$U = \begin{pmatrix} 1 - \frac{\lambda^2}{2} & \lambda & A\lambda^3(\rho - i\eta) \\ -\lambda & 1 - \frac{\lambda^2}{2} & A\lambda^2 \\ A\lambda^3(1 - \rho - i\eta) & -A\lambda^2 & 1 \end{pmatrix} \tag{1.4}$$

There are two interesting points about the form (1.4). We see that CP violation enters only at order λ^3; this ultimately will explain why CP violation in K^o decay is small. On the other hand in order to explain the data it turns out that η must be comparable to ρ so that the phase of U_{ub} is not small. In this sense CP violation is in some sense intrinsically large.

There are a number of reasons for looking at theories of CP violation beyond the standard (KM) model. (1) It is possible that the predictions of the KM model will prove inconsistent with experiment. This is not the case at present and difficulties, both of theoretical calculation and experiment, make it unlikely that the KM model will be disproved very soon. (2) In order to have a feeling as to what experiments may be particularly interesting it is useful to compare the predictions of different theories. (3) Finally there is a hope that possibly CP violation is a phenomenon that gives us some clue as to the physics beyond the standard model. While the KM matrix allows for CP violation it is put into the standard model by hand; perhaps ultimately there is a more fundamental understanding to be found.

For purposes of illustration we consider two well-studied examples both of which were suggested when there were only two quark generations known.

A: Weinberg Higgs Model. Here we stick to SU(2) x U(1) but assume there are three Higgs boson doublets instead of only one. It is possible to introduce CP violation either as a complex coupling term linking the three bosons or via spontaneous CP violation. The latter case is particularly interesting since then even with three generations it can be shown that the resulting KM matrix is real. As a result CP violation occurs only as a result of the exchange of scalar bosons.

B: $SU(2)_L x SU(2)_R x U(1)$ Models. This is an example of a model in which the gauge group is enlarged. There are new gauge bosons W_R which couple to right-handed currents. The $SU(2)_L$ refers to the usual $SU(2)$ associated with the usual $W^\pm$, Z. The new bosons we are assumed to have a larger mass than $W^\pm$ so that right-handed currents are suppressed. If the W_R masses are not too large they may, however, be of interest in further experiments as well as playing a role in CP violation.

Mohapatra and Pati pointed out that with two generations this model could allow CP violation. The mass matrix for quarks is in general diagonalized by a biunitary transformation

$$U_L^u M_u U_R^{u+} = M_u^{diag}$$

$$U_L^d M_d U_R^{d+} = M_d^{diag}$$

where $M_u(M_d)$ are the up (down) mass matrices (We do not assume these to be hermitean since we can no longer arbitrarily redefine the right-handed quarks.) The KM matrix determining the couplings to W is

$$U = (U_L^u)^+ U_L^d$$

There is a similar matrix U_R for the coupling to W_R. With two generations a phase convention can be chosen so as to make U real but then U_R will contain complex elements. As a result CP violation will occur as a result of W_R exchanges. Of course with three generations there will also be the KM mechanism associated with W_L exchange. However, if $m(W_R)$ is not too large the W_R effect may dominate because the W_L exchange is suppressed by the small mixing angles of the third generation.

C: Superweak models. The previous two examples as well as the KM model are called *milliweak*. This means that the CP violation is a small piece of the normal weak interaction, in particular all tree-level exchanges have $\Delta S = 0$ or $|\Delta S| = 1$ where S is strangeness. The effective CP-violating interaction, at least for K^o decays, is of the order 10^{-3} times the normal in order to explain the K^o data. We shall discuss later a class of theories in which the CP-violating effective interaction is 10^{-9} times normal but allow $\Delta S = 2$ at tree level. These are often called *superweak* theories.

2. CP VIOLATION IN K^o DECAY

The violation of CP invariance has been observed in three decays of the K_L meson and nowhere else. These observations are summarized in two complex parameters η_{+-} and η_{oo} and the charge asymmetry δ defined by

$$\eta_{+-} = \frac{A(K_L \to \pi^+\pi^-)}{A(K_s \to \pi^+\pi^-)} = | \eta_{+-} | e^{i\phi_{+-}}$$

$$\eta_{oo} = \frac{A(K_L \to \pi^o\pi^o)}{A(K_s \to \pi^o\pi^o)} = | \eta_{oo} | e^{i\phi_{oo}} \qquad (2.1)$$

$$\delta = \frac{\Gamma(K_L \to \pi^- l^+ \nu) - \Gamma(K_L \to \pi^+ l^- \bar{\nu})}{\Gamma(K_L \to \pi^- l^+ \nu) + \Gamma(K_L \to \pi^+ l^- \bar{\nu})}$$

where A stands for amplitude and l is either e or μ. The experimental results are

$$| \eta_{+-} | = 2.274 \pm .022 \times 10^{-3}$$

$$\phi_{+-} = (44.6 \pm 1.2)^o \tag{2.2}$$

$$\phi_{oo} = (54 \pm 5)^o$$

$$| \eta_{+-}/\eta_{oo} | = .992 \pm .02$$

$$\delta = (3.30 \pm 0.12) \times 10^{-3}$$

In any theory the CP violation may arise either in the decay amplitude $K^o \rightarrow 2\pi$ or in the mass matrix which mixes K^o and $\bar{K}^o$. The decay amplitude is associated with the weak $| \Delta S | = 1$ transition whereas the mass matrix involves an effective $| \Delta S | = 2$ transition. If, as in the standard model, the Hamiltonian contains no $| \Delta S | = 2$ term then the mass matrix is a second-order effect. Because CP invariance is only broken a little it is convenient to start with the CP eigenstates

$$K_1 = (K^o + \bar{K}^o)/\sqrt{2} \quad CP = +$$

$$K_2 = (K^o - \bar{K}^o)/\sqrt{2} \quad CP = -$$

where $\bar{K}^o = (CP)K^o$. In the $K_1 - K_2$ representation the complex mass matrix takes the form

$$M - i\frac{\Gamma}{2} = \begin{pmatrix} M_1 & im'+\delta' \\ -im'+\delta' & M_2 \end{pmatrix} - \frac{i}{2}\begin{pmatrix} \gamma_1 & i\gamma' \\ -i\gamma' & \gamma_2 \end{pmatrix} \tag{2.3}$$

The off-diagonal terms which mix K_1 and K_2 are the result of CP violation. Since these are small, to a good approximation the diagonal values are equal to the eigenvalues

$$M_1 - M_2 + i(\gamma_1 - \gamma_2)/2 = (M_s - M_L) + i(\Gamma_s - \Gamma_L)/2 = -\Delta M + i\Gamma_s/2$$

where ΔM is the mass difference between K_L and K_s and in the last approximation we use $\Gamma_L << \Gamma_s$. The experimental data (9) on ΔM and the widths are

$$\Gamma_s^{-1} = 0.892 \pm 0.022 \times 10^{-10} sec.$$

$$\Gamma_s/\Gamma_L = 581 \pm 4.5$$

$$\Delta M/\Gamma_s = .477 \pm .003$$

The factor i and the antisymmetry indicate that the term m' violates not only CP but also T as expected from the CPT theorem. On the other hand the term δ' is T-invariant and violates CPT; we set it equal to zero for this reason, but discuss it later. As an exercise show that if you write M in the $K^o - \bar{K}^o$ representation the quantity $2\delta'$ represents the difference between the masses of K^o and $\bar{K}^o$ before mixing is included. The terms m' and $(M_1 - M_2)$ then are the $\Delta S = 2$ pieces that mix K^o and $\bar{K}^o$.

To describe the $K \to 2\pi$ decays we use the isospin decompositions of the 2π states which must be either $I = 0$ or 2 since the final state has $L = 0$. The transition amplitudes are written

$$< I \mid T \mid K^o >= A_I e^{i\delta_I}$$

$$< I \mid T \mid \bar{K}^o >= \bar{A}_I e^{i\delta_I}$$

where δ_I is the $\pi\pi$ phase shift in the final state with isospin I. From CPT invariance and unitarity it can be shown to lowest order in the weak interactions and assuming the $\pi\pi$ scattering conserves isospin that (the proof is given in Sect. 2 of my 1968 Erice lectures)

$$< I \mid T \mid K^o >^* = e^{-2i\delta_I} < I \mid T \mid \bar{K}^o >$$

It then follows that

$$< I \mid T \mid K_1 > = \sqrt{2} ReA_I e^{i\delta_I}$$

$$< I \mid T \mid K_2 > = \sqrt{2} ImA_I e^{i\delta_I} \tag{2.4}$$

Experimentally we know that $ReA_o >> ReA_2$ in accordance with the empirical $\Delta I = \frac{1}{2}$ rule governing non-leptonic strange particle decays. On the basis of the $\Delta I = \frac{1}{2}$ rule one can use the rate of $K^+ \rightarrow \pi^+\pi^o$ to determine the small parameter

$$w = ReA_2/ReA_o = .045$$

The phase shifts δ_I are deduced from reactions of the form $\pi + N \rightarrow \pi + \pi + N$; the quantity of interest to us is

$$\theta' = \delta_2 - \delta_o + \frac{\pi}{2} = 48 \pm 8^o \tag{2.5}$$

The quantity γ' in Eq. (2.3) represents a contribution associated with real on-mass-shell transitions. Just as the widths γ_i are the squares of transition amplitudes so from unitarity we have to a good approximation

$$\frac{i\gamma'}{\Gamma_s} = \frac{< 0 \mid T \mid K_2 >< 0 \mid T \mid K_1 >^*}{|< 0 \mid T \mid K_1 >|^2} = i\frac{ImA_o}{ReA_o} \tag{2.6}$$

The approximation involves neglecting all final states except the $\pi\pi$ state with $I = 0$.

The phenomenology is seen to contain three CP-violating quantities m', ImA_o and ImA_2. We now express the observables in terms of these. As a result of the CP violation in the mass matrix the mass eigenstates differ from the CP eigenstates

$$K_s = (K_1 + \tilde{\epsilon} K_2)/(1 + |\tilde{\epsilon}|^2)$$

$$K_L = (K_2 + \tilde{\epsilon} K_1)/(1 + |\tilde{\epsilon}|^2) \tag{2.7}$$

$$\tilde{\epsilon} = i\frac{m' - i\gamma'/2}{\Delta M + i\Gamma_s/2} \tag{2.8}$$

After a little algebra one finds

$$\eta_{+-} = \epsilon + \epsilon'/(1 + w/\sqrt{2})$$

$$\eta_{oo} = \epsilon - 2\epsilon'/(1 - \sqrt{2}w) \tag{2.9}$$

$$\epsilon = \tilde{\epsilon} + i(ImA_o/ReA_o) \simeq \frac{1}{\sqrt{2}} e^{i\theta} (\frac{m'}{\Delta M} + \frac{ImA_o}{ReA_o}) \tag{2.10}$$

$$\epsilon' = \frac{1}{\sqrt{2}} e^{i\theta'} w (\frac{ImA_2}{ReA_2} - \frac{ImA_o}{ReA_o}) \tag{2.11}$$

$$\theta = tan^{-1}(2\Delta M/\Gamma_s) = (43.67 \pm 0.14)^o \tag{2.12}$$

We can now use the data summarized in Eq. (2.2). This gives

$$|\epsilon| = (2.27 \pm .02) \times 10^{-3}$$

$$\epsilon'/\epsilon = |\eta_{+-}/3\eta_{oo}| - 1 = (-3 \pm 6) \times 10^{-3}$$

Note that the theory tells us (Eq.s 2.5 and 2.12) that the phases of ϵ and ϵ' are almost the same which means that the measurement of $|\eta_{+-}/\eta_{oo}|$ provides a sensitive measure of ϵ'. Given the small value of (ϵ'/ϵ) the phase of ϵ is essentially the phase of η_{+-} and we see from Equation (2.2) that the experimental value of ϕ_{+-} agrees perfectly with the value θ given in Equation (2.12). While this prediction for ϕ_{+-} was first made in the superweak theory we see here that it follows to a good approximation (the main approximation is Equation (2.6)) from CPT invariance once we know $\frac{\epsilon'}{\epsilon}$ is small. We also expect the phase ϕ_{oo} to be almost the same as ϕ_{+-}; the experimental value taken literally cannot be understood without invoking CPT violation but allowing for a two standard deviation error it is consistent with ϕ_{+-}.

Let us turn to the other observable, the charge asymmetry δ. In nearly all theories the semi-leptonic decays arise from the quark transition

$$s \to u + e^- + \bar{\nu}$$

Before quarks this was expressed as the $\Delta Q = \Delta S$ rule which said that when the hadrons had $\Delta S = +1$ they also had $\Delta Q = +1$. From CPT invariance we expect the same amplitude for $\bar{s} \to \bar{u} + e^+ + \nu$. As a result

$$A(K^o \to \pi^- + e^+ + \nu) = A(\bar{K}^o \to \pi^+ + e^- + \bar{\nu}) \tag{2.13}$$

(This holds separately for the two cases (e_L^+, e_R^-) and (e_R^+, e_L^-).) Thus CP violation arises only because of the mass matrix mixing effect. From Eqs. (2.7) and (2.13) (together with the definitions of K_1 and K_2) we find

$$\delta = 2Re\tilde{\epsilon} = 2Re\epsilon \qquad (2.14)$$

Check that the experimental value of δ indeed agrees with this.

While our results were expressed in terms of m', ImA_o, and ImA_2 only the combinations ϵ and ϵ' enter and the phases of each of these is determined. In fact, as first emphasized by Wu and Yang, the parameters m' , ImA_o, and ImA_2 cannot be determined unambiguously because it is possible to redefine the s quark by transforming its phase: $s \rightarrow se^{i\alpha}$. This is possible because the strong and electromagnetic interactions conserve strangemenss and so are not affected by this transformation. Writing this as an infinitesimal transformation

$$s \rightarrow s(1 - i\alpha) \qquad (2.15a)$$

As a result

$$ImA_I \rightarrow ImA_I - \alpha ReA_I \qquad (2.15b)$$

$$m' \rightarrow m' + \alpha(\Delta M) \qquad (2.15c)$$

$$\tilde{\epsilon} \rightarrow \tilde{\epsilon} + i\alpha \qquad (2.15d)$$

Thus any of the three original parameters may be set equal to zero, Wu and Yang chose $ImA_o = 0$. Most theoretical models are expressed in terms of a convenient phase convention such that in general none of the three turns out to be zero.

The result of the analysis can be summarized as follows: given CPT invariance and the $\Delta S = \Delta Q$ rule all the present observations on CP violation in the K^o system depend to a good approximation on two parameters, which may be chosen as $|\epsilon|$ and $|\epsilon'|$. The parameter $|\epsilon'|$, which unambiguously depends on CP violation in the decay amplitude, is consistent with zero. Our only measure of CP violation therefore is $|\epsilon|$, which could arise solely from the K^o mass matrix term m'; $|\epsilon|$ also contains a contribution from the decay amplitude A_o but these two contributions cannot be separated on the basis of experiment because of the phase ambiguity (2.15).

So far we have assumed CPT invariance. This is considered very fundamental and a violation of CPT invariance would be a great discovery. However, it is important to emphasize that we know CP is violated at a level of parts per thousand and searches for other CP-violating variables are devoted to explicating a known phenomenon. If CPT is violated we have no idea at what level and random searches for CPT violation are extreme shots in the dark. In fact there does exist one limit on CPT violation which is far better than any other. This is the limit on the parameter δ' in Eq. (2.3), which as we discussed represents the $K^o - \bar{K}^o$ diagonal mass difference. If δ' were not zero this would change the theoretical phase of ϵ given by Eq. (2.12). Since ϵ' is empirically small this would change the phase of η_{+-}. The fact that $\phi_{+-} = \theta$ within 2° can be interpreted as giving

$$\frac{|M(K^o) - M(\bar{K}^o)|}{M(K^o)} < 10^{-18}$$

Clearly it is valuable to improve this limit if possible; in general this requires not only measuring ϕ_{+-} but also limits on CP violation in different K^o decays. For a recent analysis see the work of Bardin et al. What about ϕ_{oo}? From CPT invariance $\phi_{oo} = \phi_{+-}$ to within a fraction of a degree given the small value of ϵ'. Thus the measured difference (2 S.D.) between ϕ_{+-} and ϕ_{oo}, if correct, would imply CPT violation. However, to explain this one would have to assume CPT violation in two different places which conspire to give the correct value for ϕ_{+-} but the wrong value for ϕ_{oo}. A direct measurement of $(\phi_{oo} - \phi_{+-})$ is clearly desirable.

In the KM model there are three $\Delta S = 1$ amplitudes due to W exchange

$$A(s \rightarrow u + \bar{u} + d) \sim U_{us}U^*_{ud} \approx \lambda \qquad (2.16A)$$

$$A(s \rightarrow c + \bar{c} + d) \sim U_{cs}U^*_{cd} \approx -\lambda + i\eta A^2\lambda^5 \qquad (2.16B)$$

$$A(s \to t + \bar{t} + d) \sim U_{ts}U_{td}^{*} \approx -A^2\lambda^5[(1-\rho) + i\eta] \qquad (2.16C)$$

where we have kept the leading power of λ for the real and imaginary terms. We expect that (2.16A) makes the main contribution to the decay $K^o \to 2\pi$ since the (2.16B and C) yield heavy quark pairs at tree level. As a result with this phase convention the decay amplitude is approximately CP-conserving. On the other hand to calculate the $\Delta S = 2$ quantities Δm and m' we need to go to second-order in the weak interaction so that virtual intermediate states involving $c\bar{c}$ and $t\bar{t}$ are naturally included. Thus we expect that the main CP-violating contribution in the KM model is m'.

The second-order calculation is described by a so-called "box diagram" involving the exchange of two W's. With either c or u in the legs of the box the value of Δm is proportional to λ^2 as seen from (2.16A) and (2.16B). On the other hand in order to get CP violation needed for m' we must pick up one factor proportional to $\eta\lambda^5$ so that $m' \sim \lambda^6\eta$. It follows that

$$m'/\Delta m \sim \lambda^4\eta$$

Since this is the major source of CP violation $\epsilon \sim \lambda^4(\lambda^4 \approx 2.5 \times 10^{-3})$ and so the small value of ϵ is explained in this way in the KM model.

Unfortunately the quantitative calculation of ϵ is very difficult. The basic problem involves going from a quark calculation to a meson calculation. The result also depends on the value of m_t. A reasonable approximation to the result of the calculation (for m_t between 35 and 60 Gev) is

$$\epsilon e^{-i\theta} = 3.1 \times 10^{-3}A^2\eta B[1 + \frac{1}{2}(1-\rho)(m_t/42Gev)^2]$$

The main uncertainty is in the factor B which is proportional to the matrix element

$$< \bar{K}^o \mid \bar{s}\gamma_\lambda(1-\delta_5)d\bar{s}\gamma^\lambda(1-\delta_5)d \mid K^o >$$

The result $B = 1$ corresponds to the original evaluation by Gaillard and Lee using a vacuum insertion method; a value of $B = \frac{1}{3}$ was derived using $SU(3)$ plus soft pion approximations. For the value $B = \frac{1}{3}$ and $m_t < 60Gev$ the experimental value of ϵ cannot be fit. However, with $B = 1$ and $m_t = 45Gev$ there is no difficulty provided $\eta \geq 0.4$. This value of η provides a lower limit on the decays of the type $b \to ue\nu$ and also shows that the phase of U_{ub} is larger than 30°.

To determine ϵ' we need to use the CP-violating amplitudes (2.16B) and (2.16C). It was first emphasized by Gilman and Wise that (2.16B) and (2.16C) may not be neglected in calculating the decay amplitude even though they yield $c\bar{c}$ and $t\bar{t}$ because they contribute to the transition $s \to d+$ gluons via a loop diagram referred to as a penguin. These diagrams were first discussed as a possible explanation of the $\Delta I = \frac{1}{2}$ rule since the transition $s \to d + gluons$ automatically has $\Delta I = \frac{1}{2}$. At the present time it is doubtful if this explains the $\Delta I = \frac{1}{2}$ rule but it still seems likely that penguin diagrams give the major contribution to a CP-violating $K \to 2\pi$ amplitude. Since these obey the $\Delta I = \frac{1}{2}$ rule we have, ignoring any isospin violation, only a contribution to ImA_o with $ImA_2 = 0$. We then have from Eq. (2.11)

$$\sqrt{2}\epsilon' e^{-i\theta'} = -.045 ImA_o/ReA_o \tag{2.17}$$

A recent summary of the calculation of the penguin, graphs then gives

$$\sqrt{2}\epsilon' e^{-i\theta'} = -A^2\lambda^4\eta(.045)(.017ln[m_t^2/m_c^2])P$$

where the matrix element P is estimated to lie between 0.7 and 2.6. The main uncertainty here as in the case of B lies in going from quarks to mesons. Using the experimental value of ϵ and the lower limit on η needed to fit ϵ this yields

$$7.0 \times 10^{-3} > \mid \epsilon'/\epsilon \mid > 1.0 \times 10^{-3}$$

The analysis also yields a positive sign for ϵ'/ϵ.

The discussion of the KM model has assumed three generations. Adding a fourth generation introduces three new mixing angles and two new phases. If these new parameters are not constrained by theoretical assumptions the experimental value of ϵ can be fit without any significant limits on the magnitude of ϵ' for either sign of ϵ'/ϵ.

Let us now return briefly to superweak models. It is assumed that some new interaction, most simply a scalar boson exchange, allows $\Delta S = 2$ and violates CP and so contributes to m' at tree level. Since $\epsilon \sim m'/\Delta m$ and Δm is second-order in the usual weak interaction the new interaction can be very weak. Of course, this interaction may also cause $\Delta S = 1$ and so contribute to ϵ' but the interaction is so weak that ϵ'/ϵ would only be of the order 10^{-8}. Of course, it is possible that the major contribution to ϵ is superweak but that ϵ' may come from another source such as the KM mechanism. There are many possible realizations of the superweak model since they require the introduction of some relatively arbitrary new physics. None of these realizations at the moment seems especially attractive. Superweak theories are discouraging for experimentalists since in general they predict that no new CP violation effects are likely to discovered. As long as no such effects are discovered, however, superweak models represent a viable possibility.

In the Weinberg Higgs model with spontaneous CP violation there is no phase in the KM matrix and all CP violation comes from Higgs boson exchange. As far as K decays are concerned CP violation results from diagrams like those in the standard model with the W boson replaced by charged Higgs bosons. (Since the theory starts with three Higgs doublets there remain two physical charged Higgs bosons while the third charged Higgs becomes the longitudinal part of the W.) In order to get a large enough effect, since Higgs bosons have very small couplings, it is necessary that these charged Higgs bosons be lighter than the W boson, although exactly how light depends on various arbitrary parameters in the model.

Detailed calculations indicate that the box diagram involving changed Higgs bosons is very small and with reasonable choices of parameters can be neglected. The main CP violation is then believed to come from the penguin diagram with the W replaced by the charged Higgs. This contributes directly to ImA_o/ReA_o but also contributes to m' as a result of virtual transitions of the form $K_o \rightarrow I \rightarrow \bar{K}_o$ where I is some fairly low-mass intermediate state. (Such contributions are often referred as long-distance contributions to $\Delta S = 2$). The value of ϵ' is given by Eq. (2.17) and of ϵ by (2.10). Thus ϵ'/ϵ depends on the relative importance of $m'/\Delta m$ and ImA_o/ReA_o in Eq. (2.10) and thus on a detailed analysis of the virtual transitions.

The most important states I are probably 2π and η' which contribute with opposite signs. A rough estimate yields the result $\epsilon'/\epsilon = -.01$, but it is very uncertain.

In the $SU(2)_L \times SU(2)_R \times U(1)$ model the major contribution to ϵ comes from the box diagram involving the exchange of one W_L and one W_R. This box diagram gives CP violation already in the two generation case due to the phases in the right-handed quark mixing matrix U_R. In general the model has too many parameters to allow one to relate the value of ϵ to other observables. There exists a particularly simple version of the model discussed by Darwin Chang in which all CP violation is spontaneous and there is one CP-violating phase. Grimus and Ecker have found in this version that $|\epsilon'/\epsilon| \sim .005$ with either sign possible. This model necessarily has flavor-changing neutral Higgs bosons which contribute to m' at tree level as in the superweak model. Unless these bosons are assumed to have a very large mass they will also contribute to ϵ but not to ϵ' and so the value of $|\epsilon'|\epsilon|$ may be reduced.

3. OTHER OBSERVABLES

3.1 Electric Dipole Moment

As discussed in the lecture of Calaprice time-reversal violation can be detected by the measurement of a non-zero electric dipole moment of the neutron. The present limit on the neutron electric dipole moment d_n is

$$d_n < 4 \cdot 10^{-25} e - cm$$

$$d_n/\mu_n < 10^{-11}$$

where μ_n is the neutron magnetic moment. The significance of this result may be assessed from a rough theoretical order-of-magnitude estimate for milliweak models

$$d_n/\mu_n \sim 10^{-6} \times 10^{-3} = 10^{-9}$$

Here 10^{-6} is needed for weak P violation and the extra factor 10^{-3} (of order ϵ) is the price paid for T violation. Thus the present limit already seems quite significant.

In the KM model d_n is zero in lowest order. To obtain a non-zero result one needs to consider second-order contributions of the form

$$n \rightarrow S = 1 \; state \; \rightarrow n$$

where one of the transitions involves T violation. Estimates of these contributions by Gavela et al and the Khriplovich group give a value of 10^{-32} for d_n. Similarly in the superweak model $d_n < 10^{-30}$. In contrast most milliweak models would be expected to give values of the order of the present limit. In the Weinberg Higgs model calculations give values of the order 10^{-24} to 10^{-25} although the result really depends on parameters not constrained by the value of ϵ. In the one-parameter version of $SU(2)_L \times SU(2)_R \times U(1)$ Ecker and Grimus find

$$| d_n | = | \epsilon'/\epsilon | \times 10^{-22} to 10^{-23}$$

This implies that present limits on d_n and $| \epsilon'/\epsilon |$ are approximately equal in sensitivity.

It must be noted, however, that a non-zero value of d_n could always be blamed on "strong CP violation". This is the P-odd T-odd Δ S=0 interaction expected in QCD if CP is violated anywhere. Unfortunately its magnitude (measured by the parameter θ_{QCD}) is fundamentally incalculable in most models. In practice the neutron electric dipole model is by far the most sensitive probe of θ_{QCD} so that the present limit on d_n already assures us that this type of CP violation will not show up anywhere else. A non-zero value of d_n therefore cannot be unambiguously related to the weak CP violation that shows up in K^o decay and so does not necessarily rule out the KM or superweak explanations. However, if prospective experiments should lower the limit on d_n by an order of magnitude, milliweak models like the Weinberg Higgs model may well be ruled out. As noted in the talk of Calaprice an alternative to the measurement of d_n is the measurement of the electric dipole moment of a nucleus. It is conceivable that if a non-zero moment were measured both for the neutron and a nucleus it might be possible to distinguish whether the cause was strong or weak CP violation.

3.2 Semi-leptonic Decays

Calaprice presented results on the parameter D that measures the correlation between the nuclear spin and the normal to the plane of the decay products in nuclear beta-decay. A non-zero result would be a sign of time-reversal invariance and could be interpreted as a phase θ_{AV} between the weak V and A currents. The results of all experiments was given as

$$\theta_{AV} = -(0.3 \pm 1.1) \times 10^{-3}$$

In the KM model there is obviously no phase entering the normal beta-decay diagram and D is zero. In the Higgs model there is CP violation associated with the charged Higgs exchange; however this is small because it is proportional to m_e and contributes to an effective scalar not a vector interaction. In the $SU(2)_L \times SU(2)_R \times U(1)$ a value of θ_{AV} of the order 10^{-3} is conceivable as noted by Herczeg. However, this only takes into account the limit on $m(W_R)$ imposed by Δm_K and not the limit on CP violation given by ϵ. In the one-parameter model discussed above one finds $\theta_{AV} = 10^{-5}$ to 10^{-6}

In the decay $K \to \pi\mu\nu$ a similar correlation, that of the muon spin with respect to the decay plane normal, has been sought as a sign of time-reversal. The result is expressed in terms of Im ξ where ξ is the ratio of the two form factors f_+ and f_-. The present result is $Im\xi = -.02 \pm .02$. Again in the KM model no effect is expected. In the $SU(2)_L \times SU(2)_R \times U(1)$ there also no effect because only vector and scalar interactions contribute to this decay so that a relative phase of V and A does not contribute. However, here an effect is expected in the Higgs model, a reasonable estimate is $Im\xi < 10^{-3}$.

3.3 Other K^o Decays

It is natural to look for CP violation in non-leptonic strange particle decays besides $K \to 2\pi$. The goal is to find an effect that cannot be blamed on $K^o - \bar{K}^o$ mixing but is associated with a CP-violating $\Delta S = 1$ decay amplitude. We can consider decays like K^+ or hyperon decays or look in other K^o decays for observables analogous to ϵ'. In a sense any such experiment has to be judged in comparison to ϵ' because one is looking at the same basic weak transition in all cases. We already have a limit on ϵ' of $2 \cdot 10^{-5}$ and ongoing experiments discussed in the lectures of Manelli hope to go down at least to $4 \cdot 10^{-6}$. This provides stiff competition. Some hope for other experiments is provided by the observation that ϵ' is suppressed by the $\Delta I = \frac{1}{2}$ rule; if that rule were exact then the ratio $\pi^+\pi^-/\pi^o\pi^o$ would have to be 2 for all decays and ϵ' would equal zero. In fact we know for the CP-conserving amplitude that the $\Delta I = \frac{1}{2}$ rule is violated by the factor $w = .045$ so that the suppression of ϵ' is expected to be a factor of 20 to 25. Taking this into account an experiment to be competitive with the present limit on ϵ' should be looking for a CP-violating effect of order $2 \cdot 10^{-5} \times$ (20 to 25) $= 5 \cdot 10^{-4}$. And that's not easy.

We look first at the decay $K^o \to 3\pi$. Analogous to η_{+-} we can define the parameters η_{ooo} and η_{+-o} where

$$\eta_{ooo} = \frac{A(K_s \to 3\pi^o)}{A(K_L \to 3\pi^o)}$$

$$\eta_{+-o} = \frac{A(K_s \to \pi^+\pi^-\pi^o)_{I=1}}{A(K_L \to \pi^+\pi^-\pi^o)}$$

The subscript $I = 1$ in the second equation means that we must consider the final $I = 1$ CP-odd states of $\pi^+\pi^-\pi^o$ which is the only one which can interfere with K_L. There is also a CP-even $I = 2$ state of $\pi^+\pi^-\pi^o$ which can occur without CP violation although it is suppressed by the $\Delta I = \frac{1}{2}$ rule and angular momentum barriers. (For a more detailed discussion of $K \to 3\pi$ see my 1968 Erice lectures.) For either of these cases we can write

$$\eta_{3\pi} = \tilde{\epsilon} + i\frac{ImA_{3\pi}}{ReA_{3\pi}}$$

and thence from Eq. (2.10)

$$\eta_{3\pi} = \epsilon + i\left(\frac{ImA_{3\pi}}{ReA_{3\pi}} - \frac{ImA_o}{ReA_o}\right) = \epsilon + \epsilon'_{3\pi} \tag{3.1}$$

Note that $\epsilon'_{3\pi}$, like ϵ', is the difference between two CP-violating decay amplitudes (more accurately, two phases that indicate CP violation) and so a non-zero value indicates that CP violation is not simply in the mass matrix.

In the KM model it is possible using PCAC soft-pion techniques to relate $\epsilon'_{3\pi}$ and ϵ' giving

$$\epsilon'_{3\pi} \approx 2\epsilon' \tag{3.2}$$

Very recently Donoghue and Holstein (U Mass preprint UM HEP-25 (1986)) have claimed that this result is too pessimistic because of corrections arising from the next order in chiral perturbation theory. They obtain as an estimate a value of $\epsilon'_{3\pi}$ equal to $10\epsilon'$. H. Y. Cheng using PCAC obtains the same results as Eq. (3.2) for the Weinberg Higgs model. In the one parameter $SU(2)_L \times SU(2)_R \times U(1)$ model Darwin Chang found a value of $\epsilon'_{3\pi}$ of the same order or less than ϵ'.

There does exist a class of models for which ϵ' vanishes but $\epsilon'_{3\pi}$ does not. Such models have the form

$$H(CP\ odd) = H_- + e^{i\alpha} H_+ \tag{3.3}$$

where H_- is P-odd and H_+ is P-even. In such a model there can be no CP-violating phase difference between a set of amplitudes all of which are P-violating (or P-conserving). Thus ϵ' must vanish since it is just the phase difference between the transition to $I = 2$ and the transition to $I = 0$ $\pi\pi$ states. On the other hand as seen in Eq. (3.1) $\epsilon'_{3\pi}$ compares P-odd and P-even transitions and so is approximately equal to $i\alpha$. An example is the original version of the CP-violating $SU(2)_L \times SU(2)_R \times U(1)$ model due to Mohapatra and Pati. In their case CP violation was introduced only in the Yukawa interaction eventually showing up in the mixing matrix U_R for the coupling of quarks to W_R. Even in this case, however, the value of $\epsilon'_{3\pi}$ is only expected to be of the order of 10^{-4} to 10^{-5}.

Another possibility mentioned here is the decay $K^o \to \gamma\gamma$. The only decay of this sort observed is $K_L \to \gamma\gamma$ with a branching ratio of 5×10^{-4}. There are two possible J=0 states of two photons, corresponding in helicity representation to (LL + RR) and (LL - RR), with opposite CP eigenvalues. It is assumed that the K_L decay goes primarily to the CP-odd state allowed by CP invariance. It is expected, but not yet observed (see Manelli's lectures) that K_s decays to the CP-even $\gamma\gamma$

state with a comparable rate. A study of interference effects in $K^o(\bar{K}^o) \rightarrow \gamma\gamma$ could be used to measure

$$\eta_- = \frac{A(K_s \rightarrow \gamma\gamma,\ CP\ odd)}{A(K_L \rightarrow \gamma\gamma,\ CP\ odd)} = \epsilon + \epsilon'_{\gamma\gamma}$$

An estimate by Golowich gives $\epsilon'_{\gamma\gamma} = 30\epsilon'$ for the KM model.

3.4 Hyperon decays

Leaving K^o decays we will consider one other case of non-leptonic strange particle decays, namely hyperon decays. The same general formalism can be applied to $\Lambda \rightarrow N\pi$, $\Sigma \rightarrow N\pi$, and $\Xi \rightarrow \Lambda\pi$. We shall use the Λ decays as an example. The idea is to compare the decays $\Lambda \rightarrow p\pi^-$ or $n\pi^o$ with the decays $\bar{\Lambda} \rightarrow \bar{p}\pi^+$ or $\bar{n}\pi^o$. In general there are four amplitudes for Λ decay $A_e(I)e^{i\delta}(l, I)$ where $l = s, p$ and $I = \frac{1}{2}, \frac{3}{2}$ and $\delta(l, I)$ is the corresponding pion-nucleon phase shift in the final state. In the absence of CP violation $A_e(I)$ is real so that there are three possible CP-violating phases $\theta_1, \theta_3, \theta$ defined by

$$A_s(1)/A_p(1) = \mid A_s(1)/A_p(1) \mid e^{i\theta_1}$$

$$A_s(3)/A_p(3) = \mid A_s(3)/A_p(3) \mid e^{i\theta_3}$$

$$A_s(3)/A_s(1) = \mid A_s(3)/A_s(1) \mid e^{-i\phi}$$

One can consider three possible observables for which to compare Λ and $\bar{\Lambda}$: (1) the branching ratio; that is the $p\pi^+/n\pi^o$ ratio, (2) the decay asymmetry, which measures the correlation between the Λ polarization $\vec{\sigma}_\Lambda$ and the final N or π momentum, (3) final proton polarization transverse to the plane defined by $\vec{\sigma}_n$ and $\vec{p}$, which determines a parameter called β. The branching ratio difference is expected to be very small because it is suppressed both by the $\Delta I = \frac{1}{2}$ rule and by the small value of the final phase shifts. In the absence of CP violation we expect $\alpha(\Lambda) = -\alpha(\bar{\Lambda})$

since then P violation implies C violation. A measure of CP violation is then given by

$$D_\alpha = \frac{\alpha(\Lambda) + \alpha(\bar{\Lambda})}{\alpha(\Lambda) - \alpha(\bar{\Lambda})} = -tan[\delta(s, \frac{1}{2}) - \delta(p, \frac{1}{2})]tan\theta_1 \approx 0.1tan\theta_1$$

In a recent analysis of different hyperon decays, Donoghue and Pakvasa (Phys. Rev. Lett. 55, 162 (1985)) give the following estimates:

$$KM\ model \qquad D_\alpha = 4 \cdot 10^{-3}\epsilon'/\epsilon$$

$$Weinberg\ Higgs\ model \qquad D_\alpha \sim 10^{-4}$$

$$SU(2)_L \times SU(2)_R \times U(1)\ model \qquad D_\alpha \sim 10^{-4}\ to\ 10^{-5}$$

Donoghue and collaborators have discussed ways of measuring D_α in the process $\bar{p}+p \rightarrow \bar{\Lambda}+\Lambda$, which has the great advantage of producing Λ and $\bar{\Lambda}$ in a symmetrical fashion. One needs first to find an energy at which there is a large Λ polarization P. Of course, since the production is a strong process Λ and $\bar{\Lambda}$ have the same polarization. One can then measure the asymmetry of the p and $\bar{p}$ from the *decay* relative to the production plane:

$$A = \frac{N(p, up) - N(p, down) + N(\bar{p}, up) - N_{\bar{p}}(down)}{Sum}$$

The quantity of interest D_α is then given by

$$D_\alpha = 1.56\ A/P$$

There is a larger CP-violating effect for the parameter β, but it would appear still harder to measure.

3.5 $B^o - \bar{B}^o$ System

Analogous to the K^o system there are the $B^o(\bar{b}d)$ and the $B_s^o(\bar{b}s)$ systems. Here one can look again for mixing effects and CP violation as in the $K^o - \bar{K}^o$ case. We shall concentrate on the $B^o - \bar{B}^o$ system because in the KM model one expects large CP violation for this system in contrast to K^o. The fundamental reason is easy to see by looking at the analogues of Eqs. (2.16), now seeking transitions that convert b to d

$$A(b \to u + \bar{u} + d) \sim U_{ub}U_{ud}^* = A\lambda^3(\rho - i\eta)$$

$$A(b \to \bar{c} + \bar{c} + d) \sim U_{cb}U_{cd}^* = -A\lambda^3 \tag{3.4}$$

$$A(b \to t + \bar{t} + d) \sim U_{tb}U_{td}^* = A\lambda^3[(1-\rho) + i\eta]$$

To calculate Δm and m' for this system we again use a box diagram with the exchange of two W's and with u, c, or t in the legs. From Eqs. (3.4) you see that each side of the box is proportional to λ^3 and so both m' and Δm are proportional to λ^6. The main contribution comes from the box with t in both legs (because the GIM cancellation yields a factor m_t^2) so that by inspection

$$\frac{m'}{\Delta m} = \frac{\eta(1-\rho)}{(1-\rho)^2 + \eta^2} = \frac{1}{2} sin 2\theta_{td}$$

where (θ_{td}) is the phase of U_{td}^*.

We can consider the analogue of the Fitch-Cronin experiment for B^o decay. We choose a final state with a definite CP quantum number such as $\psi + K_s$ (which is CP odd) corresponding to a transition amplitude which is CP-conserving. For example $B \to \psi + K_s$ corresponds to the transition $b \to c + \bar{c} + s$ proportional to $V_{bc}V_{cs}^* \approx A\lambda^2$. Thus the only CP violation is due to $B - \bar{B}^o$ mixing. We then find

$$\epsilon(B^o) \equiv \frac{A(B_L \to CP\ odd)}{A(B_H \to CP\ odd)} = \frac{i\eta}{1-\rho} = itan\theta_{td}$$

where B_L, B_H are the B^o mass eigenstates and (CP-odd) refers to the final state. Notice in contrast to the K^o case $\epsilon(B^o)$ is not suppressed by any power of λ. There must, of course, be some way of distinguishing B_L and B_H. In the case of K_L and K_S the lifetime difference was the obvious difference but for the case of B^o decay there are so many final states that one expects B_L and B_H to have practically the same lifetime. Thus we must make use of the mass difference Δm; hence the subscripts L and H for light and heavy. As a result of the mass difference an initial B^o, which is a mixture of B_L and B_H, will show as a function of time interference effects just as happens in the case of K^o. Thus there will be a difference in the decays of particles that are tagged at birth as B^o or $\bar{B}^o$. Thus one finds

$$r_{\pm} \equiv \frac{\Gamma(B^o \to \pm) - \Gamma(\bar{B}^o \to \pm)}{\Gamma(B^o \to \pm) + \Gamma(\bar{B}^o \to \pm)} = \mp \frac{(\Delta m/\Gamma)}{(1+(\Delta m/\Gamma)^2} sin\ 2\theta_{td}$$

where $\pm$ is the final state CP. The value of $\Delta m/\Gamma$ is unknown but is expected to be only about 0.2. On the other hand as discussed $sin 2\theta_{td}$ may be of order unity. Thus a fairly large CP-violating effect is expected. However the experiment still looks extremely difficult because the exclusive B^o branching ratio to any simple exclusive final state is likely to be much less than 1%.

3.6 Conclusion

The best bet for a new non-zero observable is still the search for ϵ' in the decay $K^o \to 2\pi$. A non-zero value would clearly kill superweak models, although many milliweak models including the KM model would still be allowed. The next best bet is the electric dipole moment of the neutron because of the sensitivity of present and prospective experiments. The sensitivity of the measurement of d_n is far from that needed to see the order-of-magnitude predicted by the KM model; on the other hand the strong CP problem renders the interpretation of a non-zero d_n ambiguous. The KM model does predict large effects in the $B^o - \bar{B}^o$ system but the experiments are very difficult and are not likely be done in the near future if at all.

All other experiments are really long shots. However any experiment that can search for a CP-violating effect at the level 10^{-3} is probably worth doing unless there are some obvious suppression factors. Given our present limited knowledge the next advance on the CP problem will likely come in an unexpected way.

BIBLIOGRAPHY AND REFERENCES

1. General reviews

Much of the material is based on my article "Present Status of CP Violation" to be published in the Annual Review of Nuclear and Particle Science, Volume 36. I also refer to my 1968 Erice lectures published in "Theory and Phenomenology in Particle Physics (A. Zichichi, Ed.) Academic Press (1969).

2. Experimental Summaries

A good review of older experiments is given by K. Kleinknecht, Ann. Rev. Nuc. Sci. 26:1 (1976). A recent review of ϵ'/ϵ measurements is given by G. Gollin, Proc. of the Oregon Meeting, World Scientific (1986).

3. KM Matrix Elements

V_{us} : H. Leutwyler and M. Roos Z. Phys. C 25, 91 (1984).

U_{cb} : B. Grinstein, N. Isgur, and M. Wise, Phys. Rev. Lett. 56 298 (1986). Unfortunately the theoretical result is more uncertain than indicated in this paper.

General parameterization : L. Maiani, Int. Symp. on Lepton Photon Interaction p. 867 DESY Hamburg (1977). L. Wolfenstein, Phys. Rev. Lett. 51, 1945 (1983).

4. Weinberg Higgs Model

Recent analyses include J. F. Donoghue and B. R. Holstein, Phys. Rev. D32, 1152 (1985) and H. Y. Cheng, Phys. Rev. D34, 1397 (1986).

5. $SU(2)_L \times SU(2)_R \times U(1)$ Model

Most of our results are taken from the work of G. Ecker and W. Grimus, Nuc. Phys. B 258, 328 (1985); Phys. Lett. B 153, 279 (1985).

6. CPT Violation

A detailed analysis of data allowing for CPT violation is given by V. Barmin et al, Nucl. Phys. B 247, 293 (1984).

7. Quantitative Calculation of ϵ and ϵ'

J. F. Donoghue, E. Golowich, and B. Holstein, Physics Reports 131, 319 (1986) and references therein.

8. Dipole Moment of the Neutron in the KM Model

M. B. Gavela et al, Phys. Lett 109B, 215 (1982), Zeits. Phys. C23, 251 (1984); I. B. Khriplovich et al, Soviet Physics JETP 60, 873 (1985).

9. CP Violation in $p + \bar{p} \rightarrow \Lambda \times \bar{\Lambda}$

J. F. Donoghue in Proc. First Workshop on Antimatter Physics at Low Energy, p. 242 Fermilab (1986).

10. CP Violation in $K^o \rightarrow \gamma\gamma$

E. Golowich in Flavour Mixing and CP Violation (ed. J. T. T. Van) Editions Frontieres (1985).

USE OF QCD SUM RULES IN THE EVALUATION OF WEAK HADRONIC MATRIX ELEMENTS

Roger Decker

Institut für Theoretische Kernphysik
Universität Karlsruhe
D-7500 Karlsruhe 1

INTRODUCTION

The kaon system constitutes a great challenge for the standard model. Recent measurements of the B lifetime and decay channels indicate that the standard model may have some difficulties in explaining the kaon physics [1]. However, firm statements are not possible because any existing phenomenological analysis is plagued by uncertainties related to the evaluation of weak hadronic matrix elements and to the socalled long distance effects. We will not expand on the long distance problem but concentrate on the short distance predictions of the standard model.

The first step in a phenomenological analysis of the kaon system is devoted to the short distance behavior of weak interactions. It results in an effective $\Delta S=1,2$ Hamiltonian which is a sum of local four-quark operators constructed with light quarks (u,d,s) only [2].

$$H_{eff}^{\Delta S} = A(M_w, m_h, \Theta_i, \mu)\ \bar{\psi}\,\Gamma_a\psi\ \bar{\psi}\,\Gamma_b\psi \qquad (1)$$

where Γ_i are Dirac matrices; A is a function of the weak boson mass, the heavy quark mass, the quark mixing angles (Θ_i), and the overall renormalization scale μ.

The second step is devoted to the evaluation of weak hadronic matrix elements using the effective Hamiltonians. The computation is a controversial subject. We will not review all possible methods but elaborate on a method based on QCD sum rules, allowing a direct determination of matrix elements of the type $\langle p'|O|p\rangle$. This method can be applied directly for the $K^\circ\bar{K}^\circ$ mixing but must be handled with care if we want to compute the decay amplitudes: one pion has to be reduced by PCAC.

QCD SUM RULES FOR WEAK MATRIX ELEMENTS

The basic idea is to write down a dispersion relation for invariant functions appearing in the decomposition of some three-point function. The imaginary part of the function will be saturated by the desired matrix element and a description of the continuum.

On the other hand, in the short distance limit the real part is computed using the operator product expansion (OPE) which is assumed to hold in the physical vacuum.

Taking advantage of Borel transforms, the contribution of the desired matrix element is enhanced and higher dimensional operators in the OPE are suppressed. A socalled sum rule then relates the weak matrix element to fundamental quantities of the QCD Lagrangian: quark masses and some condensates $\langle\bar{\psi}\psi\rangle$, $\langle G^2\rangle$, etc. The condensates are fitted numerically by reproducing the spectrum of mesons and their decay constants [3].

Consider the three-point function $\Gamma^{\mu\nu}(p,p',q)$ [4]

$$\Gamma^{\mu\nu}(p,p',q) = -\int d^4x d^4y \exp(px-qy) \langle 0|TJ^\mu(x)O(y)J^\nu(0)|0\rangle \qquad (2)$$

with J^μ, J^ν the axial currents coupling to the states $|p\rangle$, $|p'\rangle$, O(y) the four-quark operator of eq. 1. $\Gamma^{\mu\nu}$ includes a lot of different nontrivial tensor structures:

$$\Gamma^{\mu\nu} = g^{\mu\nu}F_0(p^2,p'^2,q^2) + p^\mu p^\nu F_1(p^2,p'^2,q^2) + \ldots \qquad (3)$$

For any of these structure functions (F_i) we are allowed to write a double disperson relation in p^2 and p'^2 at fixed q^2. We will restrict ourselves to $|p\rangle$, $|p'\rangle$ being pseudoscalar particles, in which case the structure function proportional to $p^\mu p'^\nu$ is the interesting one

$$F(p^2,p'^2,q^2) = \int dsds' \frac{\rho(s,s',q^2)}{(s+p^2)(s+p'^2)} + \text{sub} \qquad (4)$$

The quantity $-4\pi^2\rho(s,s',q^2)$ equals the double discontinuity of F on the cuts $m^2_p<s<\infty$, $m^2_{p'}<s'<\infty$. The subtractions are polynomials in p^2 and p'^2. In order to get rid of the subtractions we perform a Borel transform:

$$B\,B'\;F(p^2,p'^2,q^2) = \frac{1}{M^2M'^2}\int dsds'\rho(s,s',q^2)\;\exp-(\frac{s}{M^2}+\frac{s'}{M'^2}) \qquad (5)$$

$$\text{with}\quad B\,g(p^2) = \lim_{\substack{n\to\infty \\ p^2\to\infty \\ p^2/n=M^2}} \left[\frac{(p^2)^n}{n!}\left(\frac{-d}{dp^2}\right)^n g(p^2)\right] \qquad (6)$$

In equation 5 the advantage of Borel transform comes out clearly: the exp function allows an enhancement of a particular contribution to $\rho(s,s',q^2)$. In our case of interest ρ has a particularly simple form:

$$\rho(s,s',q^2{=}0) = f_p f_{p'}\,\langle p'|O|p\rangle + [1-\theta(s_0-s)\theta(s_0-s')]\rho^q(s,s',o) \qquad (7)$$

We assumed that the contribution of the continuum is given by the free quark result obtained as the imaginary part of the Feynman diagram. The new parameter will thus be fixed by asking for best stability of the sum rule (explained later).

Next we have to discuss the lefthand side of equation 5. In the limit of large p^2, p'^2 we apply the OPE to compute F. Aside from the free quark result we have to compute power corrections to F. Due to the complicated structure of the QCD vacuum, condensates of quark and gluon fields will arise. Our calculations are performed in the background field method. Taking operators up to dimension 6 we transform eq. 5 to

$$f_p f_{p'} \langle p'|O|p\rangle \exp-\left[\frac{m_p^2}{M^2} + \frac{m_{p'}^2}{M'^2}\right]$$

$$= \int_o^{s_o} ds \int_o^{s_o} ds'\, \rho^q(s,s',o) \exp-\left(\frac{s}{M^2} + \frac{s'}{M'^2}\right) + M^2M'^2[F_\psi(M^2,M'^2)\langle\bar\psi\psi\rangle$$

$$+ F_G(M^2,M'^2)\langle\frac{\alpha}{\pi}G^2\rangle + F_\sigma(M^2,M'^2)\langle g\bar\psi\sigma^{\mu\nu}\psi G^{\mu\nu}\rangle] \tag{8}$$

In equation 8 we took advantage of combining the free quark contribution of the lefthand and righthand side of equation 5 together. Eq. 8 is our master equation and will provide us with the value of the matrix element. Plotting eq. 8 as a function of M^2 and M'^2 the value of the physical matrix element is obtained in the region where it is independent of M^2 and M'^2. The parameter s_0 has to be chosen such that the stability region is as large as possible and the contribution of the free quark contribution is about 70% to 80% [5].

PREDICTIONS FOR SOME HADRONIC MATRIX ELEMENTS

A direct application is the $K^o\bar{K}^o$ mixing. The short distance contribution includes

$$O^{\Delta s=2} = \bar{s}_L \gamma_\mu d_L \bar{s}_L \gamma^\mu d_L \tag{9}$$

Here it is usual to introduce the bag parameter B defined by

$$\langle\bar{K}^o|O^{\Delta s=2}|K^o\rangle = 2/3B\, f_K^2 m_K^2 , \quad f_K = 156\text{MeV} \tag{10}$$

B=1 corresponds to the historical vacuum insertion assumption made by Gaillard and Lee [6].

Our master equation reads:

$$\frac{2B}{3} f_K^4 m_K^2 \exp\frac{-m_K^2}{M^2}\left(1+\frac{1}{\alpha}\right) = \int_{m_s^2}^{s_o} ds \int_{m_s^2}^{s_o} ds'\, \rho(s,s') \exp\frac{-1}{M^2}\left(s+\frac{s'}{\alpha}\right)$$

$$\frac{-1}{36\pi^2} m_s\left(\alpha+\frac{1}{\alpha}\right) \langle 0|g\bar{d}\sigma_{\mu\nu} d\, G^{\mu\nu}|0\rangle$$

$$+ \frac{M^2(1-\alpha)^2}{108\pi^2}\left(1+\frac{1}{\alpha}\right)\left[\left[\langle\frac{\alpha_s}{\pi}G^2\rangle + 6m_s\langle\bar{s}s\rangle\right]\right] \tag{11}$$

with $\alpha = M'^2/M^2$. With a conservative error estimate our result [4] is:

$$B = (0.58 \pm 0.16)\, [\alpha(\mu)]^{2/9} \tag{12}$$

From the phenomenological point of view, this indicates that the experimental value for ϵ can be reproduced without a too large mass for the top quark [1]. The real part of ΔM cannot be reproduced but it is known to be plagued by big long distance contributions.

Soft pion techniques are used to relate the $K \to 2\pi$ amplitude to $K \to \pi$ matrix elements. For the $\Delta I=3/2$ amplitude we obtain [6]

$$a^{3/2}(K^+ \to \pi^+\pi^0) = 3/2\ f_\pi\ A_{\Delta I=3/2}\ \langle \pi^+ | :O_4: | K^+ \rangle$$

with

$$O_4 = (\bar{s}_L \gamma_\mu d_L)(\bar{u}_\ell \gamma^\mu u_\ell - \bar{d}_L \gamma^\mu d_L) + \bar{s}_L \gamma^\mu u_\ell \bar{u}_\ell \gamma_\mu d_L \tag{13}$$

The $\Delta I=1/2$ amplitude is more delicate. Indeed for penguin operators an anomalous commutator occurs. No safe predictions with this method can be obtained; we are unable to make a conclusive statement about the $\Delta I=1/2$ rule. In case of the $\Delta I=3/2$ amplitude the sum rule reads as

$$\langle \pi^+ | O_4 | K^+ \rangle\ f_K f_\pi\ \exp - \frac{1}{M^2} \left[m_K^2 + \frac{m_\pi^2}{\alpha} \right]$$
$$= \int_{m_s^2}^{s_o} ds \int_0^{s_o} ds'\ \rho(s,s')\ \exp - \frac{1}{M^2}\ (s + \frac{s'}{\alpha}) + \frac{m_s}{18\pi^2} \langle \bar{s}s \rangle\ M^2 \alpha (1-\alpha)$$
$$+ \frac{m_s}{108\pi^2} \langle g\bar{s}\sigma_{\mu\nu} s G^{\mu\nu} \rangle\ \alpha(5\alpha-2) + \frac{M^2}{216\pi^2} \langle \frac{\alpha_s}{\pi} G^2 \rangle\ (1-\alpha)^2\ (1 + \frac{1}{\alpha}) \tag{14}$$

We obtain

$$\langle \pi^+ | O_4 | K^+ \rangle = 1.0 \pm 0.2\ 10^{-3}\ (\mathrm{GeV})^4 \tag{15}$$

Our result is compatible with experimental results. Furthermore, an SU(3) breaking is observed, i.e. the relation derived in reference [7] does not hold. This is not so surprising after all, because during the calculation our parameters were SU(3) broken: $m_s \neq m_u, m_d$; $\langle \bar{d}d \rangle \neq \langle \bar{s}s \rangle$.

CONCLUSIONS

Our conclusion is that weak hadronic matrix elements can be calculated in the frame of QCD sum rule method. We are unable to make a firm prediction for the $\Delta I=1/2$ rule because of the presence of anomalous commutators relating the $K \to \pi\pi$ amplitude to the $K-\pi$ amplitude.

The same method could be used to calculate in a pole dominance model the $K_s \to 2\gamma$ decay.

We would like to inform the reader that QCD sum rules in a quite different context (use of an effective chiral Lagrangian) are used by the authors of reference [8] to calculate the hadronic matrix elements.

REFERENCES

[1] For instance: L. Wolfenstein, these proceedings.

[2] F.J. Gilman and M.B. Wise, Phys. Lett. 93B:129 (1980); Phys. Rev. D27:1128 (1982).

[3] L.J. Reinders, H. Rubinstein and S. Yazaki, review in: CERN-TH 4079/84.

[4] R. Decker, Nucl. Phys. B277:660 (1986).

[5] B.L. Ioffe and A.V. Smilge, Nucl. Phys. B216:373 (1983).

[6] R. Decker, Z. Physik C29:131 (1985).

[7] J.F. Donaghue, E. Golowich and B.R. Holstein, Phys. Lett. 119B:412 (1982).

[8] A. Pich and E. de Rafael, in: "Proc. Workshop on Nonperturbative Methods", ed. S. Narison, World Scientific Publ., Singapore (1986) p.51.

SHORT-DISTANCE EFFECTS IN THE K^o-$\bar{K}^o$ MIXING IN THE STANDARD MODEL

I. Picek *

DESY, Hamburg, FR. Germany and
Rudjer Bošković Institute, Zagreb
Croatia, Yugoslavia

From the accelerator experiments in the 1950s up to the CP experiments of the present time the neutral kaon complex has remained one of the most interesting systems in nature. The breakthroughs which it has provided in our knowledge of fundamental physics have been crucial in establishing the standard SU(3)xSU(2)xU(1) model (SM) of electroweak and strong interactions. More recently there have been claims that the existing CP-violation measurements represent a serious threat to the standard model. In this paper I would like to illustrate that such announcements are premature in view of the degree of accuracy in theoretical predictions. Therefore, I would stick to the theoretically more tractable short-distance contributions and then I shall gradually show that, on the one hand, their present treatment is incomplete and, on the other hand, they have to be supplemented at various stages (which I intend to explicate) by some sort of long-distance (LD) calculation.

Since this meeting is intended to be a school, it might be appropriate to start with a short history. Strange particles are produced in a large number in strong interactions, and thus it has been natural to introduce the strong-interaction (and strangeness) eigenstates

$$|K^o\rangle \ (S=-1) \ \text{and} \ |\bar{K}^o\rangle \ (S=+1) \ .$$

However, it has been discovered that these states oscillate (mix) owing to the weak interaction. The resulting removal of degeneracy

$$\Delta m_K \simeq 3.5 \times 10^{-15}\,\text{GeV} \tag{1}$$

represents the most precisely measured quantity in particle physics. The corresponding weak-interaction (and CP-symmetry) eigenstates

$$|K_1\rangle = (|K^o\rangle + |\bar{K}^o\rangle)/\sqrt{2} \quad (CP=+) \quad \text{and} \quad |K_2\rangle = (|K^o\rangle - |\bar{K}^o\rangle)/\sqrt{2} \quad (CP=-)$$

are still not the last word in the story. The small impurity discovered in these states,

$$\varepsilon = 2.27 \times 10^{-3} \tag{2}$$

* A. v. Humboldt fellow at DESY, Hamburg, F.R. Germany

represents the only measured CP-violating effect. Thus, our present knowledge stops at short- and long-lived eigenstates,

$$|K_S\rangle = \frac{1}{\sqrt{2}}(|K_1\rangle + \varepsilon_S|K_2\rangle) \quad , \quad |K_L\rangle = \frac{1}{\sqrt{2}}(|K_2\rangle + \varepsilon_L|K_1\rangle)$$

and we take $\varepsilon_S = \varepsilon_L = \varepsilon$, assuming CPT invariance.

Although the derivation of hadronic states is still beyond our capabilities, we can hope to get a better understanding of the inner machinery of the $\Delta S=2$ mixing, i.e. to reproduce the values in (1) and (2) theoretically. Our hope resides in the SM. From the outset there are two possible contributions within the SM, namely, the short-distance /SD/ contribution $\Delta S=2$ (Fig. 1) and the long-distance /LD/ dispersive contribution $(\Delta S=1)^2$ (Fig. 2). Accordingly,

$$\Delta m_K = \Delta m_K^{SD} + \Delta m_K^{LD} \tag{3}$$

and

$$\varepsilon = \varepsilon^{SD} + \varepsilon^{LD} \quad . \tag{4}$$

In general, the calculation of the dispersive LD effects is plagued with large uncertainties. We would like to distinguish the SD from the LD contribution or at least to know their relative size for a given physical quantity. For the mass difference Δm_K the potential significance of the LD effects has been pointed out by Wolfenstein /1/ and Hill /2/, and our ignorance in its precise knowledge has been phrased /1/ as a D-parameter

$$\Delta m_K^{LD} = D\,\Delta m_K \quad . \tag{5}$$

The situation with the ε-parameter is considerably better /3/,

$$\varepsilon^{LD} < 0.2\,\varepsilon \quad , \tag{6}$$

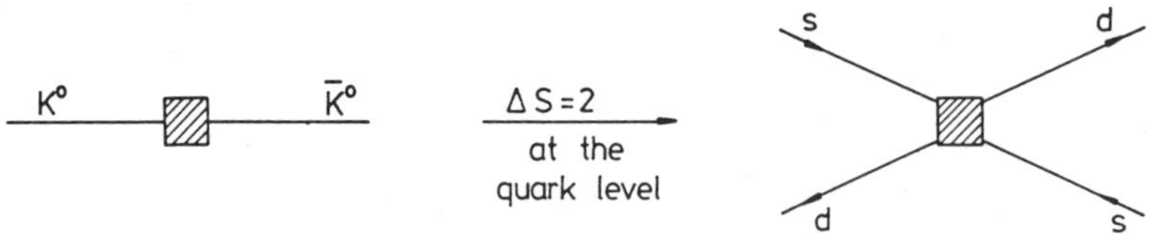

Fig. 1

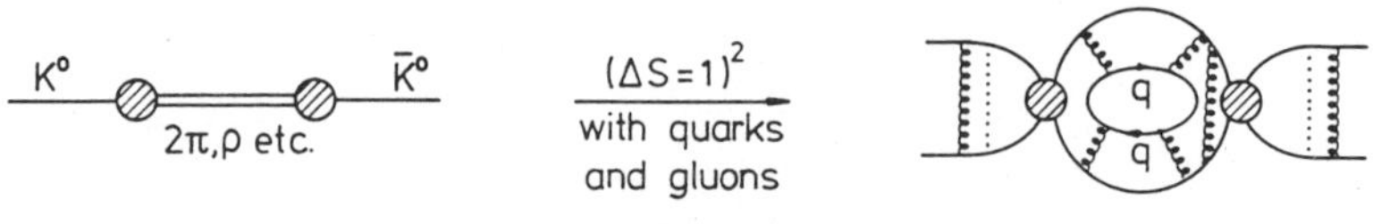

Fig 2

indicating that ε is SD dominated. Then, this motivates us to explore the SD contributions to ε in more detail. Since ε and Δm are, respectively, the imaginary and real parts of the same K^0 to $\bar{K}^0$ amplitude, we simultaneously obtain the improvement of the SD contribution to Δm. However, in view of (5), this improvement will only be of marginal importance.

The simplest SD diagram for the $\Delta S=2$ transition of interest is the standard box /4/ (Fig. 3). In principle, this diagram can give rise to both Fig. 1 and Fig. 2. Here we restrict ourselves to its contribution described by a four-quark effective hamiltonian

$$H^{\Delta S=2}_{Box} = \frac{G_F^2}{16\pi^2}\left[(\eta_1)\, m_c^2 \lambda_c^2 + (\eta_2)\, m_t^2 \lambda_t^2 + (\eta_3)\, \frac{2\lambda_c\lambda_t m_c^2}{1-m_c^2/m_t^2}\ln\frac{m_t^2}{m_c^2}\right] O^{\Delta S=2}, \qquad (7)$$

where $\lambda_i = V_{id}V^*_{is}$ are the Kobayashi-Maskawa factors constrained by the relation $\lambda_u + \lambda_c + \lambda_t = 0$, and

$$O^{\Delta S=2} = \bar{d}^i \gamma^\mu (1-\gamma_5) s^i \, \bar{d}^j \gamma_\mu (1-\gamma_5) s^j \quad . \qquad (8)$$

The real and imaginary parts of (7) account for Δm_K and ε, respectively, whereby one has to calculate the K^0 to $\bar{K}^0$ transition amplitude. Conventionally, this is done by the vacuum saturation approximation (VSA)

$$\langle K^0 | O^{\Delta S=2} | \bar{K}^0 \rangle \xrightarrow{VSA} \frac{16}{3} F_K^2 m_K^2 \;\; ; \;\; F_K \simeq 0.113\ \mathrm{GeV} \; . \qquad (9)$$

Here we face the <u>second</u> LD uncertainty, phrased as a B-value

$$B = \frac{\langle K^0 | O^{\Delta S=2} | \bar{K}_0 \rangle}{\langle K^0 | O^{\Delta S=2} | \bar{K}^0 \rangle_{VSA}} \; . \qquad (10)$$

This represents the gap in tracing the transition from the quark to the hadronic states. In particular, this uncertainty reflects for the mass

Fig. 3

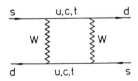

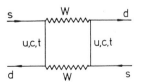

difference as

$$\Delta m_K^{Box} = B \times (\Delta m_K^{Box})_{VSA} \quad . \tag{11}$$

In the standard box, the role of the QCD part of the SM is not explicit. The QCD enters when taking into account the hard-gluon (SD) corrections to $H_{Box}^{\Delta S=2}$ /5/, namely the three terms in (7) corrected by the factors

$$\eta_1 \simeq 0.7 \quad , \quad \eta_2 \simeq 0.6 \quad , \quad \eta_3 \simeq 0.5 \quad . \tag{12}$$

The role of QCD is more explicit in considering a topologically different class of diagrams, the double penguin-like (DPL) diagrams of Fig. 4. Here one employs the induced flavour-changing (penguin) vertices, known by a "magical" property of reducing the power like GIM cancellation to the logarithmic one /6/. The role of double-penguin (DP) diagrams for Δm_K has been controversial for some time /7,8/. The solution /9/ is that the particular DP diagram displayed in Fig. 4 represents only a percentage of the standard box and thus is completely negligible. The larger class of DPL diagrams /10/ might give rise to ~ 10% of the standard box, but it is still immaterial in view of the LD contribution in (3). Note that, in the ratio, the two SD contributions in (3)

$$\Delta m_K^{SD} = \Delta m_K^{Box} + \Delta m_K^{DPL} \tag{13}$$

are free of the LD uncertainty of the B-factor given in (10) and (11). This is due to the fact that both contributions in (13) result in the same local four-quark operator.

However, the DPL diagrams turn out to be more relevant for the ε-parameter in (2) and (4) /10,11/. This is due to the negligible ε^{LD} from eq. (6). Accordingly,

$$\varepsilon \simeq \varepsilon^{SD} = \varepsilon^{Box} + \varepsilon^{DPL} \quad . \tag{14}$$

The numerical value of the RHS of (14) depends on the parameters involved, the top quark mass m_t and the infrared (IR) cut-off μ. The latter has been introduced by

$$\alpha_s(\mu^2) = 1 \quad , \tag{15}$$

in order to have a sensible perturbative QCD calculation $(\alpha_s(p^2) < 1)$. At this point we face the third uncertainty of the LD type, related to the IR cut-off μ. This cut-off indicates the border of the non-perturbative region and the LD effects in the form of bound states. Since the effects of the latter are not well controlled by the present QCD techniques, we have to check the stability of our results with respect to the variation

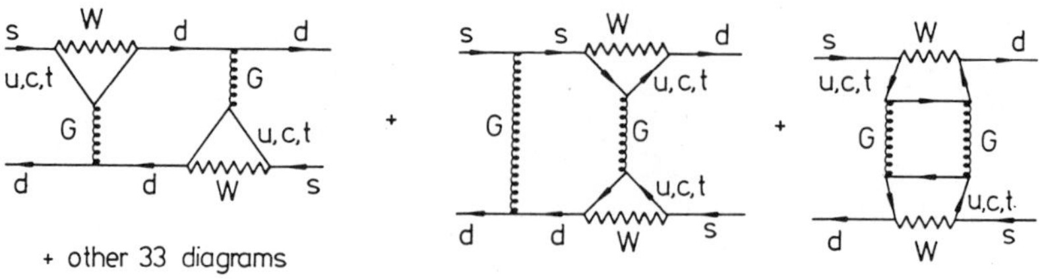

Fig. 4

in μ /11/. For m_t = 45 GeV and μ = 0.7 GeV, ε^{DPL} represents -25% of the QCD-corrected standard-box value. Varying μ in the range (0.3, 1.2) GeV gives ε^{DPL} in the interval (-15%, -40%).

Apparently, the result for ε with the DPL added, (eq. (14)), seems to be even more off the experimental value (2) than the standard box alone. The mere fact that the DPL contributions are not negligible is not surprising in view of previous experience that the extra powers of α_s might be compensated when penguin vertices are inserted in a higher loop /12/. Since our DPL class of diagrams represents just another contribution, which in no sense makes the final prediction for ε , it merely illustrates the need of more accurate calculations. First of all, this refers to the need of going beyond the leading logarithm. For example, the QCD corrections to the standard box quoted in (12) represent the result of leading logarithms summed by the renormalization-group method to all orders in α_s. There is no such simple treatment of non-leading terms. As a result, the calculations become more tedious and resemble the calculation for the g-2 (compare the lecture presented at this School by V. Hughes) where one has to procede order by order in the coupling. The point is that, in these cases, the level of accuracy required from theoretical predictions is dictated by the precise measurements performed so far. Before reaching the appropriate calculational accuracy, one is not able to infer about the crisis of the SM with respect to CP phenomenology.

References

1. L. Wolfenstein, Nucl. Phys. B160 (1979) 501
2. C.T. Hill, Phys. Lett. 97B (1980) 275
3. J.F. Donoghue, E. Golowich and B. Holstein, Phys. Reports 131 (1986) 319
4. M.K. Gaillard and B.W. Lee, Phys. Rev. D10 (1974) 897
5. F.J. Gilman and M.B. Wise, Phys. Rev. D27 (1983) 1128
6. A.I. Vainshtein, V.I. Zakharov and M.A. Shifman, ZhETF Pisma 22 (1975) 123 (JETP Lett. 22 (1975) 55);
   M.A. Shifman, A.I. Vainshtein and V.I. Zakharov, Nucl. Phys. B120 (1977) 316
7. D. Hochberg and R.G. Sachs, Phys. Rev. D27 (1983) 606
8. G. Ecker, Phys. Lett. 147B (1984) 369
9. J.O. Eeg and I. Picek, Phys. Lett. 160B (1985) 154
10. J.O. Eeg and I. Picek, Phys. Lett. 177B (1986) 432
11. J.O. Eeg and I. Picek, DESY report 86-135, submitted to Nucl. Phys. B
12. J.O. Eeg and I. Picek, Nucl. Phys. B244 (1984) 77

THE CP LEAR PROJECT: PS195

Basle, CERN, CEN Saclay, DAP Athens, Democritus, ETH Zurich, Friburg, Liverpool, SIN, Stockholm, Thessaloniki Collaboration

Presented by: P.J. Hayman
Department of Physics, Liverpool University, U.K.

1. INTRODUCTION

It has been proposed by Adiels et al to carry out tests of CP violation with K^0 and $\bar{K}^0$ at LEAR by measuring interference effects and asymmetries of the different amplitudes in the 2π, 3π and semi-leptonic decay channels. These measurements will be possible when intense $\bar{p}$ beams are available from late 1987 after construction of the $\bar{p}$ accumulator (ACOL) at CERN. The symmetrical production of K^0 and $\bar{K}^0$ in $\bar{p}p$ annihilation at rest was shown by Gabathuler and Pavlopoulos to have particular advantages in minimising systematic errors compared to other measurement techniques where K_S and K_L are used in investigating CP violation. Because of this new technique if there is new physics to be seen, the well controlled systematics of this experiment will enable it to be seen at the level of the anticipated statistical and systematic uncertainties.

2. K^0, $\bar{K}^0$ PRODUCTION

Neutral kaons are produced in fluxes of equal magnitude and at the same time by $\bar{p}$ (at 105 MeV/c) stopped in a gas target and in conjunction with charged kaons which uniquely identify (tag) the final states in the exclusive channels:

$$\bar{p}p \rightarrow \quad K^+ \pi^- \bar{K}^0 \text{ (total BR } 2\ 10^{-3})$$
$$K^- \pi^+ K^0 \text{ (total BR } 2\ 10^{-3})$$

In order to achieve the required statistical limits a beam of 2.10^6 $\bar{p}$'s per second will give a 2 KHz event rate for 25%

acceptance. This corresponds to $3.5 \cdot 10^8$ each of K^o and $\bar{K}^o$ per day. All estimates of the measureable effects have been calculated assuming a total of 10^{13} stopped $\bar{p}$.

3. CP VIOLATION

The K^o and $\bar{K}^o$ produced in strong interactions can be represented as linear conbinations of the weak interaction eigenstates K_S and K_L where the complex parameter ε specifies the CP violation in the mass matrix and ε' in the decay matrix. The advantage of initially pure K^o or $\bar{K}^o$ beams when used to investigate CP violation is that they each contain exactly equal K_S and K_L fluxes. Assuming that CPT is conserved to 10^{-19} we will study the decay modes of K_S and K_L to final states

$$
\begin{array}{lll}
K_S \rightarrow \pi^+ \pi^- & & \text{total BR} \sim 100\% \\
 \pi^o \pi^o & & \\
K_L \rightarrow \pi^+ \pi^- \pi^o & & \text{total BR} \sim 34\% \\
 \pi^o \pi^o \pi^o & & \\
K_L \rightarrow \pi^{\pm} \mu^{\mp} \nu & & \text{total BR} \sim 66\% \\
 \pi^{\pm} e^{\mp} \nu & &
\end{array}
$$

to investigate the violation of CP which is as mysterious now as was the original discovery of the 'strange' particle by Rochester and Butler which is now manifested in the K_S having $\frac{1}{\gamma_s} = \tau_s \sim 0.9 \times 10^{-8}\text{s}$.

4. THE OBSERVABLES

CP violation can be detected through the interference between the amptitudes for decay into a given final state of the particle and its antiparticle.

The observed decay rate R $(K^o \rightarrow f)$ then depends on the interference

$$A_S\, A_L\, e^{-\gamma_s t} \cos(\Delta m.\ t - \phi)$$

where $\qquad \Delta m = m_2 - m_1$

is the mass difference of K^o and $\bar{K}^o$ and ϕ = the relative phase of A_S and A_L.

For a final state f where

$$\eta_f = \frac{A(K_L \rightarrow f)}{A(K_S \rightarrow f)} = \eta\ e^{i\phi}$$

the time development asymmetry factor

$$A_f(t) = \frac{R(\bar{K}^o \to f)\ (t) - R(K^o \to f)\ (t)}{R(\bar{K}^o \to f)\ (t) + R(K^o \to f)\ (t)}$$

or the difference from unity

$$U_f(t) = 1 - \frac{R(K^o \to f)\ (t)}{R(\bar{K}^o \to f)\ (t)}$$

then depend only on the interference term and measurements of these quantities will enable the phase ϕ and amplitude η to be obtained.

If the time dependance is not required, the integral rate asymmetry I_f can be evaluated by fast on line event summation where

$$I_f = \int_o^T A_f(t)\ dt.$$

<u>2π Final States</u> $(f = \pi^+ \pi^-, \pi^o \pi^o)$

For the $\pi^+ \pi^-$ $(\pi^o \pi^o)$ final state

$$\eta_{2\pi} = \frac{A(K_L \to 2\pi)}{A(K_S \to 2\pi)}$$

so that $\eta_{+-} = \varepsilon + \varepsilon'$, $\eta_{oo} = \varepsilon - 2\varepsilon'$

For $T \geq 15\ \tau_S$

$$I_{+-} = 2Re\varepsilon + 4Re\varepsilon', \quad I_{oo} = 2Re\varepsilon - 8Re\varepsilon'$$

so that $\varepsilon'/_\varepsilon = \frac{1}{6}\ [I_{+-}/I_{oo} - 1]$

The measurement of $\varepsilon'/_\varepsilon$ using the integral asymmetry method and depending only on a ratio of I_{+-}/I_{oo} is then independant of systematic errors such as the K^o vertex resolution, efficiences, and solid angles all of which are independent of the initial state of K^o or $\bar{K}^o$ so long as measurements are carried out over a large range of τ_S. This thus sets the size of the particle tracking chamber region in the experiment to a radius of ~ 55cm and also corresponds to the region of interest for the interference term which is shown in figure 1. Below 15cm (~ $5\tau_S$) the interference amplitude is negligible. The intercepts when A_{+-} (t) = 0 allow the measurement of ϕ and the amplitude η_{+-} can be obtained from $A_{+-}(t)$.

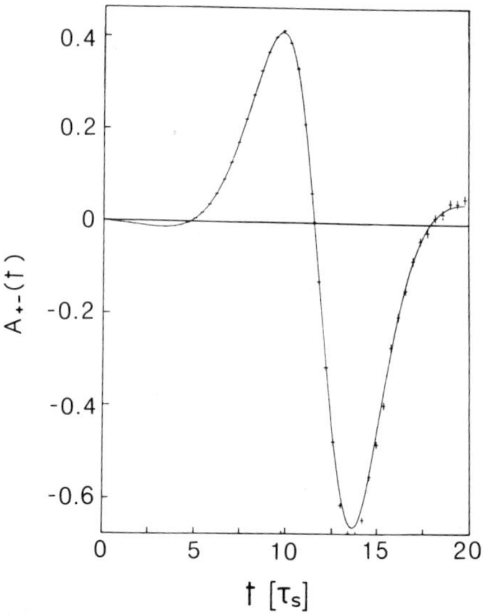

Figure 1. The asymmetry factor A_{+-} for the decay mode $\pi^+ \pi^-$.

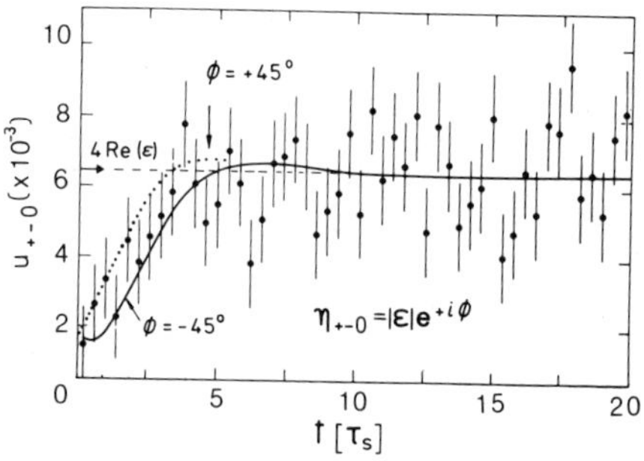

Figure 2. The difference from unity U_F of the decay mode $\pi^+ \pi^- \pi^0$.

<u>3π Final States</u> $(f = \pi^+ \pi^- \pi^0,\ \pi^0 \pi^0 \pi^0)$

For the 3π final states

$$\eta_{3\pi} = \frac{A(K_S \to 3\pi)}{A(K_L \to 3\pi)}$$

and $\qquad U_{\pi\pi\pi} = 4[\mathrm{Re}\,\varepsilon - \text{interference term}].$

The $\pi^+ \pi^- \pi^0$ final state variation of $U_{\pi^+\pi^-\pi^0}$ is shown in figure 2. Here the interference effect in U_{+-0} from the value $4\mathrm{Re}\varepsilon$ takes place in the region $< 5\tau_S$ (15cm from the decay point)and allows the value of η_{+-0} to be obtained. Detection of a change in U_{+-0} from $4\mathrm{Re}\varepsilon$ is a clear indication of CP violation and its magnitude is appreciable for either sign of the expected value of $\phi \simeq 45^\circ$.

The integral rate asymmetry

$$I_{\pi\pi\pi}(T) = \frac{4\tau_S}{T} \mathrm{Re}\ \eta_{3\pi} - 2\mathrm{Re}\varepsilon$$

is a function of the integral limit T.

From $U_{\pi^+\pi^-\pi^0}$ and $I_{\pi^+\pi^-\pi^0}$ the phase of η_{+-0} can be obtained which is expected to be better for this channel than the $3\pi^0$ channel because of better decay vertex definition.

Detailed consideration of the channels involving π^0 measurement and its effect on vertex resolution are given in the accompanying paper by A. Schopper. The simultaneous measurements of the two 3π decay modes which we propose are unique to this experiment.

Semi-leptonic Decays

Measurement of the semi leptonic decays

$K^0 \to \pi^- l^+ \upsilon$, A ; $\bar{K}^0 \to \pi^+ l^- \bar{\upsilon}$, B ; $\Delta S = \Delta Q$, or

$K^0 \to \pi^+ l^- \bar{\upsilon}$, C ; $\bar{K}^0 \to \pi^- l^+ \upsilon$, D ; $\Delta S = -\Delta Q$

will allow a check on the $\Delta S = -\Delta Q$ decay contributions which are forbidden in the standard model.

If

$$x = \frac{A(\Delta S = -\Delta Q)}{A(\Delta S = +\Delta Q)}$$

then CP violation is indicated by $Im(x) > o$.

The asymmetry

$$A_{13} = \frac{B + D - (A+C)}{B + D + (A+C)}$$

is shown in figure 3(a) together with an amplitude of Im(x) ten times the present limit, figure 3(b) and a clear indication of the measurement sensitivity.

Other quantities which can be studied using the time distributions of A, B, C, and D are

$$A_T = \frac{B - C}{B + C} \sim 2Re\varepsilon$$

$$R_1 = A+B+C+D \;\alpha\; Rex$$

$$R_2 = A-C \;\alpha\; f(\Delta m)$$

5. THE DETECTOR

The detector is required to

1. tag K^0 and $\bar{K}^0$
2. identify decay channels of K^0 and $\bar{K}^0$
3. track decay products and reconstruct the vertex to get the decay time
4. have high efficiency to detect as many K^0 and $\bar{K}^0$ as possible.

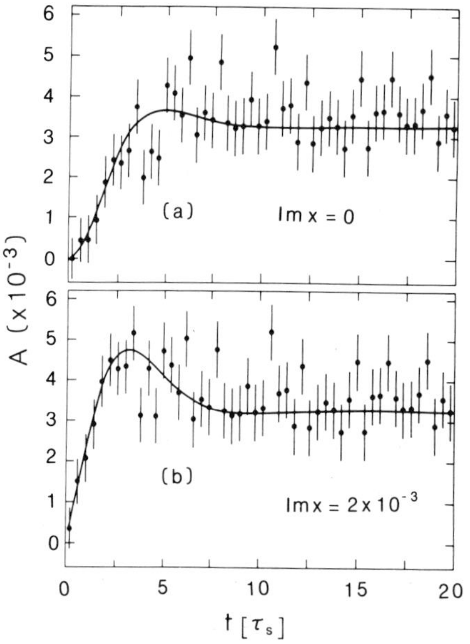

Figure 3. The asymmetry for semi-leptonic decays.

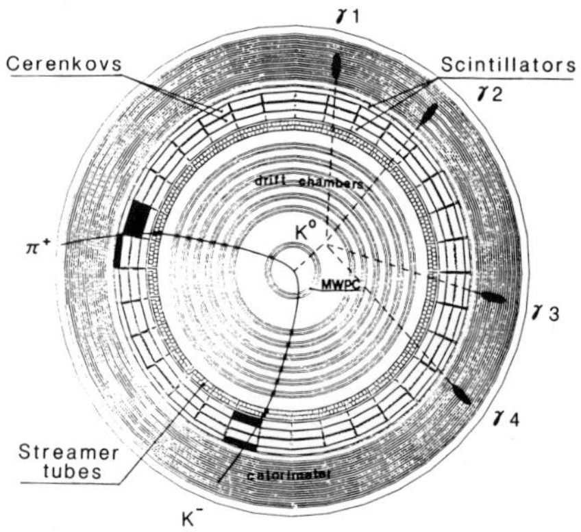

Figure 4. The CP Lear detector cross-section.

The detector is shown in figure 4, with a typical $\bar{p}p \rightarrow \pi^+ K^- K^0$ event having charged multiplicity two. Two MWPC's and six drift chambers carry out the magnetic analysis and a threshold Cerenkov detector is used to tag the decay channel from the identity of the charged kaon. A double layer (water or FC72) conventional Cerenkov detector viewed at each end by a photomultiplier has been investigated and an alternative design with a thin FC72 radiator having a quartz window and wire chamber with TMAE to detect the radiation is being considered. A double layer streamer tube array provides a fast z-coordinate to enable rapid readout of the traversed regions of the drift chambers. The double scintillation counter hodoscope is used to provide additional time of flight information for πK separation and also the multiplicity to identify each decay channel trigger. The outer section of the detector contains an integrated assembly of the electromagnetic calorimeter and magnet solenoid. The calorimeter is a 19 layer gas sampling detector having a total thickness of 6 radiation lengths and is described in detail in the following paper. The segmentation and layer design give good resolution in K^0 vertex reconstruction. Monte Carlo simulation of showers show that the resolution in the reconstruction of the decay vertex affects not only the statistical but also the systematic errors especially for $K^0 \rightarrow 2\pi$ where the decay curve is very steep. The solenoid which is derived from the former DM2 detector is 3.3m in length and 2m internal diameter and has a central field value of 0.45T constant to 2%.

6. THE TRIGGER

Purpose built hard wired trigger processors will be used to select the different final state channels and reduce the event rate from the beam value (2MHz) to give master triggers less than a few KHz, figure 5. Processor KT1 in 40ns uses multiplicity information from the scintillator, and Cerenkov and TOF information to give particle identication and a reduction factor of 4. Processor KTA in a further 250ns uses the MWPC's to get the charge of the kaon from the track curvature and from the momentum apply a cut to increase the Cerenkov efficiency giving a further reduction factor of 10. Two further processors, HWP1 and 2 within a further 1μs each, carry out full momentum reconstruction using look-up tables and then calculate missing mass.

A separate processor handles the calorimeter data. It uses a simple algorithm counting the number of hit streamer tubes to give the coordinates of the centroids of the γ-rays and values of their energies from which the K^0 vertex position and energy are obtained.

7. BACKGROUND AND BIASES (SYSTEMATICS)

2π Final States

High precision measurements are required for

$$\frac{\varepsilon'}{\varepsilon},\ \phi_{+-} - \phi_{00},\ \text{and}\ \Delta m$$

The time dependant asymmetries are used to get $\eta_{\pi\pi}$, $\phi_{\pi\pi}$ and Δm and the ratio I_{+-}/I_{00} from the integral asymmetry method to get ε'/ε.

The main contribution to the background comes from pions misidentified as kaons through the reaction

$$\bar{p}\ p \rightarrow \pi^+\pi^-(\pi^+\pi^-)$$

and through the semi leptonic decay,

$$\bar{K}^0 \rightarrow (\pi^+ l^-)\bar{\nu}$$

see figure 6. The semi leptonic background is also symmetric for the K^0 channel. Since K^0 and $\bar{K}^0$ are equally affected these systematics cancel to high order. The effect of the material in the detectors (mainly in the region of the MWPC's) can produce changes in the decay rates through regeneration. This effect is

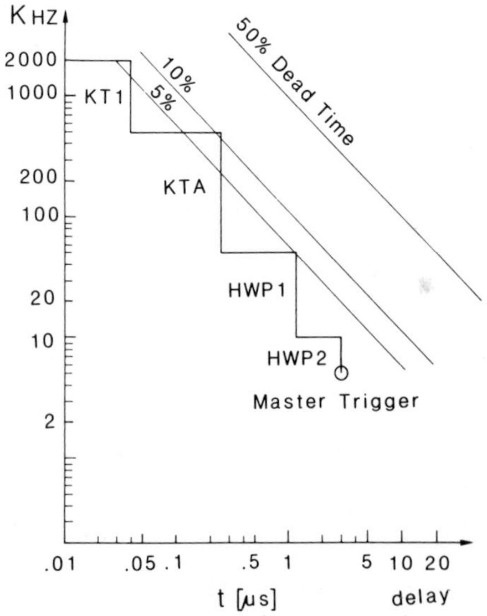

Figure 5. The effect of the hard processors on the trigger rates.

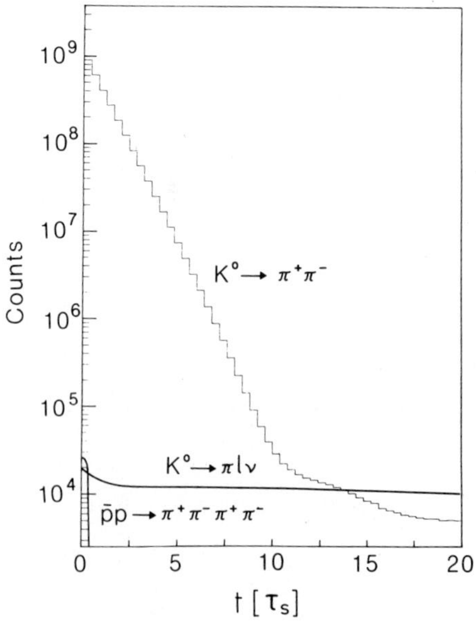

Figure 6. The contributions of bias to the $\pi^+\pi^-$.

shown in figure 7 for a medium of average density 0.04gm cm^{-3} in comparison to a vacuum. The effect can be corrected.

3π Final States

Measurement of $U_{\pi\pi\pi}(t)$ and $I_{\pi\pi\pi}(T)$ can be used to get $\eta_{3\pi}$ and $\phi_{3\pi}$. The decays allow an extremely clean selection of the $\pi^+ \pi^- \pi^0$ or $3\pi^0$ events but the misidentification of $K^{\pm}$ remains in the backgrounds see figure 8.

$$\bar{p}\,p \rightarrow \pi^+ \pi^- (3\pi).$$

The effect is reduced by good time resolution ($\sim 0.07\ \tau_s$) in the affected region $<0.5\tau_s$.

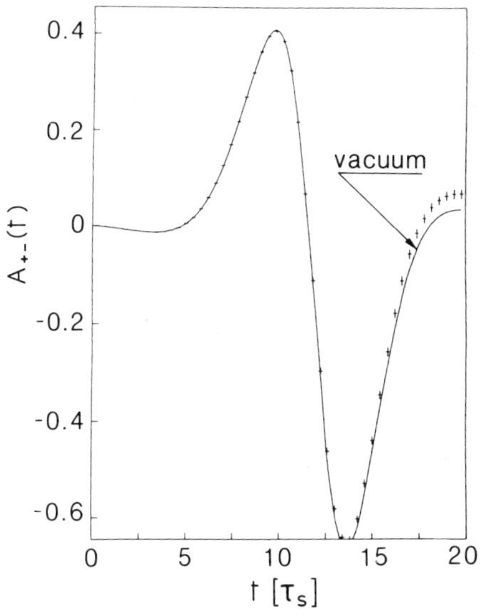

Figure 7. The effect of regeneration on the asymmetry factor A_{+-}.

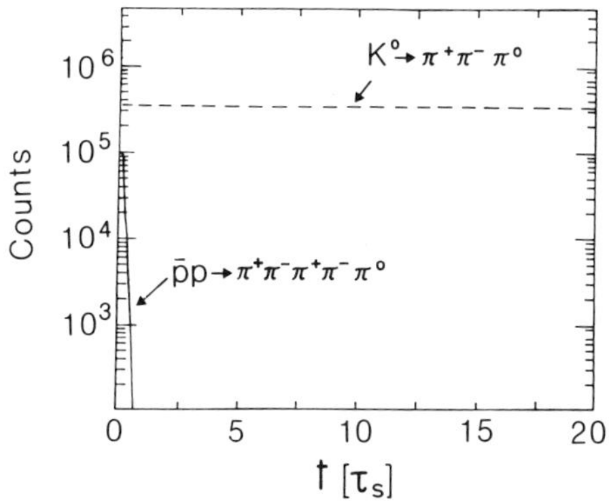

Figure 8. The contributions of bias to the $\pi^+ \pi^- \pi^0$ decay mode.

Semi-leptonic States

The electron can be identified by momentum-energy balance using the magnetic analysis and the calorimeter. The three body decays also have a continuous angular distribution and a cut on the angle of the charged particles can separate $\pi e \nu$ decays from the $\pi^+ \pi^-$ decays especially where separation of the electron is not required.

Table

CP parameter	Present precision	Standard errors, σ ($10^{13}\bar{p}$) Statistical	Total
ϵ'/ϵ	$7 \cdot 10^{-3}$	$1{\cdot}5 \cdot 10^{-3}$	$\leqslant 2 \cdot 10^{-3}$
$\Phi_{+-} - \Phi_{00}$	5°	$0{\cdot}3^{\circ}$	$\leqslant 1^{\circ}$
η_{+-0}	$<0{\cdot}12$	$6 \cdot 10^{-4}$	$\sim 6 \cdot 10^{-4}$
η_{000}	$<0{\cdot}1$	$8 \cdot 10^{-4}$	$\sim 8 \cdot 10^{-4}$
Re(x)	$<2 \cdot 10^{-2}$	$6 \cdot 10^{-4}$	$\sim 6 \cdot 10^{-4}$
Im(x)	$<2{\cdot}6 \cdot 10^{-2}$	$7 \cdot 10^{-4}$	$\sim 7 \cdot 10^{-4}$
T		$6 \cdot 10^{-2}$	$\sim 6 \cdot 10^{-2}$
$\gamma(\Delta m/m)$	$4 \cdot 10^{-3}$	$1{\cdot}2 \cdot 10^{-3}$	$1{\cdot}2 \cdot 10^{-3}$

8. RATES AND STATISTICS

It is estimated that 10^{13} $\bar{p}p$ interactions at 2.10^{6} incident $\bar{p}$ per second will give in 100 days 2.10^{10} tagged reactions for K^{o} and the same for $\bar{K}^{o}$.

The precision of the measured CP parameter which can be attained is given in the Table where they are compared with the present values.

9. ACKNOWLEDGEMENTS

The work contained in this paper is the work of many members of the collaboration. Their help and discussions are gratefully acknowledged.

10. REFERENCES

Adiels et al CERN/PSC/85-6,PSCCC 86-34

E. Gabathuler and P. Pavlopoulos in Proc Workshop in Physics at LEAR with Low-energy cooled Antiprotons, Erice 1982 (eds U Gastaldi and R Klapish) Plenum Press New York 1984, p 747

G. D. Rochester and C.C. Butler, Nature, 160, 855, (1947)

A. Schopper, following presentation at this school.

THE MEASUREMENT OF THE PHASE DIFFERENCE $(\phi_{+-} - \phi_{00})$

WITH THE CPLEAR EXPERIMENT AT CERN

The CPLEAR Collaboration*)

Presented by A. Schopper**)

Physikalisches Institut der Universität Basel
Basle, Switzerland

The phenomenology of CPT violation in the neutral kaon system is presented and the relevance of the measurement of the phase difference $(\phi_{+-} - \phi_{00})$ briefly discussed. The method, its advantages, and the limits expected of the measurements of the CPLEAR experiment are presented.

1. INTRODUCTION

Following several studies of the problem, Barmin et al.[1] pointed out in 1984 the inconsistency of the experimental data with CPT invariance and extensively discussed the importance of measuring the phase difference of the CP violating parameters[2] η_{+-} and η_{00}. The parameters of CP and CPT violation can be found in Ref. 1 and only a brief outline will be presented here for the clarity and completeness usually required at a summer school.

In discussing the validity of CP and CPT symmetries in the neutral kaon system, one has to distinguish between two possible cases where symmetry violation can appear; i.e. CP or CPT can be violated either at the level of the $\Delta S = 2$ particle–antiparticle transitions (mass matrix) or at the level of the $\Delta S = 1$ decays (decay matrix).

In order to separate the CP from a possible CPT violation in the mass matrix, one can introduce the parameters ϵ and δ measuring respectively the CPT-conserving CP violation and the CPT violation. Then the strong eigenstates $|K^0\rangle$ and $|\bar{K}^0\rangle$ can be expressed through the physical states $|K_{S,L}\rangle$ according to Table 1.

In order to describe a symmetry violation in the decay matrix, it is convenient to decompose the decay amplitudes into the amplitudes corresponding to the different final-state isospins of the 2π system. The amplitudes A_I and B_I of Table 2 describe the CPT-conserving and CPT-violating amplitudes of isospin I; Δ_I is the strong interaction phase[1]. The parameters ϵ' and a defined in Table 2 are the measures corresponding respectively to CP and CPT violation in the decay matrix. In order to relate the parameters ϵ' and a to the physical states $|K_{S,L}\rangle$, it is convenient to introduce the parameters ϵ_0 and ϵ_2 defined in Table 2, which describe CP and CPT violation, separately for the $I = 0$ and $I = 2$ final states. These parameters can be expressed by

* Basle - CERN - Demokritos - Fribourg - Liverpool - Saclay (CEN) - Stockholm - Thessaloniki - Villigen (SIN) - Zurich (ETH)

** Visitor at CERN, Geneva, Switzerland

Table 1

The parametrization of the neutral kaon states and their decays into two pions for different symmetry assumptions

CPT invariance	CPT violation in the mass matrix	CPT violation in the decay matrix
$K^0 = \frac{1}{\sqrt{2}}\left[(1-\varepsilon)K_S^0 + (1-\varepsilon)K_L^0\right]$	$K^0 = \frac{1}{\sqrt{2}}\left[1-\varepsilon+\delta)K_S^0 + (1-\varepsilon-\delta)K_L^0\right]$	$K^0 = \frac{1}{\sqrt{2}}\left[(1-\varepsilon)K_S^0 + (1-\varepsilon)K_L^0\right]$
$\bar{K}^0 = \frac{1}{\sqrt{2}}\left[(1+\varepsilon)K_S^0 - (1+\varepsilon)K_L^0\right]$	$\bar{K}^0 = \frac{1}{\sqrt{2}}\left[1+\varepsilon-\delta)K_S^0 - (1+\varepsilon+\delta)K_L^0\right]$	$\bar{K}^0 = \frac{1}{\sqrt{2}}\left[(1+\varepsilon)K_S^0 - (1+\varepsilon)K_L^0\right]$
$A(K^0) = A_I\, e^{i\Delta_I}$	$A(K^0) = A_I\, e^{i\Delta_I}$	$A(K^0) = (A_I+B_I)\, e^{i\Delta_I}$
$A(\bar{K}^0) = A_I^*\, e^{i\Delta_I}$	$A(\bar{K}^0) = A_I^*\, e^{i\Delta_i}$	$A(\bar{K}^0) = (A_I^*-B_I)\, e^{i\Delta_I}$
$a = i\,\frac{\mathrm{Im}\,A_0}{\mathrm{Re}\,A_0} \equiv 0$	$a = i\,\frac{\mathrm{Im}\,A_0}{\mathrm{Re}\,A_0} \equiv 0$	$a = \frac{\mathrm{Re}\,B_0}{\mathrm{Re}\,A_0 + i\,\mathrm{Im}\,B_0}$
$\varepsilon_0 = \varepsilon$	$\varepsilon_0 \simeq \varepsilon - \delta$	$\varepsilon_0 \simeq \varepsilon + a$
$\varepsilon' = \frac{1}{\sqrt{2}}\, i\, \frac{\mathrm{Im}\,A_2}{\mathrm{Re}\,A_0}\, e^{i(\Delta_2-\Delta_0)}$	$\varepsilon' = \frac{1}{\sqrt{2}}\, i\, \frac{\mathrm{Im}\,A_2}{\mathrm{Re}\,A_0}\, e^{i(\Delta_2-\Delta_0)}$	$\varepsilon' = \frac{1}{\sqrt{2}}\, \frac{\mathrm{Re}\,B_2 + i\,\mathrm{Im}\,A_2}{\mathrm{Re}\,A_0 + i\,\mathrm{Im}\,B_0}\, e^{i(\Delta_2-\Delta_0)}$

Table 2

The definition of the CP-violating parameters ϵ_0, ϵ_2, a, ϵ' and the $\Delta I = {}^1\!/_2$ parameter ω for the case of CPT violation

$$\varepsilon_0 = \frac{A(K_L^0 \to 2\pi,\ I=0)}{A(K_S^0 \to 2\pi,\ I=0)} \ ; \quad a = \frac{A(K_2^0 \to 2\pi,\ I=0)}{A(K_1^0 \to 2\pi,\ I=0)} = \frac{\mathrm{Re}\,B_0 + i\,\mathrm{Im}\,A_0}{\mathrm{Re}\,A_0 + i\,\mathrm{Im}\,B_0}$$

$$\varepsilon_2 = \frac{1}{\sqrt{2}}\frac{A(K_L^0 \to 2\pi,\ I=2)}{A(K_S^0 \to 2\pi,\ I=0)} \ ; \quad \varepsilon' = \frac{1}{\sqrt{2}}\frac{A(K_2^0 \to 2\pi,\ I=2)}{A(K_1^0 \to 2\pi,\ I=0)} = \frac{1}{\sqrt{2}}\frac{\mathrm{Re}\,B_2 + i\,\mathrm{Im}\,A_2}{\mathrm{Re}\,A_0 + i\,\mathrm{Im}\,B_0}\, e^{i(\Delta_2-\Delta_0)}$$

$$\omega = \frac{1}{\sqrt{2}}\frac{A(K_S^0 \to 2\pi,\ I=2)}{A(K_S^0 \to 2\pi,\ I=0)} \approx \frac{1}{\sqrt{2}}\frac{A(K_1^0 \to 2\pi,\ I=2)}{A(K_1^0 \to 2\pi,\ I=0)} = \frac{1}{\sqrt{2}}\frac{\mathrm{Re}\,A_2 + i\,\mathrm{Im}\,B_2}{\mathrm{Re}\,A_0 + i\,\mathrm{Im}\,B_0}\, e^{i(\Delta_2-\Delta_0)}$$

$$\epsilon_0 \simeq a + (\epsilon - \delta) \tag{1}$$

$$\epsilon_2 \simeq \epsilon' + (\epsilon - \delta)\omega \tag{2}$$

and thus depend on symmetry violation in both the mass and the decay matrix. The parameters ϵ_0 and ϵ_2 are related to the experimental observables[2] η_{+-} and η_{00} by:

$$\epsilon_0 = 2/3\,\eta_{+-}\,(1+\omega) + 1/3\,\eta_{00}\,(1-2\omega) \tag{3}$$

$$\epsilon_2 = 1/3\,(\eta_{+-} - \eta_{00}) + 1/3\,(\eta_{+-} + 2\eta_{00})\omega\ . \tag{4}$$

In the following discussion the Wu–Yang phase convention[3] is adopted, which implies that the CPT-conserving isospin-zero decay amplitude into two pions is real ($\mathrm{Im}\, A_0 = 0$). Table 1 summarizes the above-defined parameters for the case of CPT invariance and CPT violation.

2. THE EXPERIMENTAL CONSISTENCY WITH CPT INVARIANCE

For the case of CPT invariance in the decay amplitude, the vanishing value of B_I implies

$$\arg a = \pm \pi/2 \quad \text{and} \quad |a| = 0 \quad \text{with the Wu–Yang phase convention} \tag{5}$$

$$\arg \epsilon' = \Delta_2 - \Delta_0 \pm \pi/2 \tag{6}$$

$$\arg \omega = \Delta_2 - \Delta_0 \;(\text{or } \Delta_2 - \Delta_0 + \pi)\,. \tag{7}$$

Since the world average of the $\pi\pi$ scattering phase shift[4] is $(\Delta_2 - \Delta_0) = -45.3° \pm 4.6°$, it follows that CPT conservation in the decay matrix implies a phase of ϵ' approximately equal to $+45°$ or $-135°$. Any significant deviation from this direction would correspond to a failure of CPT invariance in the neutral kaon decays. According to Eqs. (2) and (4), ϵ' is given, as a function of η_{+-}, η_{00}, ω, ϵ, and δ, by

$$\epsilon' = 1/3\,(\eta_{+-} - \eta_{00}) + [1/3\,\eta_{+-} + 2/3\,\eta_{00} - (\epsilon - \delta)]\omega\,. \tag{8}$$

Taking into account the present experimental values[5] of

$$|\eta_{+-}| = (2.275 \pm 0.021) \times 10^{-3}\,, \qquad \phi_{+-} = (44.6 \pm 1.2)°$$
$$|\eta_{00}| = (2.299 \pm 0.036) \times 10^{-3}\,, \qquad \phi_{00} = (54 \pm 5)°\,,$$

and omitting the second term of Eq. (8) ($|\omega| \ll 1$), one finds a value of $\arg \epsilon' \approx -47°$ and

$$\mathrm{Re}\,\epsilon' = (0.9 \pm 0.5) \times 10^{-4}\,, \qquad \mathrm{Im}\,\epsilon' = (1.0 \pm 0.4) \times 10^{-4}\,. \tag{9}$$

The one- and two-standard-deviation contours for ϵ' are shown in Fig. 1. The obtained value of ϵ' is in agreement with CPT invariance in the decay matrix only within two standard deviations. The difference arises exclusively from the $(\phi_{+-} - \phi_{00})$ phase difference (Ref. 1). The significance of this discrepancy is relatively weak, because of the large uncertainty of the measured value of ϕ_{00} and because of the recently obtained values[5] of $|\epsilon'/\epsilon|$, which indicate $\mathrm{Re}\,\epsilon'$ to be much smaller than in Eq. (9).

For the case of CPT invariance in the mass matrix, δ vanishes. In order to estimate the value of δ from the present experimental results it is convenient, using Eqs. (1) and (3), to introduce the parameter $\tilde{\delta} = (\delta - a)$, which depends on both the CPT violation in the decay amplitudes and that in the mass matrix. Defining the projections of the vectors $\tilde{\delta}$ and ϵ_0: $\tilde{\delta}_{\parallel}$, $\epsilon_{0\parallel}$, $\tilde{\delta}_{\perp}$, and $\epsilon_{0\perp}$, where the

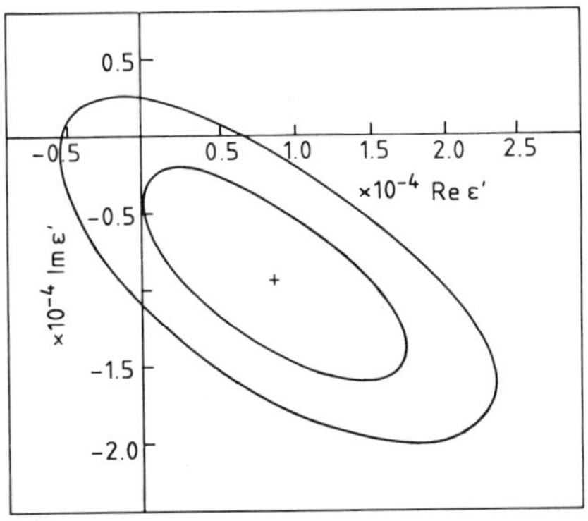

Fig. 1 One- and two-standard deviation contours[1] for ϵ'. (CPT invariance would imply $\arg \epsilon' \approx +45°$ or $-135°$.)

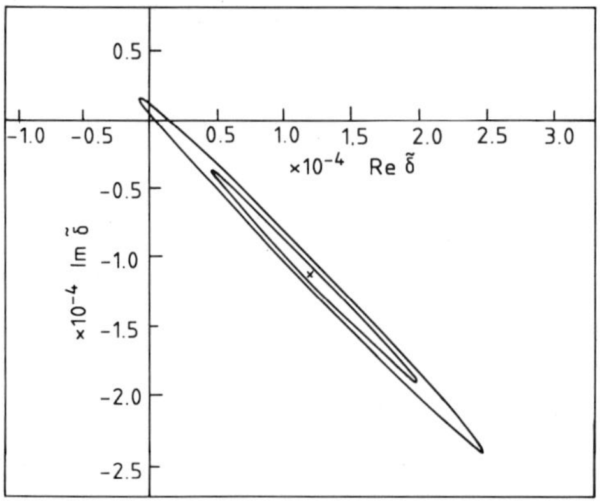

Fig. 2 One- and two-standard deviation contours[1] for $\tilde{\delta}$. (CPT invariance would imply $\tilde{\delta} = 0$.)

indexes ∥ and ⊥ refer to the direction parallel and perpendicular to the angle of superweak interaction ϕ_ϵ on the complex plane[1], one gets from the present experimental data (see also Fig. 2):

$$\tilde{\delta}_{\parallel} \ll \tilde{\delta}_{\perp} \simeq -\epsilon_{0\perp} = (-0.16 \pm 0.07) \times 10^{-3} \, . \tag{10}$$

This non-vanishing value of the CPT-violating parameter $\tilde{\delta}$ is again exclusively determined by the non-vanishing difference of ϕ_{+-} and ϕ_{00}.

Therefore Barmin et al.[1] conclude that the deviation of 2σ from CPT invariance of the present experimental data in the neutral kaon system is exclusively related to the inaccurate measurement of ϕ_{00}, independent of CPT violation in the decay matrix (ϵ') or CPT violation in both the decay and the mass matrix ($\tilde{\delta}$).

3. THE MEASUREMENT OF $(\phi_{+-} - \phi_{00})$ IN THE CPLEAR EXPERIMENT

In the CPLEAR experiment[2,6], the phase difference in the decay of the kaon into the final states $\pi^+\pi^-$ and $\pi^0\pi^0$ can be measured through the time-dependent asymmetries $A_{+-}(t)$, $A_{00}(t)$ requiring the definition of the eigentime of each decay. The study of the time dependence of the tagged K^0 and $\bar{K}^0$'s decaying into two pions needs high-quality reconstruction of the decay vertex.

For the channel of two charged pions, we use a set of drift chambers. A good vertex resolution of the order of 2 mm is expected. For the detection of the two neutral pions, an electromagnetic calorimeter, measuring the impact position and the energy of the four photons produced, is used. The vertex of the neutral pions is reconstructed from the measured positions and energies of the photons via a constraint fit, using energy and momentum conservation.

In order to keep the systematic effects related to the eigentime resolution of the phase measurement at an acceptable level, a vertex resolution of the order of $0.5\tau_S$ or better ($\tau_S \equiv$ lifetime of the K_S), corresponding to a spatial resolution of $\sigma \approx 1.5$ cm, is required, as discussed in Section 5.

4. THE ELECTROMAGNETIC CALORIMETER

As shown in Fig. 3, the vertex resolution on the photon impact point has a considerable effect on the resolution of the decay length, whereas for a spatial resolution of a few millimetres the energy resolution of the calorimeter is rather uncritical. This allows us to use, instead of an expensive crystal calorimeter with extremely good energy resolution and medium modularity, a less expensive calorimeter of high modularity, with good spatial resolution but with relatively poor energy resolution. To achieve a decay length resolution of the order of $\sigma \approx 1.5$ cm, one has to

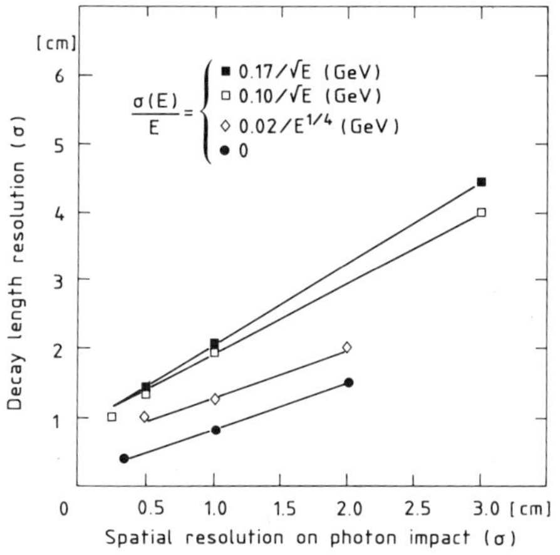

Fig. 3 Decay length resolution of $K^0 \rightarrow 2\pi^0$ as a function of the spatial and the energy resolution of the calorimeter.

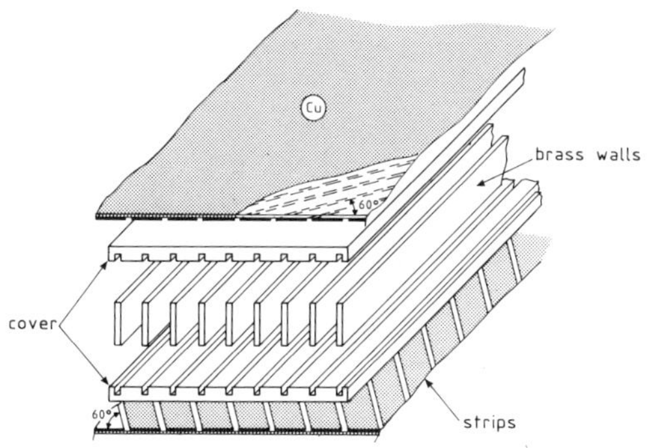

Fig. 4 Exploded view of a chamber and strips.

measure the position of the photons with an accuracy of ~ 5 mm, assuming an energy resolution of about $\sigma(E)/E \approx 15\%/\sqrt{E(GeV)}$.

The electromagnetic calorimeter is a lead sampling calorimeter placed inside the magnet coil with a total thickness of 6.2 radiation lengths. Each of the 19 layers consists of a lead converter plate, one-third of a radiation length, followed by a set of two sheets of strips placed above and below the streamer tubes.

One chamber consists of 8 rectangular tubes with an internal cross-section of 4.5 mm × 4 mm. The chamber is assembled from two fibre-glass plates, coated with graphite of a resistivity $R \approx 1$ MΩ per square on the internal side. Nine brass walls are then glued, with conductive epoxy, in slots on the plates (Fig. 4). A sheet of strips consists of 5.5 mm wide strips, interspaced by 1 mm, inclined at 30° with respect to the wire axis.

The tubes are operated in a limited streamer mode. This is a saturated mode, where the charge collected on the wires is independent of the primary ionization, leading to large pulse height, low noise, and a relatively simple electronic readout. The energy of a converted photon is measured by merely counting the charged track segments in the shower, allowing the use of a digital readout.

The streamer tubes have an efficiency of ≳ 90%, measured with a β-particle source. We aim both at a good strip efficiency and, in order to simplify the pattern recognition, at a low strip multiplicity per wire hit. These requirements imply the following restrictions on the construction of the chamber:
i) As shown in Fig. 5, the resistivity on the inner side of the plates forming a chamber should be high. This guarantees a good transparency for the induced strip signal, leading to a considerable but localized strip charge.
ii) The tubes must be separated by low-resistivity material. Figure 6 shows the effect of the resistivity of the tube separation walls on the strip charge in logarithmic scale. Even though there is a loss of a factor of three in the maximum of the charge distribution using metallic walls, the dynamical range of the distribution is much larger than with more resistive walls (R ~ kΩ per square), because the wings of the distribution are suppressed by a factor of ten. This leads to a low strip multiplicity and therefore to less ambiguities. A typical strip-multiplicity distribution with a strip efficiency of ≈ 95% is shown in Fig. 7.

The azimuthal spatial resolution of the chambers is only determined by the wire spacing of 5 mm, which leads to $[\Delta(R\phi)]_{rms} \approx 1.5$ mm. The longitudinal spatial resolution Δz is determined

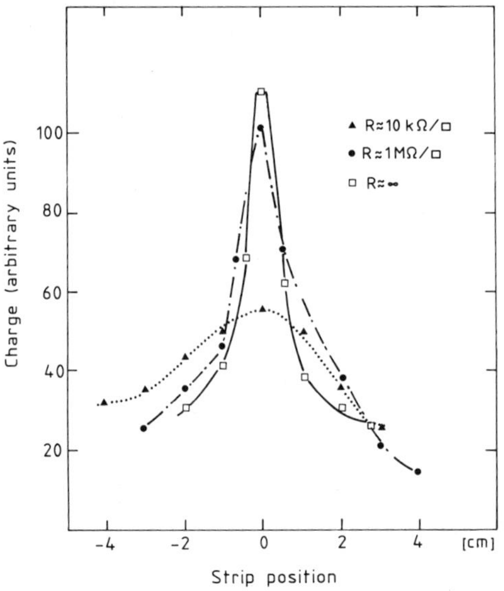

Fig. 5 Effect of the resistivity of the cover of a chamber on the spatial distribution of the strip charge.

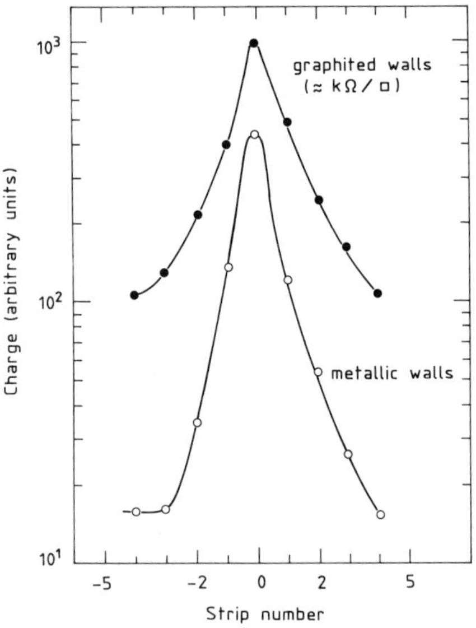

Fig. 6 Effect of the resistivity of the separation walls of the tube on the spatial distribution of the strip charge.

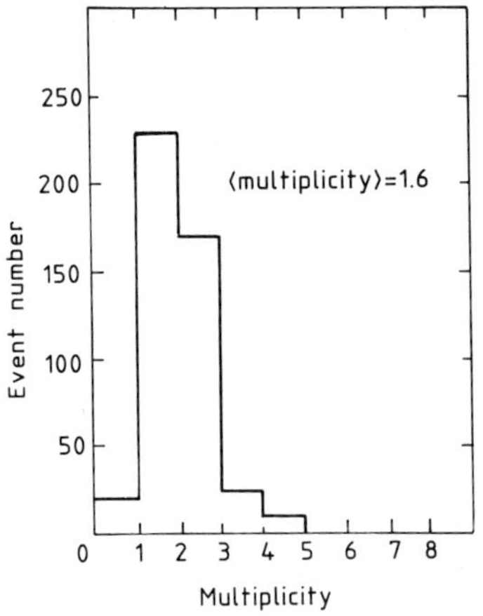

Fig. 7 Typical multiplicity distribution of the strip charge.

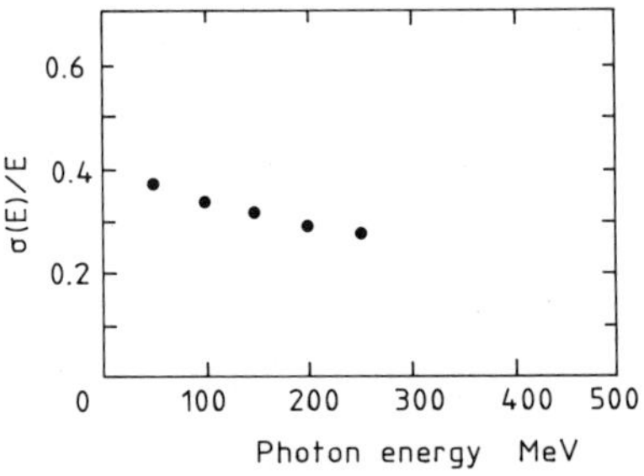

Fig. 8 The simulated energy resolution of the calorimeter.

by the resolution of the position measurement provided by the strips. This position is given by the mean value of the number of strips which gave a signal. Tests showed that with the strips at a step of 6.5 mm and an inclination of 30° the longitudinal resolution is $[\Delta z]_{rms} \approx 3.5$ mm.

Further properties of the calorimeter were investigated using Monte Carlo simulations which include the exact geometry of the calorimeter, the effect of the magnetic field, and the effect of other detectors in front of the calorimeter. The strip response was simulated according to the measured multiplicity distribution and to the efficiency obtained from the above tests. The simulated energy resolution of the calorimeter is $\sigma(E)/E \approx 12\%/\sqrt{E(GeV)}$ (Fig. 8). The efficiency for the photon detection as a function of the energy, assuming a minimal threshold of three hit tubes, is shown in Fig. 9. The detection efficiency of a $2\pi^0$ event in which all four photons enter the calorimeter is $\approx$ 48%. The spatial resolution on the photon impact point, obtained by averaging the reconstructed hit coordinates of the lowest layer hit by each individual photon, is

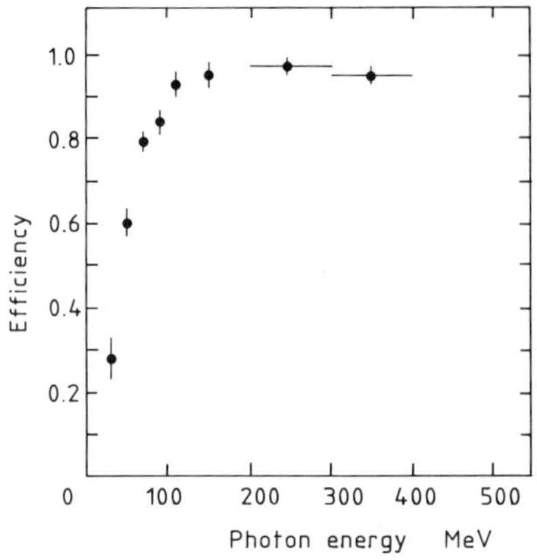

Fig. 9 The simulated efficiency of the photon detection.

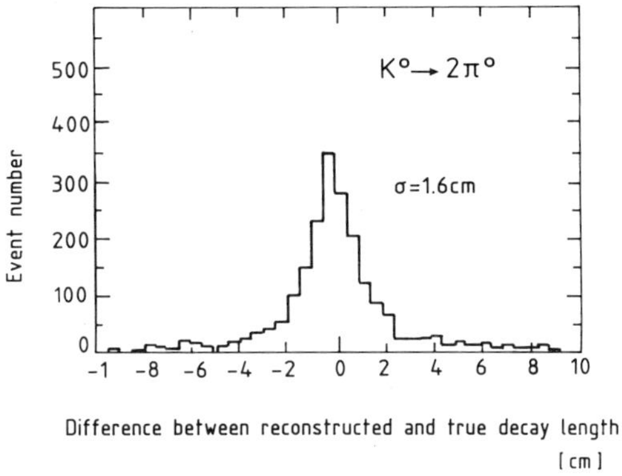

Fig. 10 The simulated decay length resolution of the calorimeter for the $K^0 \rightarrow 2\pi^0$ decays.

$[\Delta(R\phi)]_{rms} \approx 3$ mm for the azimuthal and $[\Delta z]_{rms} \approx 5$ mm for the longitudinal position. The achieved vertex resolution is $\sigma = 1.6$ cm (Fig. 10), which corresponds to $\sigma \approx 0.5\tau_S$.

5. THE SYSTEMATIC UNCERTAINTIES IN THE MEASUREMENT OF $(\phi_{+-} - \phi_{00})$

Assuming $\sim 10^{13}$ stopped antiprotons, the statistical error on $(\phi_{+-} - \phi_{00})$ will be of the order of 0.2°.

In order to discuss the influence of the different systematic effects on the measurement of $(\phi_{+-} - \phi_{00})$, one should point out the following advantages of our method:

i) By using pure K^0 and $\bar{K}^0$ beams we directly measure the interference of K_S and K_L, without the background of the direct K_S and K_L contributions.

ii) The tagging provides separate normalizations for K^0 and $\bar{K}^0$.

iii) The acceptance and detection efficiencies of the decay products for K^0 and $\bar{K}^0$ are identical. Therefore the measurement of the time-dependent asymmetry leads to the cancellation in first order of the systematic effects of the apparatus.

The time-dependent asymmetry measurements are affected by uncertainties arising from the vertex resolution. In order to extract the relevant physical quantities ϕ_{+-} and ϕ_{00}, the asymmetries $A_{+-}(t)$ and $A_{00}(t)$ have to be unfolded with the experimental resolutions of the decay vertex. These resolutions therefore have to be measured separately with sufficient precision. For the neutral pions, the resolution σ_{00} could be measured and monitored at the origin ($t = 0$) through the reaction $p\bar{p} \rightarrow \pi^+\pi^-\pi^0\pi^0$, and its time dependence could be calibrated from the $K^+ \rightarrow \pi^+\pi^0\pi^0$ decays (branching ratio of 1.7%) over the whole decay length. A σ_{00} of $0.5\tau_S$ has to be known at the 10% level, which leads to an uncertainty of $\Delta\phi_{00} \leq 0.8°$. With a vertex resolution of $\sigma_{+-} = 0.07\tau_S$ for the charged pions, the error on the phase ϕ_{+-} is negligible.

Further contributions to the error on ϕ_{00} are due to different deviations of the vertex resolution from Gaussian shapes and to its time dependence.

- Generating the asymmetry $A_{00}(t)$ with a time-dependent vertex resolution of $\sigma(t=0) = \sigma_{00}$, which increases linearly up to 50% for $t = 20\tau_S$, and fitting it with the fixed resolution σ_{00}, gives an idea of the maximum error on ϕ_{00} for this effect, i.e. $\Delta\phi_{00} = 0.6°$.
- If one makes a simulation with an asymmetric vertex resolution σ of $1\sigma_{00}$ before the maximum of the distribution and $1.5\sigma_{00}$ after, one finds $\Delta\phi_{00} = 0.5°$.
- If one uses a vertex resolution with very long tails for the simulation of $A_{00}(t)$ (superposition of two Gaussian distributions with $\sigma_1 = \sigma_{00}$ and $\sigma_2 = n\sigma_{00}$ with a percentage P of events belonging to σ_2) the following upper limits for the error on ϕ_{00} are obtained:

P (%)	n	$\Delta\phi_{00}$ (°)
5	4	2.2
5	2	0.9
2	4	0.4

An uncertainty in the definition of the eigentime can be due to either a systematic error on the kaon momentum, which has exactly the same effect on the $\pi^+\pi^-$ and $\pi^0\pi^0$ final states, or to a systematic shift in the definition of the vertex in the transition $\overset{(-)}{K}{}^0 \rightarrow \pi^0\pi^0$ which has to be calibrated with $p\bar{p} \rightarrow \pi^+\pi^-\pi^0\pi^0$ for $t = 0$ and $K^+ \rightarrow \pi^+\pi^0\pi^0$ for $t > 0$.

Fitting the time-dependent asymmetry, systematic errors due to uncertainties on the different input parameters are also introduced. Although the K_S, K_L mass difference introduces an error of 1° for each individual phase measurement, it is cancelled out by measuring $(\phi_{+-} - \phi_{00})$. The uncertainty on $\gamma_S = 1/\tau_S$ leads to an equal deviation of the time scale for the $\pi^+\pi^-$ and $\pi^0\pi^0$ final states. Therefore, the error on the phase difference vanishes. The error on the phase difference obtained by simultaneously decreasing or increasing the actual experimental value of both $|\eta_{+-}|$ and $|\eta_{00}|$ by 5% is $\Delta\phi = 0.3°$. The difference in the absolute values $|\eta_{+-}|$ and $|\eta_{00}|$ is expected to be very small, since a limit is given by using Eqs. (1) to (4) and the present experimental value[5] of $|\epsilon'/\epsilon|$:

$$(\eta_{+-} - \eta_{00}) = \frac{3(\epsilon' - a\omega)}{(1+\omega)(1-2\omega)} \approx 3\epsilon' \qquad \text{for } |\omega| \ll 1 \tag{11}$$

From this discussion and the expected quality obtained in monitoring the resolution via the above reactions, we are confident of being able to measure the phase difference with an accuracy better than or equal to 1°.

Backgrounds can also influence the precision of the phase difference. The transitions $\overset{(-)}{K}{}^0 \rightarrow \pi^0\pi^0\pi^0$ give a background for the $2\pi^0$ final state and the $\overset{(-)}{K}{}^0 \rightarrow \pi^\pm\ell^\mp\overset{(-)}{\nu}$ decays (ℓ = leptons) for the $\pi^+\pi^-$ final state. These decays can be reduced by a factor of ten by kinematical constraints. Nevertheless, since these decays originate from the K_L component, they represent a nearly constant background which has no effect on the measurement of the phases.

6. CONCLUSION

The CPLEAR experiment is particularly suited[6] for measuring the phase difference $(\phi_{+-} - \phi_{00})$ with a precision less than or equal to 1°, which would clarify the present discrepancy of the experimental data with CPT invariance.

Referring to the work of Barmin et al.[1] we stress that for the case of $|\epsilon'| \neq 0$, the measurement of this phase difference provides the best possible limit for the CPT violating parameters. However, for the case of very small or vanishing values of $|\epsilon'|$, the best test of CPT invariance[7] is the phase difference of η_{+-} relative to the superweak angle, i.e. $(\phi_{+-} - \phi_\epsilon)$.

REFERENCES

1. V.V. Barmin et al., Nucl. Phys. B247:293 (1984).
2. P. Hayman, See lecture in these Proceedings.
3. T.T. Wu and C.N. Yang, Phys. Rev. Lett. 13:380 (1964).
4. T.J. Devlin and J.O. Dickey, Rev. Mod. Phys. 51:237 (1979).
5. Particle Data Group, Review of Particle Properties, Phys. Lett. 170B (1986).
6. L. Adiels et al., Proposal for experiment PS195 at CERN, CERN/PSCC/85-6, P-82 (1985) and CERN/PSCC/86-34, M263 (Status report).
7. L. Wolfenstein, private communications.

CP VIOLATION IN THE $K_S \to \pi^+\pi^-\pi^\circ$ DECAY

N.W. Tanner and I.J. Ford

Dept. of Nuclear Physics
University of Oxford
Oxford

Dept. of Theoretical Physics
University of Oxford
Oxford

INTRODUCTION

The failure of CP-invariance observed in $K^\circ \to \pi\pi$ can be represented by a single complex number[1] $\epsilon \simeq 2.3e^{i\pi/4}$ (assuming CPT-invariance[2]), and is compatible with the "standard model". However there is no evidence for a CP failure except in the $K^\circ \to \pi\pi$ channel, and the known failure could just as well be caused by a super-weak, $\Delta s = 2$, interaction as by some pecularity of the standard model.

The observation of a difference, ϵ', between the CP failure in the $\pi^+\pi^-$ and $\pi^\circ\pi^\circ$ channels would clarify the situation, but ϵ' is expected to be small ($\sim 10^{-3}\epsilon$, present limit[3] $< 10^{-2}\epsilon$) and may not be experimentally accessible. It is unlikely that any upper limit for $|\epsilon'/\epsilon|$ which might be obtained could embarrass the standard model.

The position of the neutron electric dipole moment (strictly a test of T- rather than CP-invariance) is similar[4]. Powerful experiments have been done, but have yielded only upper limits from which no conclusion can be drawn about the validity of the standard model.

The decay $K^\circ \to \pi\pi\pi$ is firmly predicted to exhibit a failure of CP-invariance with a ratio of K_S CP-non-conserving to K_L CP-conserving amplitudes which is similar (identical for superweak) to the ratio for the $\pi\pi$ channel. However the phase space factors and lifetimes, which are favourable for the observation of a CP-violation in the $\pi\pi$ channels, are correspondingly unfavourable for the $\pi\pi\pi$ channel. The branching ratio of K_S to the CP-odd state of $\pi\pi\pi$ is expected to be $\sim 3 \times 10^{-9}$, c.f. the branching ratio of $\sim 3 \times 10^{-3}$ for K_L to $\pi\pi$. In addition there are CP-even states of $\pi^+\pi^-\pi^\circ$ (but not of $\pi^\circ\pi^\circ\pi^\circ$) of higher angular momentum which allow CP-conserving transitions $K_S \to \pi^+\pi^-\pi^\circ$ and which are expected to be stronger than the CP-non-conserving transition[5].

Any measurement of $K_S \to \pi\pi\pi$ is certain to be difficult, but even a qualitative demonstration of a failure of CP-invariance in the $\pi\pi\pi$ channel would be of interest. A measurement which might distinguish between the magnitudes of the CP-violations in the $\pi\pi$ and $\pi\pi\pi$ channels is more remote.

The problem of identifying $K_S \to \pi\pi\pi$ has previously been considered[6] for sources of neutral kaons in the states K° and $\bar{K}^\circ$ i.e. linear combinations of K_S and K_L. In this paper attention is directed towards the pure K_S state using the two body reaction

$$\bar{p}p \to K_S K_L$$

and the advantages that might accrue from using regeneration to add a small coherent K_L amplitude to the dominant K_S amplitude.

K_S SOURCES

Usually neutral kaons are produced in the states K° or $\bar{K}^\circ$ by primary production by protons, and at large distances from the source become pure K_L beams as the K_S component dies out. Cleaner but less intense K° and $\bar{K}^\circ$ beams can be obtained from $\pi^- p \to K^\circ \Lambda$ and $K^- p \to \bar{K}^\circ n$, but for precise comparisons between K° and $\bar{K}^\circ$ decays it is necessary to use[2,6,7] anti-proton annihilation in hydrogen

$$\begin{array}{lll} \bar{p}p \to & K^\circ K^- \pi^+ + \bar{K}^\circ K^+ \pi^- & 0.4\% \text{ of stopped } \bar{p} \\ & +K^\circ K^- \pi^+ \pi^\circ + \bar{K}^\circ K^+ \pi^- \pi^\circ & 0.9\% \text{ of stopped } \bar{p}. \end{array}$$

Lipkin[10] has considered the properties of a two body $K\bar{K}$ state with spin-parity $J^\pi = 1^-$. This might be prepared by e^+e^- at the ϕ resonance (1.02 GeV) or more usefully by the s-state annihilation[8] of stopped anti-protons in liquid hydrogen. The possible $J^\pi = 1^-$ sources are

$$\begin{array}{rl} e^+e^- \to \phi \to K\bar{K} & 35\%\ K^\circ \bar{K}^\circ,\ p_K = 110 \text{ MeV/c},\ K_S \text{ decay length 6mm} \\ \bar{p}p \to K\bar{K} & \sim 0.1\%\ K^\circ \bar{K}^\circ,\ p_K = 800 \text{ MeV/c},\ K_S \text{ decay length 43 mm} \end{array}$$

The interesting part of the state function at proper time $t = 0$ is[10]

$$\begin{aligned} \psi(t_+ = t_- = 0) &= \frac{1}{\sqrt{2}}[K^\circ(+)\bar{K}^\circ(-) - K^\circ(-)\bar{K}^\circ(+)] \\ &= \frac{1}{\sqrt{2}}[K_1(-)K_2(+) - K_1(+)K_2(-)] \\ &\simeq \frac{1}{\sqrt{2}}[K_L(+)K_S(-) - K_L(-)K_S(+)] \end{aligned} \tag{1}$$

where $K(+)$ indicates a particle propagating in the positive hemisphere and $K(-)$ a collinear particle in the negative hemisphere. Interchanging + and − reverses the sign of ψ thus verifying the odd parity. The full time dependence $\psi(t_+, t_-)$ is obtained by multiplying $K_S(\pm)$ and $K_L(\pm)$ by $e^{-\gamma_S t_\pm}$ and $e^{-\gamma_L t_\pm}$ respectively, where

$$\gamma_S = iM_S + \frac{1}{2}\Gamma_S \qquad \gamma_L = iM_L + \frac{1}{2}\Gamma_L$$

are the eigenvalues and $t_\pm$ is the proper time of the particle in the $\pm ve$ hemisphere. It follows that the detection of a K_L, identified as a strongly interacting neutral particle at a time $t \gg \Gamma_S^{-1}$, specifies the existence of a pure K_S in the diametrically opposite direction. A hadron "calorimeter", say, $\sim$ 100g/cm^2 of lead and streamer tubes, might serve as a K_L detector with $\sim$ 50% efficiency at 800 MeV/c, identifying direction reasonably but having poor energy resolution.

It is of some interest to consider the effects of regeneration on the state function equation (1) and the purity of the K_S produced. Particles entering a regenerator in the state K_S (or K_L) will re-emerge after a proper time t with a small admixture of the state K_L (or K_S). Following Kabir's[11] development, regeneration gives

$$\begin{aligned} K_S &\to e^{-\gamma'_S t}\left\{K_S - \rho(1 - e^{-(\gamma'_L - \gamma'_S)t})K_L\right\} \\ K_L &\to e^{-\gamma'_L t}\left\{K_L + \rho(1 - e^{+(\gamma'_L - \gamma'_S)t})K_S\right\} \end{aligned} \tag{2}$$

where the regeneration parameter

$$\rho = \pi\nu \cdot \frac{\hbar}{M_K c} \cdot (f - \bar{f})[i\{i(M_L - M_S) + \frac{1}{2}(\Gamma_L - \Gamma_S)\}]^{-1}$$

and f, $\bar{f}$ are the forward scattering amplitudes for K°, $\bar{K}^\circ$

ν is the number density of the scatterers in the regeneration material

γ'_S, γ'_L are the eigenvalues within the regenerator material and are such that $\gamma'_L - \gamma'_S \simeq \gamma_L - \gamma_S$ and $|e^{-\gamma'_S t}|^2 = e^{-\Gamma_S t} e^{-\bar{\sigma}\nu x}$, $|e^{-\gamma'_L t}|^2 = e^{-\Gamma_L t} e^{-\bar{\sigma}\nu x}$ where x is the thickness of the regenerator and $\bar{\sigma}$ is the mean total cross-section for K° and $\bar{K}^\circ$.

For a point source of $K_S K_L$ and a spherical regenerator the state function at time t is obtained by substituting the regenerated amplitudes of equations (2) into equation (1). To first order in the small regeneration parameter ρ and $t_+ = t_- = t$,

$$\psi(t) = \frac{1}{\sqrt{2}} e^{-(\gamma'_S + \gamma'_L)t} [K_L(+)K_S(-) - K_L(-)K_S(+)] \tag{3}$$

No regeneration occurs for a spherical regenerator or indeed any regenerator with inversion symmetry about the source point. That is to say, the amplitude for $K_S K_S$ cancels to zero*. It is possible to confine a $\bar{p}p$ annihilation source to $< 1\text{mm}^3$ of liquid hydrogen[7], which for present purposes approximates to a point.

For a hemispherical regenerator, regeneration does occur adding a small amplitude for K_L (or K_S) to the dominant K_S (or K_L). In the case of a thin regenerator, $\Gamma_S t \ll 1$, the ratio of small to large amplitudes is the same for both K_S and K_L, viz.,

$$\rho(\gamma_S - \gamma_L)t = i\pi\nu\frac{\hbar}{p_K}(f - \bar{f})x \tag{4}$$

where x is thickness of the regenerator, ν the number density, and f, $\bar{f}$ the forward scattering amplitudes for K°, $\bar{K}^\circ$ as before.

The magnitude of the regenerated amplitude can be estimated using the forward scattering amplitudes of Baillon *et al*[13] for $K^\pm N$ noting that

$$K^+(n+p) = K^\circ(p+n), \quad K^-(n+p) = \bar{K}^\circ(p+n). \tag{5}$$

For p_K = 800 MeV/c and 1cm of Be, putting the forward scattering amplitude for $KBe \simeq 4K(n+p)$, the regeneration amplitude $\pi\nu\hbar|f - \bar{f}|x/p_K$, equation (4), is 0.9×10^{-2}.

K_S AND K_L DECAYS TO $\pi^+\pi^-\pi^\circ$

It is convenient to consider the general neutral kaon state

$$\psi(t) = a_S K_S e^{-\gamma_S t} + a_L K_L e^{-\gamma_L t} \tag{6}$$

with spin-parity $J^\pi = 0^-$ and isospin $I = \frac{1}{2}$. To order ϵ K_S and K_L are assumed to be eigen states of CP with eigenvalues $+1$ and -1 respectively.

We can construct the states of $\pi^+\pi^-\pi^\circ$ by expanding in terms of the orbital angular momentum of the $\pi^+\pi^-$ pair, as in Fig.1.

For $J = 0$, $L = \ell$

CP $(\pi^+\pi^-) = (-1)^\ell(-1)^\ell = +1$ for all ℓ

CP $(\pi^\circ) = (+1)(-1) = -1$

so CP $(\pi^+\pi^-\pi^\circ) = -1$ (or $+1$) according to whether $L = \ell$ =even (or odd), and to satisfy Bose statistics I =odd (or even). As there is very little energy released in the decay, $Q = 84$ MeV, it is only the lowest angular momentum states which are of interest:-

$$(\pi^+\pi^-\pi^\circ)_s \quad \ell = 0 \quad J^\pi = 0^- \quad I = \text{odd} \quad \text{CP} = -1$$

---

* The observation, outside a spherical regenerator, of diametrically opposite $\pi^+\pi^-$ decays from a pair of neutral kaons in excess of the small probability allowed by CP-violation[10] would represent a failure of the non-separability property of quantum mechanics[9,12]. However, non-separability does not forbid incoherent (scattering) regeneration which per unit solid angle at $0^\circ \simeq (\nu\lambda^2 x)^{-1}\times$ coherent (transmission) regeneration, where λ is the K wavelength. (For 1cm Be and an angular resolution of 1° scattering regeneration is about one third of transmission regeneration). There is also an incoherent background of $K_S K_S$ events from p-state annihilation[8] of $\bar{p}p$.

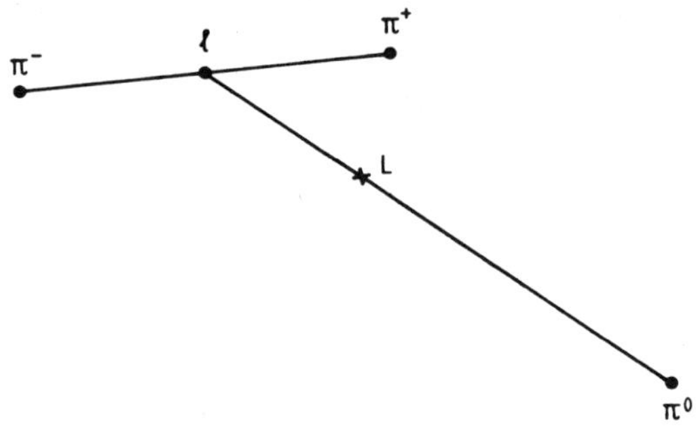

Fig.1. Orbital angular momenta of $\pi^+\pi^-\pi^\circ$.

$$(\pi^+\pi^-\pi^\circ)_p \quad \ell = 1 \quad J^\pi = 0^- \quad I = \text{even} \quad \text{CP} = +1$$

We can represent a non-relativistic $|\pi^+\pi^-\pi^\circ>$ state as

$$|\pi^+\pi^-\pi^\circ >= |(\pi^+\pi^-\pi^\circ)_s > +|(\pi^+\pi^-\pi^\circ)_p >$$

There is of course a representation of these angular momentum states in terms of isospin states, but which is not necessary here.

There are three interesting decay amplitudes for K_S and K_L.

(a) The CP allowed decay of K_L to $(\pi^+\pi^-\pi^\circ)_s$

$$< (\pi^+\pi^-\pi^\circ)_s|T|K_L > \tag{7}$$

which gives rise to the 12.4% branch of K_L decay to $\pi^+\pi^-\pi^\circ$;

$$\Gamma_L(\pi^+\pi^-\pi^\circ) \propto | < (\pi^+\pi^-\pi^\circ)_s|T|K_L > |^2$$

(b) The CP forbidden decay of K_S to the same final state $(\pi^+\pi^-\pi^\circ)_s$ for which

$$\eta_{+-0} =< (\pi^+\pi^-\pi^\circ)_s|T|K_S > / < (\pi^+\pi^-\pi^\circ)_s|T|K_L > \tag{8}$$

is expected to be similar to $\eta_{+-} \simeq 2.3 \times 10^{-3} e^{i\frac{\pi}{4}}$.

(c) The CP-allowed but angular momentum inhibited decay of K_S to $(\pi^+\pi^-\pi^\circ)_p$ which is expected[5] to have a rate relative to the $K_L \to \pi^+\pi^-\pi^\circ$ rate of

$$|A|^2 = \Gamma_S[(\pi^+\pi^-\pi^\circ)_p]/\Gamma_L[(\pi^+\pi^-\pi^\circ)_s] \simeq 10^{-3} \tag{9}$$

with an uncertainty of $\sim$ 50%. There is no CP-allowed decay of K_S to $\pi^\circ\pi^\circ\pi^\circ$ (B.-E. statistics).

The amplitudes (a) and (b) for K_L and K_S decays to $(\pi^+\pi^-\pi^\circ)_s$ give rise to a time dependent interference term similar to that observed for K_S and K_L decays to $\pi\pi$.

For decay products integrated over all angles there are no interference terms between the amplitude for (c) and (a) or (b) since the angular momenta involved are different. Within the angular distributions there are interference terms (a)/(c) and (b)/(c), and the (a)/(c) term is time dependent.

The $\pi\pi\pi$ state in the K centre of mass can be treated non-relativistically to a good approximation, in which case the Dalitz plot[14] is a circle inscribed in an equilateral triangle as shown in Fig.2. The angular momentum state $(\pi^+\pi^-\pi^\circ)_s$ is kinematically identical with the state involved in the decay $K^+ \to \pi^+\pi^+\pi^-$. The density of points (events) is uniform over the circle.

The form of the Dalitz plot for the $(\pi^+\pi^-\pi^\circ)_p$ state is dramatically different[14]. There are zeroes along the axes of Fig.2, where $T_- = T_0$, $T_0 = T_+$, and $T_+ = T_-$, and a strong radial

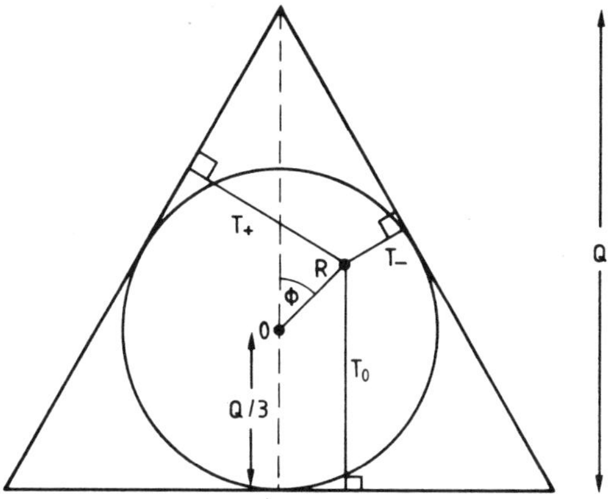

Fig.2. Dalitz plot for a non-relativistic $\pi\pi\pi$ state. The circle is the kinematic limit. T_+, T_- and T_0 are the kinetic energies of π^+, π^- and π°.

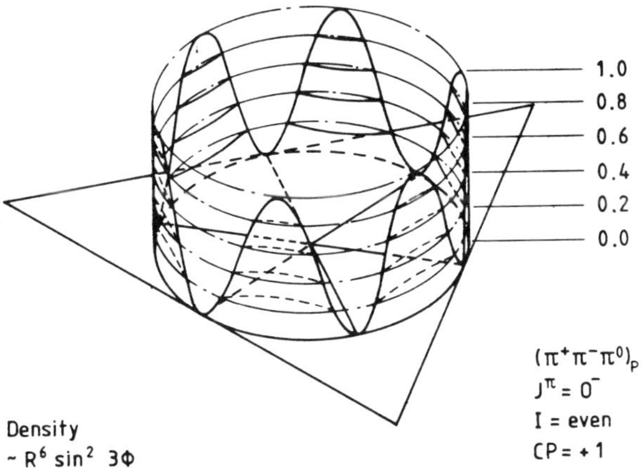

Fig.3. The Dalitz plot density distribution for the state $(\pi^+\pi^-\pi^\circ)_p$. (From Perkins[14]).

dependence. Fig.3 gives a picture of the density distribution which has the algebraic form

$$\sim R^6 \sin^2 3\phi \tag{10}$$

in terms of the coordinates (R, ϕ) defined in Fig.2. The amplitude is

$$\sim R^3 \sin 3\phi \tag{11}$$

and so alternates in sign, sector by sector, around the circle.

Adding together the three decay amplitudes from (7), (8) and (9), modified to conform with the Dalitz plot distributions, the decay rate for the general neutral kaon state, equation (6), to some area of the Dalitz plot (specified by the integrations) becomes

$$\pi^{-1}\Gamma_L[(\pi^+\pi^-\pi^\circ)_s] \int d\phi \int Z dZ \left| a_S(2\sqrt{2}Ae^{i\alpha}Z^3 \sin 3\phi + \eta_{+-0})e^{-\gamma_S t} + a_L e^{-\gamma_L t}\right|^2 \tag{12}$$

where $Z = R/R_0$, $R_0 = Q/3$ is the kinematic limit of the Dalitz plot, and α is the phase of A. Equation (12) is normalized to one K_S.

PURE K_S SOURCE

In principle a pure K_S beam can be obtained from $\bar{p}p$ annihilation in the s-state and the detection of a long lived neutral hadron at a large distance from the source. Equation (12) for the Dalitz plot density then becomes

$$\pi^{-1}\Gamma_L[(\pi^+\pi^-\pi^\circ)_s]e^{-\Gamma_S t}\int d\phi\int ZdZ|2\sqrt{2}\cdot Ae^{i\alpha}Z^3\sin 3\phi+\eta_{+-0}|^2 \qquad (13)$$

which includes an interference term which alternates in sign in the six sectors of the Dalitz plot. (Fig.2). Taking the difference of the even and odd sectors gives a ratio

$$\frac{\text{Sector Differences}}{\text{Sector Sum}}=\frac{16\sqrt{2}}{5\pi}\cos(\phi_{+-0}-\alpha)\left|\frac{\eta_{+-0}}{A}\right| \qquad (14)$$

where ϕ_{+-0} and α are the phases of η_{+-0} and A, and $|\eta_{+-0}|^2$ has been neglected c.f. $|A|^2$. The statistical error (standard deviation) is

$$\sigma=(\text{Sector Sum})^{-\frac{1}{2}}=[N_L\Gamma_L(\pi^+\pi^-\pi^\circ)|A|^2/\Gamma_S]^{-\frac{1}{2}} \qquad (15)$$

where N_L is the number of K_L triggers, and the integrals have been taken over the whole area of the Dalitz plot and over time out to $t\gg\Gamma_S^{-1}$.

It is noted that A and α are not at present known experimentally (see next section).

K_S SOURCE AND REGENERATION

A thin hemispherical regenerator of thickness x adds a small K_L amplitude

$$a_L=i\pi\nu\frac{\hbar}{p_K}(f-\bar{f})x, \qquad |a_L|\simeq 10^{-2}$$

to the pure K_S amplitude $a_S\simeq 1$ of the previous section (see equation (4)). Inserting these amplitudes into equation (12) and integrating over the Dalitz plot gives a $\pi^+\pi^-\pi^\circ$ decay rate per K_L trigger of

$$\Gamma_L(\pi^+\pi^-\pi^\circ)\{|A|^2e^{-\Gamma_S t}+|a_L|^2e^{-\Gamma_L t}+2|\eta_{+-0}||a_L|e^{-\bar{\Gamma}t}\cos(\delta Mt+\phi_{+-0}-\arg a_L)\} \qquad (16)$$

where $\delta M=M_L-M_S$ and $\bar{\Gamma}=\frac{1}{2}(\Gamma_S+\Gamma_L)$. The CP-conserving K_S decay, $|A|^2\simeq 10^{-3}$, swamps the interference term $\sim 2|\eta_{+-0}||a_L|\simeq 5\times 10^{-5}$, and a measurement would determine only $|A|^2$ and $|a_L|^2$, not η_{+-0}.

Taking the difference of the even and odd sectors of the Dalitz plot, Fig.2, gives a net $\pi^+\pi^-\pi^\circ$ rate per K_L trigger of

$$\Gamma_L(\pi^+\pi^-\pi^\circ)\frac{16\sqrt{2}}{5\pi}|A|^2\{e^{-\Gamma_S t}\left|\frac{\eta_{+-0}}{A}\right|\cos(\phi_{+-0}-\alpha)+e^{-\bar{\Gamma}t}\left|\frac{a_L}{A}\right|\cos(\delta Mt+\alpha-\arg a_L)\} \quad (17)$$

which is plotted in Fig.4 for an arbitrary choice of phase. Except close to the $\bar{p}p$ source, where the background is unfavourable, equation (16) is dominated by the second term and a measurement would determine the amplitude, A, and phase, α, characterizing the CP-conserving K_S decay to $\pi^+\pi^-\pi^\circ$. This information is required for the determination of η_{+-0} from equation (14). The ratio of the sector difference, equation (17), to the sector sum, equation (16), for $\Gamma_S t\gtrsim\frac{\pi}{2}$ is

$$\frac{\text{Sector Difference Rate}}{\text{Sector Sum Rate}}=\frac{16\sqrt{2}}{5\pi}\left|\frac{a_L}{A}\right|\cos(\delta Mt+\alpha-\arg a_L)\times e^{-\bar{\Gamma}t}/(e^{-\Gamma_S t}+\left|\frac{a_L}{A}\right|^2e^{-\Gamma_L t}) \qquad (18)$$

The regeneration amplitude a_L and its phase can be measured experimentally. For a thin regenerator the regeneration of K_S from K_L is the same as the regeneration of K_L from K_S, see equation (2). A measurement of events with back to back $\pi^+\pi^-$ decays, with the

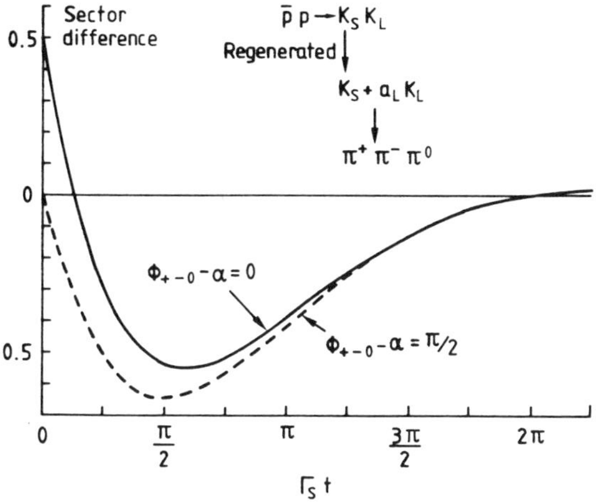

Fig.4. The proper time dependence, equation (17), for the *difference* of the sectors of the Dalitz plot for a state $K_S + a_L K_L$ at $t = 0$ decaying to $\pi^+\pi^-\pi^\circ$. Phases have been chosen arbitrarily,

$$\phi_{+-0} - \alpha = 0 \text{ and } \frac{\pi}{2}, \quad \alpha - arg a_L = \frac{\pi}{2}$$

and $|a_L| = 10^{-2}$, $|\eta_{+-0}| = 2.3 \times 10^{-3}$.

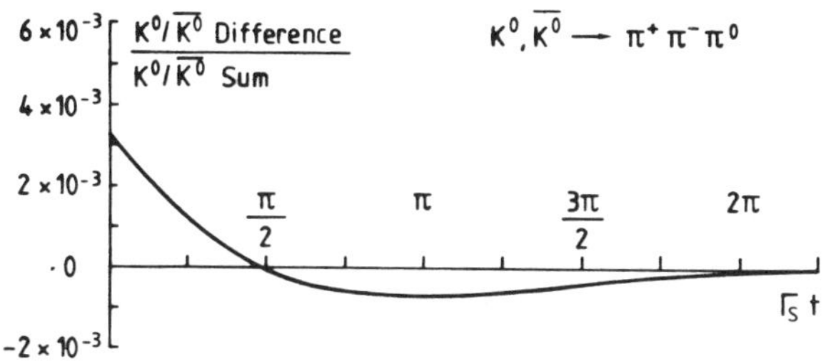

Fig.5. The difference of the decay rates for $K^\circ \to \pi^+\pi^-\pi^\circ$ and $\bar{K}^\circ \to \pi^+\pi^-\pi^\circ$ divided by the sum of those rates, equation (22). The curve is drawn for $\eta_{+-0} = \eta_{+-}$.

hemispherical regenerator, determines the properties of the regenerator in terms of the $\pi\pi$ CP parameter η_{+-}.

$K^\circ/\bar{K}^\circ$ COMPARISON

This method for determining η_{+-0} has been considered previously[6] and is summarized here for the sake of completeness.

For K° or $\bar{K}^\circ$ the neutral kaon state function, equation (6), has $a_S = 1/\sqrt{2} = \pm a_L$ to the order ϵ. Integrating equation (12) over the Dalitz plot gives a $\pi^+\pi^-\pi^\circ$ decay rate per neutral kaon

$$\frac{1}{2}\Gamma_L(\pi^+\pi^-\pi^\circ)\{(|A|^2 + |\eta_{+-0}|^2)e^{-\Gamma_S t} + e^{-\Gamma_L t} \pm 2|\eta_{+-0}|e^{-\bar{\Gamma} t}\cos\delta M t + \phi_{+-0}\} \tag{19}$$

The difference of the rates for K° and $\bar{K}^\circ$ is

$$2\Gamma_L(\pi^+\pi^-\pi^\circ)e^{-\bar{\Gamma} t}|\eta_{+-0}|\cos(\delta M t + \phi_{+-0}) \tag{20}$$

and the sum of the rates

$$\Gamma_L(\pi^+\pi^-\pi^\circ)\{(|A|^2 + |\eta_{+-0}|^2)e^{-\Gamma_S t} + e^{-\Gamma_L t}\} \simeq \Gamma_L(\pi^+\pi^-\pi^\circ) \tag{21}$$

giving

$$\frac{\text{Difference Rate}}{\text{Sum Rate}} = 2|\eta_{+-0}|e^{-\bar{\Gamma} t}\cos(\delta M t + \phi_{+-0}) \tag{22}$$

which is plotted in Fig.5 for $\eta_{+-0} = \eta_{+-}$.

PRACTICALITIES

The difficulty of observing CP-violation in the decay of K_S to $\pi^+\pi^-\pi^\circ$ may be illustrated by the numerical fact that $10^{13}\bar{p}p$ annihilations would give a signal only three or four standard deviations different from zero if $\eta_{+-0} \simeq \eta_{+-}$, both for the $K^\circ/\bar{K}^\circ$ comparison (equation (22)) and the $K_S \to \pi^+\pi^-\pi^\circ$ angular distribution (equation (14)). Furthermore the backgrounds due to $\bar{p}p \to \pi^+\pi^-$ etc and $K_S \to \pi^+\pi^-$ are severe.

The prospects of observing the CP-conserving decay of $K_S \to \pi^+\pi^-\pi^\circ$, equation (18), are much more favourable. A few hundred events might be collected from $10^{13}\bar{p}p$ and since the time dependence, equation (18), is quite gross the K_S decay contribution should be recognizable. The backgrounds are not expected to give rise to a "sector difference" in the Dalitz plot.

REFERENCES

1. J.W. Cronin, *Rev. Mod. Phys.* **53**:373 (1981).
2. N.W. Tanner and R.H. Dalitz, *Annals of Physics* **171**:463 (1986).
3. J.K. Black *et al*, *Phys. Rev. Lett.* **54**:1628 (1985).
   R.H. Bernstein *et al*, *Phys. Rev. Lett.* **54**:1631 (1985).
4. N.F. Ramsay, *Rep. Progr. Phys.* **45**:95 (1982).
5. C. Bouchiat and P. Meyer, *Phys. Lett.* **25B**:282 (1967)
   R.S. Chivukula and A.V. Manohar, preprint HUTP-86/A050.
6. P. Pavlopoulos *et al* in "Physics with Antiprotons at LEAR in the Acol Era" (Ed. U.Gastaldi *et al*). Editions Frontiéres, (1985), p.467.
7. N.W. Tanner, *ibid.*. p.483.
8. C. Amsler *et al*, *ibid.*, p.353.
   J.A. Niskanen, *ibid.*, p.193.
9. J. Six, *ibid.*, p.511.
10. H.J. Lipkin, *Phys. Rev.* **176**:1715 (1968).
11. P.K. Kabir, "The CP Puzzle", Academic Press (1968) pp.65–69.
12. F. Selleri, *Lett. Nuovo Cimento* **36**:521 (1983).
13. P. Baillon *et al*, *Nucl. Phys.* **105**:418 (1976); **134**:43 (1978).
14. D.H. Perkins, "Introduction to High Energy Physics", Addison-Wesley, 2nd Ed., (1982), pp.155, 156.

ON THE CONNECTION BETWEEN CP-VIOLATION, MUON-NEUTRINO LIFETIME, MUONIUM CONVERSION AND K^0-DECAYS: AN EXPLICIT MODEL

A.O. Barut*

International Centre for Theoretical Physics

P.O. Box 586, Trieste, Italy

The purpose of this note is to present an explicit model of CP-violation. There is a need for intuitive models of CP-violation, because most studies in this area in the last 20 years have been phenomenological and "the present theories of CP-violation are rather unsatisfactory" [1]. There is still no clear understanding why CP-violation only occurs in K^0-system (and perhaps in B^0-system) and what it means. The phenomenon is still characterized by two small experimental numbers ε and ε', of the order of 10^{-3} and 10^{-6}, respectively. The second one is consistent with being zero.

The model is based on the composite nature of the K^0-state, hence CP-violation will be connected with the internal structure of K^0. It is also interesting that it can be connected to the muon-neutrino lifetime, together with the connection to muonium conversion which of course has been already discussed[2]. It provides an explicit mechanism of "mixing" of K_0 and $\bar{K}_0$, hence a mechanism for the parameter ε. Among the interesting testable consequences of the model is the vanishing of the neutron electric dipole moment, the prediction of the sign of K_L-K_S mass difference, a relation of the mass difference to the decay rate Γ_S of K_S, and a rough estimate for the lifetime of ν_μ.

We begin with muon neutrino ν_μ and muon conversion. At the present time the lifetime of ν_μ is not precisely known (the lower limit is $\tau_{\nu_\mu} > 1.1 \times 10^5\, m_{\nu_\mu}$ sec, m_{ν_μ} in MeV), nor its precise mass (the upper limit is $m_{\nu_\mu} \leqslant 0.25$ MeV), nor possible decay modes. However, if we make the hypothesis that ν_μ is unstable with a very long lifetime (that we shall à posteriori estimate), than a number of interesting testable consequences appear.

Suppose that ν_μ decays eventually into $(\nu_e \nu_e \bar{\nu}_e)$ (that is, experimentally, it disappears). Consider now the process $e^-\mu^+ \leftrightarrow e^+\mu^-$ in scattering, or in bound state (muonium conversion), or the related process $e^-e^- \to \mu^-\mu^-$. We may imagine μ^- to be virtually composed of $e^-\bar{\nu}_e\nu_\mu$, similarly μ^+. (μ^- is a resonance in the channel $e^-\bar{\nu}_e\nu_\mu$.) Then the process $e^-\mu^+ \leftrightarrow e^+\mu^-$ will look like

*Permanent address: Department of Physics, University of Colorado, Boulder, CO 80309, USA.

$$e^-(e^+\nu_e\bar{\nu}_\mu) \leftrightarrow e^+(e^-\bar{\nu}_e\nu_\mu) \quad .$$

If the two neutrinos were the same, then the above would be just a rearrangement of constituents, or a $\nu\bar{\nu}$-exchange diagram (Fig.1(a)). However, since the two neutrinos are different, we need in order to achieve the muonium conversion, an additional process

$$\nu_e\bar{\nu}_\mu \leftrightarrow \bar{\nu}_e\nu_\mu \quad . \tag{1}$$

This in turn could proceed if ν_μ were composite or unstable as assumed. Then we have the virtual process

$$\nu_e\bar{\nu}_\mu \Rightarrow \nu_e(\bar{\nu}_e\bar{\nu}_e\nu_e) \rightarrow \bar{\nu}_e(\nu_e\nu_e\bar{\nu}_e) \Rightarrow \bar{\nu}_e\nu_\mu \quad .$$

We propose the reaction (1), or the additional vertex interaction C in Fig.1(b), as the model for a superweak interaction in CP-violation.

We now discuss how the same interaction appears in K^0 decay and causes CP-violation there.

There is a quark-lepton transformation rule[3-5] which converts quark quantum numbers and quark contents of hadrons into lepton quantum numbers and lepton contents. For the system $K^0 = (u\bar{s})$ and $\bar{K}^0 = (\bar{u},s)$, this rule gives $K^0 = (e^-\mu^+)$ and $\bar{K}^0 = (e^+\mu^-)$. In the magnetic resonance model of hadrons, K^0 and $\bar{K}^0$ are visualized as tightly bound leptons, bound due to strong short ranged magnetic forces[3,6]. The mixing mechanism we give below can also be discussed in terms of quarks, but we prefer to use observable particles.

We can visualize the $K^0 \leftrightarrow \bar{K}^0$ virtual transition or mixing as in the mixing of the two NH_3-levels in atomic masers. The $\nu\bar{\nu}$-system sees a double-well potential with minima at the positions of e^- and e^+ (in Born-Oppenheimer approximation) as shown in Fig.2, and tunnels or oscillates back and forth.

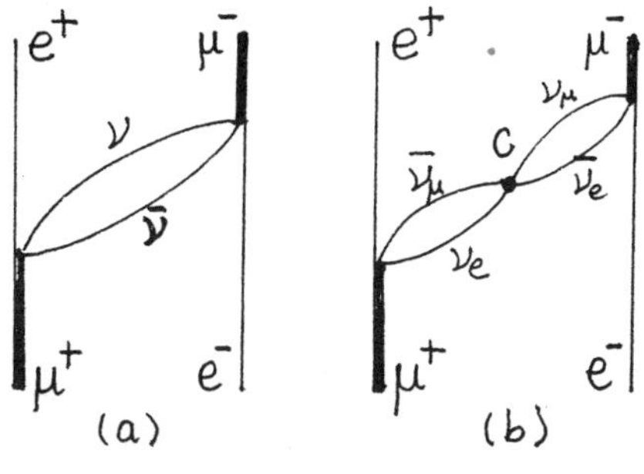

Fig.1 (a) $\nu\bar{\nu}$-exchange
(b) $\nu\bar{\nu}$-exchange with the conversion of eq.(1)

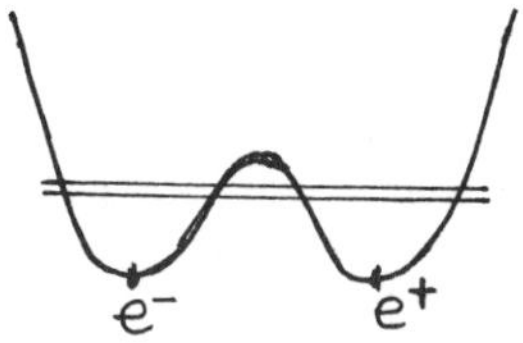

Fig.2 Double well potential for K^0-$\bar{K}^0$ and the symmetric and antisymmetric states representing K_L and K_S.

If $\nu\bar{\nu}$ is on the left-hand potential well we have a $\mu^- e^+ = K^0$ state, if $\nu\bar{\nu}$ is on the right-hand well, we have $e^- \mu^+ = \bar{K}^0$ state. But the eigenstates of the Hamiltonian are the mixtures

$$K_{\substack{1\\2}} = \frac{1}{\sqrt{2}}\left(e^-\mu^+ \pm \mu^- e^+\right) = \frac{1}{\sqrt{2}}\left(K^0 + \bar{K}^0\right)$$

corresponding to symmetric and antisymmetric wave functions of the double well which are also eigenstates of CP. Now if the two neutrinos are not the same, the double well behaviour is not entirely symmetric. Beside the tunnelling we have to introduce also the interaction (1) and this asymmetry leads to the new states

$$K_{\substack{L\\S}} = \frac{1}{\sqrt{2}}\left(\left(1 + \varepsilon\right)K^0 \mp \left(1 - \varepsilon\right)\bar{K}^0\right) = \frac{1}{\sqrt{2}}\left[\left(K^0 \mp \bar{K}^0\right) + \varepsilon\left(K^0 \pm \bar{K}^0\right)\right]$$

(assuming CPT-invariance).

According to a general theorem about the double well problem, the antisymmetric state is higher than the symmetric state by a mass difference (Dennison-Uhlenbeck formula)

$$\Delta_m / m = 1/\pi \; A^2 \quad ,$$

where A is a barrier penetration factor. Applied to our problem, this means that

$$m(K_L) > m(K_S) \quad .$$

The sign of the mass difference is thus correctly predicted, to my knowledge for the first time[3]. For the mangitude we can relate the barrier penetration factor A to the decay rate Γ_S of $K_S \rightarrow 2\pi$, and obtain $\Delta_m = \Gamma_S/2$. The experimental value is $\Delta_m = 0.477 \;\; \Gamma_S$. There is a very familiar analog of the model in the isotropic mixing of the nucleons. The charge exchange process between neutron and proton (which can also be visualized as a double well interaction) leads to the symmetric and antisymmetric states

$$\frac{1}{\sqrt{2}}\left(np \pm pn\right)$$

which are eigenstates of isospin. The exchange can be thought to be via π (or equivalently via the $(e^-\bar{\nu}_e)$-system if n is visualized to be composed of $(pe^-\bar{\nu}_e)$, as we did above for muon.

Finaly we remark that since neutron is virtually a $(pe^-\bar{\nu}_e)$ state and does not involve ν_μ, we do not expect any CP-violation in the neutron wave function, hence no electric dipole moment (at least at low energies).

REFERENCES

1. N.W. Tanner and R.H. Dalitz, Ann. of Phys. 171:463 (1986).
2. Primakoff and S.P. Rosen, Ann. Rev. Nucl. Sci. 31:145 (1981);
   A.Halpern, Phys. Rev. Lett. 48:1313 (1982);
   L.N. Mohapatra, in: "Quarks, Leptons and Beyond", H. Fritzsch, ed., Plenum Press, 1985, p.219;
   J.J. Amato, Phys. Rev. Lett. 21:1709 (1968);
   H. Orth and V.W. Hughes, in: Procs. of the Workshop on Fundamentals of Muon Physics, Los Alamos, N.M., 20 Jan. 1980, LA-10714-C.

3. A.O. Barut, Surveys in High Energy Physics 1:113 (1980).
4. A.O. Barut and S.A. Basri, Lett. N.C. 35:200 (1982).
5. A.O. Barut and D.-Y. Chung, Lett. N.C. 33:225 (1983).
6. A.O. Barut, Ann. der Phys. 43:83 (1986) and references therein.

EXPERIMENTS ON TIME REVERSAL INVARIANCE IN LOW ENERGY NUCLEAR PHYSICS

F.P. Calaprice

Department of Physics
Joseph Henry Laboratories
Princeton University, Princeton , NJ 08540

I. Introduction

Since the famous experiment of Cronin, Fitch, Turlay, and Christenson[1] we know that CP invariance is violated somewhere in the weak decay process of the $K^o{}_L$, though to be sure, the details are not clear. Unlike parity violation, which is maximal and ubiquitous in every weak interaction process, the phenomenon of CP violation is small and, so far, is found only in the neutral kaon system. This unhappy situation has stood through more than twenty years of improving, but fruitless, searches for evidence of a violation of time reversal invariance in other systems.

I would like to review here some new developments, experimental and theoretical, that may provide, within the next few years, some new insights on time reversal invariance. The three topics I have chosen to mention are as follows.

a) The T-odd angular correlation $\mathbf{J}\cdot(\mathbf{p}_e \mathrm{x} \mathbf{p}_\nu)$ in the decay of polarized nuclei.

b) Electric dipole moments of nuclei.

c) T-odd angular correlations for p-wave neutron-nuclear resonances with polarized nuclei and polarized neutrons.

We will see that atomic and optical experimental methods are likely to play a central role in the development of all three topics.

II. The T-odd Angular Correlation $J.(p_e x p_\nu)$ in β–Decay.

The triple angular correlation $\mathbf{J}\cdot(\mathbf{p}_e \mathrm{x} \mathbf{p}_\nu)$ between the initial nuclear spin $\mathbf{J}$ and the momenta of the outgoing electron and neutrino $\mathbf{p}_e$ and $\mathbf{p}_\nu$, respectively, violates time reversal invariance but conserves parity. This angular correlation is probably the most thoroughly investigated of all T-odd angular correlations.

Neutron Experiments

At least five measurements of this correlation have been made for the decay of the free neutron

$$n \rightarrow p + e^- + \nu.$$

The measurement of the correlation is performed by detecting the electron and recoil proton (to infer the neutrino momentum) emitted from a thermal beam of polarized neutrons decaying in-flight past an array of detectors. In a typical geometry, the detectors and the polarization are arranged on orthogonal axes; the T-odd angular correlation produces a count rate asymmetry between counter pairs in which the triple vector product differs in sign.

The results of these measurements, listed chronologically, are summarized in Table 1. The quantity given is the coefficient D of the correlation term which appears in the partial decay rate for polarized nuclear decay[2]. In leading approximation it is given for spin 1/2 mirror nuclear decays by

$$D = (2/\sqrt{3})\ \{|C_V M_F||C_A M_{GT}|/(\ |C_V M_F|^2 + |C_A M_{GT}|^2)\}\ \sin\theta$$

where M_F and M_{GT} are the Fermi and Gamow-Teller matrix elements and C_V and C_A are the vector and axial-vector matrix elements, respectively, and θ is the phase angle between the vector and axial-vector coupling constants. If T-invariance is valid, the phase is either 0° or 180°, and D is zero.

This last statement is modified by the presence of the final state electromagnetic interaction between the electron and the proton, which produces a non-zero D even with perfect time reversal invariance. However, the final state term vanishes in leading order for a pure VA interaction[2] and the next order, which is dominated by the weak magnetism matrix element, is small[3]. For the neutron it is given approximately by

$$D_{FSI}(n) \approx 2x10^{-5}.$$

We observe that this is about 100 times smaller than the present experimental sensitivity (see Table 1), leaving ample room for improvement in the measurements.

Table 1. Measurements of the Angular Correlation D $\mathbf{J}\cdot(\mathbf{p}_e x \mathbf{p}_v)$ for Neutron Beta Decay.

Year	Result for D	Laboratory	Reference
1958,60	-0.09±0.20	Chalk River	4
1958,60	+0.04±0.05	Argonne	5
1968,70	-0.01±0.01	Moscow	6
1974,76	-0.0011±0.0017	Grenoble	7
1978	+0.0022±0.0030	Moscow	8
Combined:	-0.0005±0.0015		
Phase:	$\theta = -(1.1 \pm 3.4) x 10^{-3}$ rad[a]		

a) Determined from the relation D(n) = +0.439 sinθ.

The most precise of the present neutron experiments is that of Steinberg, et al.[7]. The measurement was performed by detecting electron-proton coincidences from the decay of a beam of polarized neutrons produced at the ILL reactor in Grenoble. Their detector apparatus is shown if Figure 1. The neutron beam is moving out of the diagram in the center of a chamber on which are mounted two plastic scintillator electron detectors (β_1,β_2) and two proton detectors (P_1,P_2). The protons drift freely in a central field-free region and are then accelerated through a 20kV potential difference into a detector consisting of a thin layer of NaI(Tl) scintillator and a photomultiplier tube. The detectors are arranged orthogonal

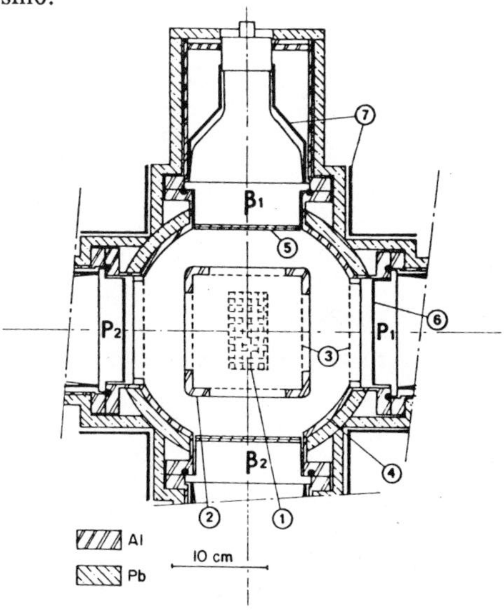

Figure 1. The Detector Arrangement for the neutron experiment on T-invariance.

to each other and to the polarization direction, which is along the beam axis. The count rate asymmetry due to the T-odd correlation is between, for example, $\beta_1 P_1$ and $\beta_1 P_2$. The detector redundancy allows for cancellation of instrumental asymmetries.

The neutron flux for the Grenoble experiment was 10^9 n/sec and the polarization was 70%. Two arrays of detectors were employed with a total coincidence count rate of 1.5 coin/sec. After 2 1/2 months of running they acquired 5×10^6 counts and obtained the result $D = -0.0011 \pm 0.0017$, consistent with time reversal invariance.

Similar experiments have been performed at Moscow by Erozolimski et al[8], with the most recent result being $D = +0.0022 \pm 0.0030$. Improvements in the experimental sensitivity are underway to achieve an error in the 10^{-4} range but details are not available.

A new experiment has been proposed at the Grenoble reactor by Bowles, et al. [9]. The detectors cover a much larger solid angle, approaching 4π for electron and proton detectors, than was used in the previous experiment. In addition, a higher flux and a higher polarization are now available. The detectors are position sensitive allowing for event reconstruction and the axial symmetry is used to reduce systematic effects. The protons are to be detected, after 35 kV preacceleration, by a thin window (40μg/cm^2) Breskin gas counter. The expected count rate is 100 coin/sec allowing for an eventual statistical error of $(1\text{-}2) \times 10^{-4}$ in the parameter D.

Neon-19 Experiments:

Several measurements of the T-odd angular correlation $\mathbf{J}\cdot(\mathbf{p}_e \times \mathbf{p}_\nu)$ have also been made on the beta decay of ^{19}Ne. The results for the correlation coefficient D are given in Table 2. The ^{19}Ne isotope decays by allowed Fermi/Gamow-Teller emission to its mirror nucleus ^{19}F.

$$^{19}\text{Ne} \rightarrow {}^{19}\text{F}^- + e^+ + \nu \qquad (t_{1/2} \approx 17 \text{ sec}, \ KE_{max} \approx 2.2 \text{ MeV})$$

The spin of this isospin doublet is 1/2 and in many ways the decay is equivalent to free neutron decay. However, we note that because of the higher atomic number the final state electromagnetic interaction is larger for ^{19}Ne than for the free neutron and therefore the final state simulation of the T-odd correlation is larger and in particular is given by

$$D_{FSI}(^{19}\text{Ne}) \approx 2 \times 10^{-4}$$

We describe the most recent experiment[13] on ^{19}Ne that employed a new and novel detector design that is much more efficient than those used in earlier designs. As in the neutron experiments, the measurement of D requires detection of the beta particle in coincidence with the recoil daughter nucleus, which in this case is the negative ion F$^-$. The recoil ion has a maximum energy of only ≈200 eV and preacceleration to several keV is therefore necessary for efficient detection. The novel feature of the new design is the realization that the recoil ion can be accelerated by an electric field along the polarization axis **J** without affecting the time reversal asymmetry. This feature makes possible essentially 4π detector solid angles for the positron and recoil ion.

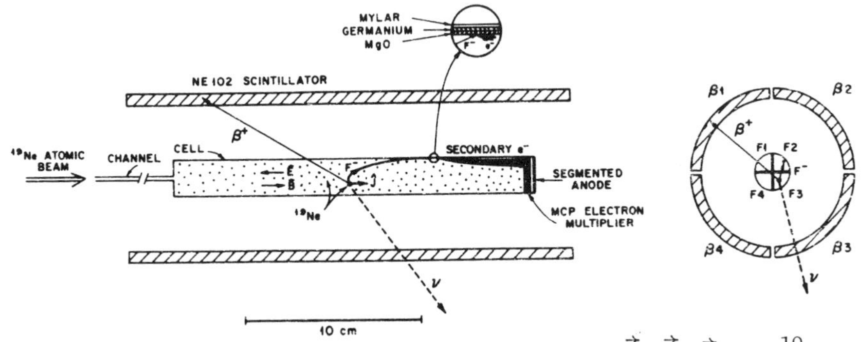

Fig. 2. The detector-cell system for measureing $\vec{J}\cdot(\vec{P}_e \times \vec{P}_\nu)$ in ^{19}Ne decay.

The detector apparatus used for the recent measurement is illustrated in Figure 2. The ^{19}Ne is produced continuosly by a beam of 12 MeV protons bombarding a ^{19}F target and polarized by atomic beam state selection. The polarized ^{19}Ne beam terminates in a special cylindrical holding cell that contains the ion detectors. The cell is 1 cm in diameter and 20 cm long and is located in a uniform magnetic field of a few Gauss such that the polarization is co-linear with the cell axis. A typical decay event is shown in Figure 3. The positron passes through the thin cell wall and is detected by one of four long plastic scintillators. The recoil ion is accelerated by the electric field along the cell axis while maintaining the component of its decay momentum tranverse to the axis. The ion collides with the wall with sufficient energy to knock out several secondary electrons which in turn are also accelerated in the same direction to the end of the cell into an array of four microchannel plate electron multipliers. The beta and ion positions are recorded for each decay. The counts in each beta-ion detector pair are combined to compute the asymmetry parameter. With one week of running time the result obtained was $D = 0.0004 \pm 0.0008$, consistent with T-invariance.

The sensitivity of the latest experiment is limited mainly by the statistical uncertainty. Development of a more intense source of polarized ^{19}Ne is underway and an eventual uncertainty of $\pm 1\text{x}10^{-4}$ is expected. At this level the final stated interactions become important and it is not very likely that the experimental test of T-invariance in beta decay can be pushed much beyond this.

Table 2. Measurements of the Time Reversal Correlation $\mathbf{J}\cdot(\mathbf{p}_e \text{x} \mathbf{p}_v)$ in ^{19}Ne Beta Decay

Year	Result for D	Laboratory	Reference
1969	+0.002±0.014	Berkeley	10
1974	+0.0020±0.0040	Berkeley	11
1977	-0.0005±0.0010	Princeton	12
1984	+0.0004±0.0008	Princeton	13
Combined:	0.0001±0.0006		
Phase:	$\theta = -(0.2 \pm 1.2)\text{x}10^{-3}$ rad[a]		

a) Determined from the relation: $D(^{19}Ne) = -0.519 \sin\theta$

Theoretical Developments

The violation of CP or T invariance can be accomodated in the standard electro-weak model by including a CP violating phase in the Kobayashi-Maskawa[(14)] mass matrix with four generations. New measurements of the CP parameters of K decay are currently putting the KM explanation of CP violation to a test but we must wait for a few years to see the results. However, even if these measurements prove to be in perfect agreement with the KM scheme there is still the possiblility of other sources of T non-conservation that would appear in other phenomena. We mention briefly some other possibilities here. The reader should consult the lecture by L. Wolfenstein for a full discussion.

Lee[(15)] and Weinberg[(16)] discussed modifications of the Higgs sector from one doublet to two or more doublets as an way to introduce CP violation. This process would produce an electric dipole moment of the neutron near the present experimental limits, whereas the KM scheme gives extremely small electric dipole moments.

Another general way to introduce CP violation is to modify the gauge boson sector, as discussed by Mohapatri and Pati[(17)]. Here they consider a model with additional right handed vector bosons that has the attraction of being left-right symmetric. Herczeg[(18)] has studied the T-violating effects in semi-leptonic decys for such a model and finds that, contrary to the standard model, in which these effects are extremely small, one might expect to have T-violating effects at resonable levels in this model. In particular, a non-vanishing triple correlation of the form D $\mathbf{J}\cdot(\mathbf{p}_e \text{x} \mathbf{p}_v)$ is expected and a value for D of $10^{-3\text{-}4}$ is not incompatible with other data. Experiments are already at this level and, as mentioned above, prospects for measurements of D at the level of $1\text{-}2\text{x}10^{-4}$ in neutron decay and in ^{19}Ne decay are reasonably good.

II. Nuclear Electric Dipole Moments

Another method fo testing time reversal invariance in nuclei is through measurements of electric dipole moments. Time reversal and parity invariance must both be violated for a nucleus to posess a non-zero electric dipole moment (EDM). A very good discussion of mechanisms by which nuclei might acquire EDM's has been written by G. Feinberg(19).

A nuclear electric dipole can arise through violation of TRI due to instanton effects in QCD. The axion was invented to prevent this violation of TRI but as the axion has not been discovered (yet) one might speculate that there is in fact a small violation of TRI in the strong interaction. Pursuing this possibility, Crewther, DiVecchia, Veneziano, and Witten(20) derived an effective P- and T-violating pion-nucleon interaction. Using this interaction Henley and Haxton(21) calculated the nuclear EDM's due to mixing of nuclear levels and find in some cases enhancements of 10 or more over the related neutron EDM.

In a study of the KM contribution to nuclear EDM's Flambaum, Khriplovich, and Sushkov(22) claim that the nuclear EDM due to KM effects is 2-3 orders of magnitude larger than it is for the neutron. This is still very small because the KM EDM of the neutron is sometimes estimated to be of the order of 10^{-31} e-cm. However, it may not be out of the question for the new generation of experiments.

The measureability of nuclear EDM's was discussed by L. Schiff in 1963(23). A fundamental problem with such experiments is that the atomic electrons are very effective at shielding the nucleus from the external electric field that must applied to measure the EDM. Schiff demonstrated, however, that the shielding of the electrons is not perfect and a small residual interaction between the external field and the EDM persists, owing to finite nuclear size and other effects. The suppression of the external field is still an impressive 10^{-7} for helium. Since the finite nuclear size effect is more favorable for heavy nuclei, it is worthwhile to explore experimental possibilities with the heaviest possible atoms.

It is well known that nuclear spin polarized rare gas atoms such as ^{3}He have very long spin relaxation times. For ^{3}He atoms the relaxation time is of the order of a day whereas for gaseous xenon isotopes the relaxation time is of the order of minutes. The long relaxation times imply that the uncertainty principle limit to the linewidth of NMR resonances is correspondingly narrow ($\Delta\nu \sim 1/\Delta\tau \approx 10^{-3-5}$/sec). The extremely narrow resonances which are possible enhance the sensitivity to the small frequency shifts associated with the interaction of the nuclear EDM with an external electric field. In fact, Schiff suggested this approach for a measurement of the ^{3}He EDM.

In the following paragraphs we summarize the experiments that have been performed on neon and xenon isotopes to search for nuclear EDM. In addition, we will discuss the development of narrow resonance NMR methods that may make it possible to perform such experiments on the heaviest possible candidate rare gas atom, radon.

The ^{19}Ne Electric Dipole Moment Experiment

The atomic beam method was used to produce a beam of polarized ^{19}Ne atoms. The beam was captured in a special holding cell that was located in a very uniform and stable magnetic field produced by a three layer mu metal shielded solenoid magnet. The holding time and the spin relaxation time were both long compared to the lifetime for decay of the isotope (τ = 25 sec) so that the expected "natural" linewidth of NMR resonances is ≈ 1/25 sec^{-1}. A resonance taken with an external field of 39.8 mG is shown in Figure 3. The absissa specifies the frequency of a tranverse rf field. The ordinate is the experimental beta asymmetry due the the angular correlation $A\mathbf{J}\cdot\mathbf{p}_e$, as measured with a pair of detectors placed along and opposite to the polarization. Note that the FWHM linewidth is approaximately 0.04Hz, as expected.

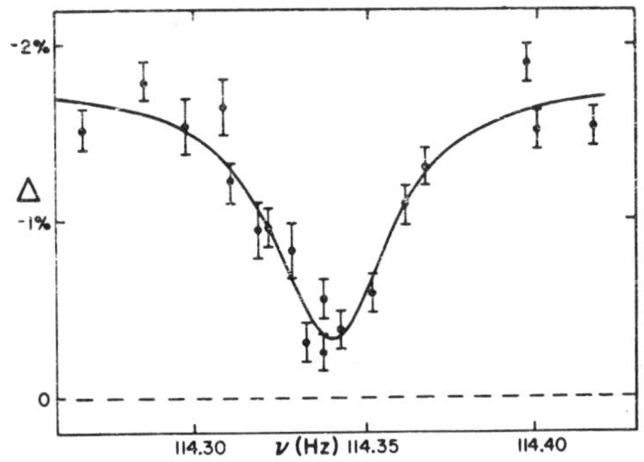

Fig. 3. Narrow NMR resonance curve for ^{19}Ne.

A preliminary search was made for a nuclear EDM. A special holding cell was constructed for this purpose consisting of a cylinder with thin-wall aluminum ends and a teflon spacer. A high voltage is applied across the ends to create a uniform electric field in the cell. Ionization detectors are built in the ends of the cell to detect the betas parallel and antiparallel to the polarization axis. Resonance curves were taken with the electric field along and opposite the magnetic field. A non-zero EDM would produce a shift in the centroid of the resonance. The centroids were found to agree within errors, the difference being 2.8±2.4 mHz. If there were no electron screening this result would imply an electric dipole moment of $(2\pm2)\times10^{-22}$e-cm, which may also be taken for the atomic EDM. A detailed calculation is needed to specify the limit for the nuclear EDM but extrapolating with a Z^2 dependence from helium we estimate an upper limit of $\sim10^{-16}$ e-cm for the nuclear moment. This limit was obtain by D.W. MacArthur in a PhD. thesis[24] and, though crude, it was the first attempt to measure a nuclear EDM. Improvement by several orders of magnitude seems possible but experiments on heavy atoms seem more favorable.

Experiments with Xenon Isotopes.- Magnetic and Electric Dipole Moments

The heavy rare gas atoms are not easily polarized by atomic beam methods. Fortunately, another method based on optical pumping is more suitable for the heavy rare gas atoms. It has been found in several studies that the stable xenon nuclei ^{129}Xe and ^{131}Xe can be polarized by spin exchange with optically pumped rubidium. A glass cell is prepared containing a few milligrams of Rb metal, a buffer gas such as N_2 at ≈ 50 torr and the xenon atoms. Circularly polarized light from a Rb lamp, or a laser tuned to the D_1 794.7 nm line of Rb, is used to polarize the Rb atoms in the cell. The polarized Rb atoms collide with the rare gas atoms and transfer spin polarization to the nucleus of the rare gas atom. For xenon and probably the heavier rare gas atom radon, the spin exchange process is enhanced by the formation of a loosely bound Rb-Xe molecule. A third body, the N_2 molecule, is needed to enhance the formation of the bound state in a collision. Large polarizations (>50%) have been obtained for the stable isotopes ^{129}Xe and ^{131}Xe and for the radioactive isotopes ^{129m}Xe, ^{131m}Xe, ^{133}Xe, ^{133m}Xe, ^{135}Xe and ^{135m}Xe.

As an illustration of the method we discuss recent measurements performed at Princeton of the magnetic moments of the radioactive isotopes [25]. The apparatus for these measurements is illustrated in Figure 4. A kryton ion laser and tuneable dye laser produce light at the Rb D_1 resonance line. The laser beam is circularly polarized with a quarter wave plate and then directed into a small glass cell containing the radioactive xenon isotopes, 50 torr of N_2 gas, and the Rb metal. The cell is heated to 150 C to obtain sufficient alkali vapor density for the spin exchange to be effective. The gamma ray spectrum is measured with a Ge(Li) detector and pulse height analysis system. The magnetic field is produced by a three lare shielded solenild magnet with active feedback control from a Cs optical pumping magnetometer. The field stability at a few gauss is better than 1 ppm.

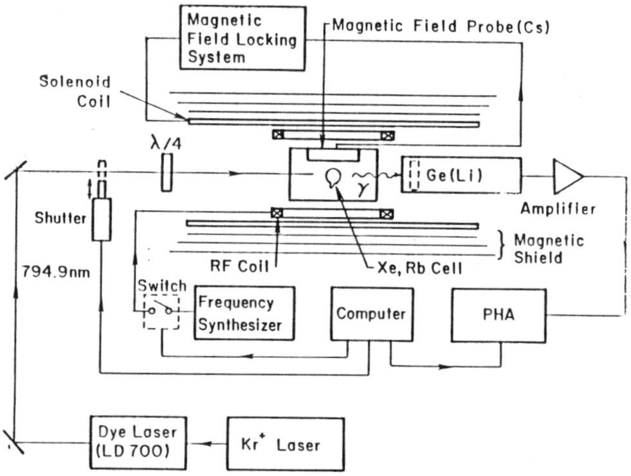

Figure 4. Apparatus for polarizing xenon isotopes

The nuclear orientation is detected by measuring the gamma ray intensity along the polarization axis. For the 11/2 to 3/2 M4 gamma ray transitions of the xenon isomers the intensity decreases as the nuclei are polarized. Magnetic resonance in the xenon nuclei is induced by a set of transverse coils that are driven by a frequency systhesizer. A typical resonance pattern is shown in Figure 5. Off resonance the xenon nuclei are polarized and the rate is low. On resonance the polarization is destroyed and the rate goes up. A characteristic linewidth for this resonance is about 0.05 Hz Hz at a frequency of 1050 Hz. The high sensitivity of the narrow resonance has made it possible to measure the magnetic moments to a few ppm.

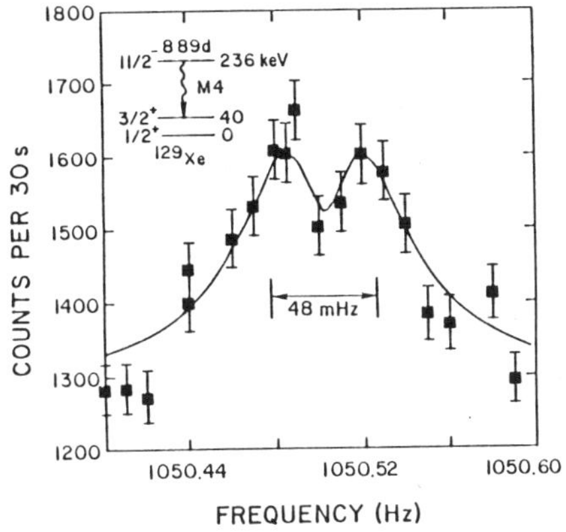

Figure 5. Typical NMR resonance for ^{129m}Xe

Vold et al[26] have applied the same optical pumping spin exchange method to search for the electric dipole moment of the stable isotope ^{129}Xe. Their apparatus is illustrated in Figure 6. A polarization of 10% was achieved and a three cell system was employed to minimize systematic effects. The polarization of the xenon was probed by observation of the re-polarization of the Rb vapor by the polarized xenon. A limit of $(-0.3\pm1.1)\times10^{-26}$ e-cm was obtained.

Figure 6. A Schematic Diagram of the Apparatus of Vold , et .al.[26].

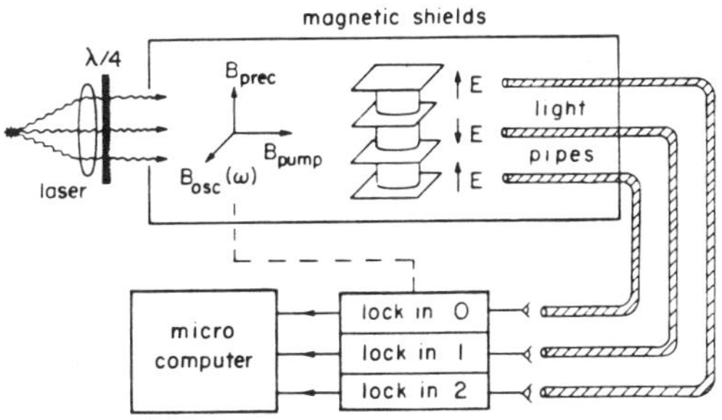

Experiments on Radon Isotopes

Radon is the heaviest rare gas element and all of its isotopes are radioactive. Measurements of the nuclear electric dipole moment is very attractive because the electron screening is least effective and because of the possibility of enhanced EDM due to mixing of states of opposite parity. The gamma ray anisotropy method employed for xenon NMR should also work for radon. For these reasons there is an effort to develop the optical pumping spin exchange method for radon polarization. Experiments are underway with separated radon isotopes at the ISOLDE facility of CERN.[27].

IV. Time Reversal Invariance in Neutron Induced Reactions

Bunakov and Gudkov have the discussed the possibility of testing time reversal invariance in neutron induced reactions[28]. The experiments involve polarized neutrons and oriented nuclear targets. A large enhancement may occur for p-wave resonances due to kinematic suppression of these resonances at thermal or epithermal energies, allowing the s-wave admixture to produce a sizeable effect. Such enhancement has been observed in parity violation in which case parity violation of several percent have been observed.

References

1. J.M. Christenson, J.W. Cronin, V.L. Fitch, and R. Turlay, Phys. Rev. Lett. **13**, 138 (1964).
2. J.D. Jackson, S.B. Treiman and H.W. Wyld, Phys. Rev. **106**, 517 (1957).
3. C. Callan and S.B. Treiman, Phys. Rev.**162**, 1494 (1967).
4. M.A. Clark and J.M. Robson, Can. J. Phys. **38**, 693 (1960).
5. M.T. Burgy, V.E. Krohn, T.B. Novey, G.R. Ringo, and V.L.Teledgi, Phys.Rev. **120,** 1829 (1960).
6. B.G. Erozolimsky, et al. Sov. J. Nuc. Phys. **11**, 583 (1970).
7. R.I. Steinberg, P. Liaud, B. Vignon, and V.W. Hughes, Phys. Rev. **D13**, 2469 (1976).
8. B. Erozolimski, et al., Sov. J. Nuc. Phys. **28,** 48 (1978).
9. T.J. Bowles, et al., J. de Physique **45**, C3-27 (1984).
10. F.P. Calaprice, E.D. Commins, H.M. Gibbs, G.L. Wick, D.A. Dobson, Phys. Rev. **184**, 117 (1969).
11. F.P. Calaprice, E.D. Commins, and D.C. Girvin, Phys. Rev. **D9**,529 (1974).
12. R.M. Baltrusaitis and F.P. Calaprice, Phys. Rev. Lett. **38**, 464 (1977).
13. A.L. Hallin, F.P. Calaprice, D.W. MacArthur, L.E. Piilonen, M.B. Schneider, and D.F. Schreiber, Phys. Rev. Lett. **52**, 337 (1984).
14. N. Kobayashi and T. Maskawa, Prog. Theor. Phys. **49**, 49 (1973).
15. T.D. Lee, Phys. Rev. **D8**, 1226 (1973).
16. S. Weinberg, Phys. Rev. Lett. **37,** 657 (1976).
17. R.N. Mohapatra and J.C. Pati, Phys. Rev. **D11,** 566 (1975).
18. P. Herczeg, Phys. Rev. **D28**, 200 (1983).
19. G. Feinberg, Trans. N.Y. Acad. Sci. Series 2 **38**, 26 (1977).
20. P.J. Crewther, P. DiVecchia, G. Veneziano, and E. Witten, Phys. Lett. **B68**, (1979).
21. H.C. Haxton and E.M. Henley, Phys. Rev. Lett. **51,** 1937 (1983).
22. V.V. Flambaum, I.B. Kriplovich, and O.P. Sushkov, Phys. Lett B, 1985
23. L.I. Schiff, Phys. Rev. **132**, 2194 (1963).
24. D.W. MacArthur, PhD. thesis, Princeton University (1982) (unpublished).
25. M.Kitano, M. Bourzutschky, F.P. Calaprice, J. Clayhold, W. Happer, and M.Musolf, Phys. Rev. **C34,** 1974 (1986).
26. T.G. Vold, F.J. Rabb, B. Heckel, and E.N. Fortson, Phys. Rev. Lett. **52**, 2229 (1984).
27. *Note added in proof.* Radon-223 and 209 were polarized in a recent run and the magnetic moment of radon-209 was measured.
    Princeton-Harvard-CERN-Mainz collaboration, to be published.
28. V.E. Bunakov and V.P. Gudkov, Nucl. Phys. **A401**, 93 (1982).

THE STRONG CP PROBLEM AND THE

VISIBILITY OF INVISIBLE AXIONS*

Wilfried Buchmüller

Institut für Theoretische Physik, Universität Hannover
and
Deutsches Elektronen-Synchrotron DESY, Hamburg

CONTENTS

INTRODUCTION

One of the remarkable successes of quantum chromodynamics (QCD), the theory of hadrons, is the solution of the U(1) problem [1]. In a quark model with three flavours one expects as a consequence of spontaneous chiral symmetry breaking nine pseudo-Goldstone bosons. Eight of them, which are related to the breaking of the nonabelian axial $SU(3)_A$ symmetry, can be identified as pions, kaons and η. The candidate for the ninth boson, however, the isoscalar meson η', is too heavy to be the pseudo-Goldstone boson of the broken axial $U(1)_A$ symmetry.

*Lectures given at the International School of Physics with Low Energy Antiprotons, Erice, September 1986.

The solution of this U(1) problem rests on two fundamental properties of gauge theories with fermions: the Adler-Bell-Jackiw anomaly [2] of the axial U(1) current and the existence of gauge field configurations with nontrivial topology, such as instantons [3], which give rise to a new parameter Θ characterizing the QCD vacuum [4]. As recent numerical QCD lattice calculations [5] show, the deviation of the η'-mass from the naively expected pseudo-Goldstone boson mass can indeed be attributed to gauge field configurations with non-zero topological charge.

The Θ-vacuum of QCD, which solves the U(1) problem, creates a new one: the strong CP problem. The new parameter Θ in the QCD lagrangian multiplies an operator which violates P, T and CP invariance. This leads to a non-zero electric dipole moment of the neutron, and from current experimental bounds one infers $\Theta < 10^{-9}$! Why should the a priori arbitrary Θ-parameter be so small? This "strong CP problem" is, contrary to the U(1) problem, to some extent a matter of taste. It is not a discrepancy between theory and experiment, and the QCD lagrangian contains also other "unexplained" small parameters such as the current quark masses. However, the small value of the Θ-parameter becomes even more miraculous when the generation of quark masses in the Glashow-Weinberg-Salam theory is taken into account. In this case the intrinsic QCD Θ-parameter and the argument of the determinant of the complex quark mass matrix have to compensate each other to less than one part in 10^9, a cancellation which demands an explanation!

The most elegant "solution" of the strong CP problem is the Peccei-Quinn mechanism [6], where a spontaneously broken chiral U(1) symmetry yields $\overline{\Theta} = 0$. An unavoidable consequence of this mechanism is the appearance of a very light pseudo-Goldstone boson, the axion [7], which has been extensively searched for experimentally. Although the standard axion, including its recent variants, has been ruled out [8], the so called "invisible" axions remain an interesting possibility.

The main subject of these lectures are general properties of axions and recent suggestions of how to detect invisible axions. After a brief review of the strong CP problem and the Peccei-Quinn mechanism in sects. 2 and 3, we will discuss axion properties by means of an effective lagrangian approach in sect. 4. Experimental bounds on axion production and decays are reviewed in sect. 5. Sect. 6 deals briefly with the recently proposed variant axion models, and in sect. 7 we discuss the possible relevance of supersymmetry to the strong CP problem. In sect. 8

we then review the different proposals for the detection of invisible axions. Some conclusions are given in sect. 9.

QCD-VACUUM STRUCTURE AND THE STRONG CP PROBLEM

The theory of mesons, baryons and their strong interactions is quantum chromodynamics which, in the case of two quark flavours, is given by the lagrangian

$$L = -\frac{1}{4} G^A_{\mu\nu} G^{A\mu\nu} + \bar{q}_L i\not{D} q_L + \bar{q}_R i\not{D} q_R$$
$$+ m_u \bar{u}_L u_R + m_d \bar{d}_L d_R + \text{c.c.}, \qquad (2.1)$$

where

$$q = \binom{u}{d}, \quad q_{L,R} = \frac{1\mp\gamma_5}{2} q, \quad D_\mu = \partial_\mu - ig_s \frac{\lambda_A}{2} G^A_\mu .$$

As the current quark masses m_u and m_d are small compared to the QCD scale parameter Λ_{QCD}, the lagrangian (2.1) possesses an approximate chiral invariance $U(2)_L \times U(2)_R$ which is spontaneously broken to the vectorial subgroup $SU(2)_V \times U(1)_V$ through the condensates

$$\langle\bar{u}u\rangle_o = \langle\bar{d}d\rangle_o = O(\Lambda^3_{QCD}) . \qquad (2.2)$$

The spontaneous breaking of the approximate axial symmetry $SU(2)_A \times U(1)_A$ gives rise to pseudo-Goldstone bosons associated with the isovector and isoscalar currents $A_{a\mu}$ and $A_{s\mu}$:

$$\pi^\pm, \pi^o \quad : \quad A_{a\mu} = \frac{1}{2} \bar{q}\gamma_\mu\gamma_5\tau_a q \qquad (2.3a)$$

$$\tilde{\eta} \quad : \quad A_{s\mu} = \frac{1}{2} \bar{q}\gamma_\mu\gamma_5 q . \qquad (2.3b)$$

In the realistic case of three flavours u, d and s the isoscalar $\tilde{\eta}$ corresponds to the pseudoscalar meson η' (up to a small mixing with the η-meson). For simplicity we will ignore in the following the existence of more than two quark flavours whenever they are unimportant for the properties of axions.

The U(1) problem of QCD is the discrepancy between the current algebra prediction [9]

$$m^2_{\eta'} < 3\, m^2_\pi \tag{2.4}$$

and the experimental observation

$$m^2_{\eta'} > 50\, m^2_\pi \;. \tag{2.5}$$

The solution of this U(1) problem was found in 1976 by 't Hooft [4] who observed that the current algebra prediction is invalidated by two subtle properties of the quantum theory based on the lagrangian (2.1): the Adler-Bell-Jackiw (ABJ) anomaly [2] and the Θ-vacuum [4].

According to the ABJ anomaly the isoscalar axial current $A_{s\mu}$ is not conserved:

$$\partial^\mu A_{s\mu} = n_f \frac{g_s^2}{32\pi^2} G^A_{\mu\nu} \tilde{G}^{A\mu\nu}, \tag{2.6}$$

$$\tilde{G}^A_{\mu\nu} = \frac{1}{2} \varepsilon_{\mu\nu\rho\sigma} G^{A\rho\sigma},$$

where n_f is the total number of quark flavours. However, this nonconservation of the isoscalar current is not sufficient to evade the current algebra prediction (2.4). What is needed in addition is the existence of gauge field configurations for which the topological charge is different from zero although the topological charge density, which appears on the r.h.s. of eq. (2.6), is a total divergence:

$$Q[G^A_\mu] = \frac{g_s^2}{32\pi^2} \int d^4x\, G^A_{\mu\nu} \tilde{G}^{A\mu\nu}$$

$$= \int d^4x\, \partial^\mu K_\mu \neq 0 \;. \tag{2.7}$$

The simplest gauge fields with non zero topological charge are the well known instantons [3].

Eqs. (2.6) and (2.7) imply that the η'-mass is larger than the current algebra prediction (2.4). Within the effective lagrangian approach for pseudo-Goldstone bosons one obtains from the anomaly (2.6):

$$\Delta L_{\tilde{\eta}} = \frac{2}{f_\pi} \tilde{\eta} \frac{g_s^2}{32\pi^2} G^A_{\mu\nu} \tilde{G}^{A\mu\nu} \;. \tag{2.8}$$

Integrating out the gluons then yields a mass term (cf. Fig. 1):

$$\Delta L_{AN} = -\frac{1}{2} m_{AN}^2 \tilde{\eta}^2 , \tag{2.9}$$

$$m_{AN}^2 = O(\Lambda_{QCD}^2) \quad .$$

The quantitative determination of the anomaly contribution m_{AN}^2 to the η'-mass is a difficult problem because it requires nonperturbative methods. In an expansion in the number of colors n_c one finds the mass formula [10,11] (f_π = 93 MeV)

$$m_{\eta'}^2 + m_\eta^2 - 2m_{K_o}^2 = \frac{2n_f}{f_\pi^2} \chi_t + O(\frac{1}{n_c}) , \tag{2.10}$$

where

$$\chi_t = \frac{1}{32\pi^2} \frac{1}{V} <\int_V d^4x \, G_{\mu\nu}^A \, \tilde{G}^{A\mu\nu}> \tag{2.11}$$

is the so called topological susceptibility. For $n_f = n_c = 3$ one obtains from eq. (2.10) and the known η'-, η- and K_o-masses

$$\chi_t^{EXP} \cong (180 \text{ MeV})^4 \quad . \tag{2.12}$$

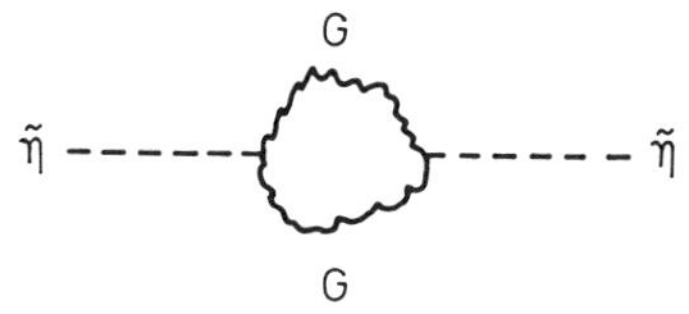

Fig. 1 Anomaly contribution to the $\tilde{\eta}$-mass

Recently χ_t has for the first time been determined numerically for the colour group SU(3) by means of lattice Monte Carlo calculations with the result [5]

$$\chi_t^{MC} = (247^{+28}_{-43} \text{ MeV})^4 , \tag{2.13}$$

where the errors are purely statistical. This result is very encouraging because it shows that topological effects are indeed large enough to explain quantitatively the η'-mass.

The existence of gauge field configurations with non zero topological charge implies that the QCD vacuum is characterized by a new quantity, the Θ-parameter. For a given value of Θ the QCD lagrangian is modified by

$$\begin{aligned} L_\Theta &= -\Theta \frac{g_s^2}{32\pi^2} G^A_{\mu\nu} \tilde{G}^{A\mu\nu} \\ &= \Theta \frac{g_s^2}{8\pi^2} \vec{E}^A \cdot \vec{B}^A, \end{aligned} \tag{2.14}$$

where $\vec{E}^A$ and $\vec{B}^A$ are the colour electric and magnetic field strengths defined in the usual way. Clearly, L_Θ violates the discrete symmetries P, T and CP.

The CP violation in strong interactions due to the Θ-term of the QCD lagrangian manifests itself in a nonvanishing electric dipole moment of the neutron. One easily estimates

$$d_N \sim \frac{e}{m_N} \left(\frac{m_u m_d}{m_u+m_d} \Theta\right) \frac{1}{\Lambda_{QCD}} . \tag{2.15}$$

Here $\frac{e}{m_N}$ is the usual dipole moment factor, $\frac{m_u m_d}{m_u+m_d}\Theta$ is the phase of a chiral rotation which eliminates the Θ-term from the lagrangian at the cost of complex quark masses, and the last factor follows on dimensional grounds. With $m_u \sim \frac{m_d}{2} \sim 5$ MeV and $\Lambda_{QCD} \sim 300$ MeV one finds

$$d_N \sim 2 \cdot 10^{-16}\, \Theta \text{ e cm}, \tag{2.16}$$

which is close to the values obtained in explicit calculations:

$$d_N = \begin{cases} 2.7 \cdot 10^{-16}\, \Theta \text{ e cm}, & \text{Baluni [12]} \\ 5.2 \cdot 10^{-16}\, \Theta \text{ e cm}, & \text{Crewther et al. [13].} \end{cases} \tag{2.17}$$

The theoretical predictions (2.17) have to be contrasted with the experimental bound [14]

$$d_N < 4 \cdot 10^{-25} \text{ e cm}, \tag{2.18}$$

which implies $\Theta < 10^{-9}$! Why should Θ be so small? One might argue that the value Θ = 0 is "natural" in the sense that the discrete symmetries are conserved in this case. Yet this argument is not possible anymore if QCD is considered as part of the standard model of strong and electroweak interactions where CP invariance cannot be imposed on the entire lagrangian. In this case Θ is replaced by $\overline{\Theta}$ which contains also a contribution from the quark mass matrix M:

$$\overline{\Theta} = \Theta + \text{argdet}\, M \,. \tag{2.19}$$

It is totally unclear why the two a priori unrelated parameters Θ and argdet M should compensate each other to 1 part in 10^9 - which is the strong CP problem.

Let us briefly mention that similar "strong CP problems" occur in many technicolour and preon models where Higgs bosons, quarks and leptons have a further substructure. For instance, if a composite SU(2)-doublet Higgs field ϕ contains coloured fermionic constituents and the hypercolour gauge interactions are not CP conserving, one expects an effective interaction of the form

$$\Delta L_{\Theta} = \frac{g_s^2}{\Lambda_c^2}\, \phi^+\phi\, G^A_{\mu\nu}\, \tilde{G}^{A\mu\nu}, \tag{2.20}$$

where Λ_c is the scale of the confining hypercolour group. After spontaneous breaking of the electroweak symmetry with $\langle\phi^+\phi\rangle_o \sim G_F^{-1}$ the effective interaction (2.20) generates the contribution to the Θ-parameter $\delta\Theta_c = 32\pi^2 \frac{v^2}{\Lambda_c^2}$. The bound $\delta\Theta_c < 10^{-9}$ then implies $\Lambda_c > 10^5$ TeV [15] which illustrates the potentially large size of CP violating effects in substructure theories with a hypercolour scale in the TeV range.

3. THE PECCEI-QUINN MECHANISM

The first step towards the Peccei-Quinn solution of the strong CP problem is to realize that in the case of at least one vanishing quark mass Θ is an irrelevant parameter. Consider, for instance, the case $m_u = o$. The QCD lagrangian (2.1) is then invariant under the chiral transformation

$$u_L \rightarrow (1-i\delta\alpha)u_L, \quad u_R \rightarrow (1+i\delta\alpha)u_R, \tag{3.1}$$

under which the lagrangian changes by

$$\delta L = -\delta\alpha\, \partial_\mu (\bar{u}\gamma^\mu\gamma_5 u)$$

$$= -\delta\alpha \frac{g_s^{\,2}}{32\pi^2} G^A_{\mu\nu} \tilde{G}^{A\mu\nu}, \tag{3.2}$$

where the second equality follows from the ABJ anomaly (2.6). Eq. (3.2) corresponds to a change in Θ:

$$\Theta \rightarrow \Theta' = \Theta + \delta\alpha . \tag{3.3}$$

Hence we can achieve $\Theta' = 0$ by means of a finite chiral rotation, i.e., we can "rotate Θ away". As the current quark masses are finite, however, the Θ-parameter cannot be rotated away in QCD.

The essential idea of the Peccei-Quinn (PQ) mechanism [6] is that the Θ-parameter can also be set to zero by means of a spontaneously broken chiral U(1) symmetry. As Weinberg and Wilczek have pointed out unavoidable consequence of this mechanism is the appearance of a very light pseudo-Goldstone boson, the axion. So far, despite remarkable experimental efforts, axions have not been discovered. Nevertheless the Peccei-Quinn mechanism appears to be the natural solution of the strong CP problem since it makes use of a chiral symmetry to which the meaning of the Θ-parameter is tied.

The simplest model, which realizes the Peccei-Quinn mechanism, contains two Higgs doublets with $SU(2)_W \times U(1)_Y$ quantum numbers

$$\phi_1 \sim (2; -\frac{1}{2}), \quad \phi_2 \sim (2; +\frac{1}{2}), \tag{3.4}$$

which couple separately to up and down quarks:

$$L_{Yuk} = \Gamma_u \bar{q}_L \phi_1 u_R + \Gamma_d \bar{q}_L \phi_2 d_R + \text{c.c.} \quad . \tag{3.5}$$

The lagrangian is invariant under the chiral $U(1)_{PQ}$ transformation

$$q_L \rightarrow e^{-i\frac{\alpha}{2}} q_L \tag{3.6a}$$

$$u_R \rightarrow e^{i\frac{\alpha}{2}} u_R, \quad d_R \rightarrow e^{i\frac{\alpha}{2}} d_R, \tag{3.6b}$$

$$\phi_1 \to e^{-i\alpha}\phi_1, \quad \phi_2 \to e^{-i\alpha}\phi_2. \tag{3.6c}$$

It is clear from (3.6c) that the $U(1)_{PQ}$ invariance is lost in the minimal standard model with a single Higgs doublet where $\phi_2 = i\sigma_2\phi_1{}^*$. As the $U(1)_{PQ}$ symmetry acts differently on left and right handed quarks the associated current has a triangle anomaly with the colour vector current (cf. Fig. 2) and the Θ-parameter can be rotated away by means of a $U(1)_{PQ}$ transformation.

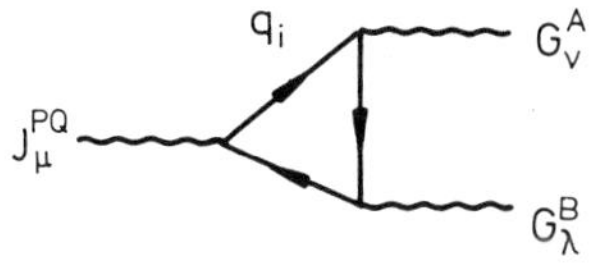

Fig. 2 Anomaly of the Peccei-Quinn current

The vacuum expectation values $(v_1{}^2 + v_2{}^2 \equiv v^2 = \frac{1}{2\sqrt{2}}G_F{}^{-1})$

$$<\phi_1>_o = \begin{pmatrix} v_1 \\ o \end{pmatrix}, \quad <\phi_2>_o = \begin{pmatrix} o \\ v_2 \end{pmatrix} \tag{3.7}$$

now break the symmetry $SU(2)_W \times U(1)_Y \times U(1)_{PQ}$ to the subgroup $U(1)_{EM}$. Hence one obtains, in addition to three massive vector bosons, a new very light pseudo-Goldstone boson, the axion, whose tiny mass $m_a = O(\frac{\Lambda^2_{QCD}}{v})$ is an effect of the anomaly. The axion mass must also vanish if at least one of the current quark masses m_q is zero, because in this case there exists an additional chiral $U(1)_A$ symmetry and hence an anomaly free $U(1)$ subgroup of $U(1)_{PQ} \times U(1)_A$ which is spontaneously broken by the vacuum expectation values (3.7). Therefore the axion mass depends on two symmetry breaking parameters, the anomaly term $m^2_{AN} \sim m^2_{\eta'} - m^2_{\pi} \sim \Lambda^2_{QCD}$, and the chiral symmetry breaking current quark masses m_q. The dependence of the masses of the pseudoscalar mesons η', π^o and a on these two symmetry breaking parameters is summarized in table I.

Table I η'-, π^o- and a-masses for different explicit symmetry breakings.

Masses	$m^2_{AN} \neq o$		$m^2_{AN} = o$	
	$m_q = o$	$m_q \neq o$	$m_q = o$	$m_q \neq o$
$m^2_{\eta'}$	>o	>o	o	>o
$m^2_{\pi^o}$	o	>o	o	>o
m^2_a	o	>o	o	o

4. AXION PROPERTIES

The couplings of axions to other particles, in particular the mixings with the neutral mesons π^o, η and η', are most easily computed by means of the effective lagrangian technique for pseudo-Goldstone bosons. Our discussion in this section follows closely the recent work of Bardeen, Peccei and Yanagida [16].

The axion field is a linear combination of the phases of the two neutral complex Higgs scalars $\phi_i{}^o$, i=1,2:

$$\phi_i{}^o = \frac{1}{\sqrt{2}} (v_i + \rho_i) e^{i \frac{\xi_i}{v_i}} \quad . \tag{4.1}$$

One easily finds for the longitudinal component of the neutral Z-boson and the axion ($v = \sqrt{v_1{}^2 + v_2{}^2}$):

$$\xi = \frac{v_1}{v} \xi_1 - \frac{v_2}{v} \xi_2, \tag{4.2a}$$

$$a = \frac{v_2}{v} \xi_1 + \frac{v_1}{v} \xi_2. \tag{4.2b}$$

In order to compute the axion couplings to quarks and leptons one may set $\xi = \rho_i = 0$ and insert into the lagrangian (3.5)

$$\phi_1{}^o = \frac{v_1}{\sqrt{2}} e^{i \frac{xa}{v}}, \quad \phi_2{}^o = \frac{v_2}{\sqrt{2}} e^{i \frac{a}{xv}} \quad , \tag{4.3}$$

where

$$x = \frac{v_2}{v_1} \quad . \tag{4.4}$$

If one defines the chiral $U(1)_{PQ}$ symmetry through the transformation of the axion field

$$a \to a + \alpha v, \tag{4.5}$$

there is still an ambiguity in the transformation of the quark field corresponding to an arbitrary admixture of the vectorial baryon number $U(1)_B$. The simplest definition of $U(1)_{PQ}$ acts trivial on the lefthanded quarks,

$$q_L \to q_L, \quad u_R \to e^{-i\alpha x} u_R, \quad d_R \to e^{-i \frac{\alpha}{x}} \quad , \tag{4.6}$$

yielding the current

$$J_\mu^{PQ} = v\partial_\mu a + x\, \bar{u}_R \gamma_\mu u_R + \frac{1}{x} \bar{d}_R \gamma_\mu d_R \tag{4.7}$$

with the ABJ anomaly

$$\partial^\mu J_\mu^{PQ} = n_f (x + \frac{1}{x}) \frac{g_s^{\,2}}{32\,{}^2} G_{\mu\nu}^A \, G^{A\mu\nu} \quad , \tag{4.8}$$

where n_f is the total number of quark-lepton families.

Inserting (4.3) into the standard model lagrangian one easily finds

$$L = L_{gauge} + \bar{q}_L i\not{\partial} q_L + \bar{q}_R i\not{\partial} q_R + \frac{1}{2} \partial_\mu a \partial^\mu a$$

$$- m_u \bar{u}_L u_R \, e^{i \frac{xa}{v}} - m_d \bar{d}_L d_R \, e^{i \frac{a}{xv}} - m_e \bar{e}_L e_R \, e^{i \frac{za}{v}} + c.c., \tag{4.9}$$

where

$$z = \frac{1}{x} \quad \text{or} \quad -x \quad , \tag{4.10}$$

depending on which Higgs field couples to leptons. By means of the local transformation

$$u_R \to e^{-i\frac{xa}{v}} u_R, \quad d_R \to e^{-i\frac{a}{xv}} d_R, \quad e_R \to e^{-i\frac{za}{v}} e_R \tag{4.11}$$

one then obtaines the result

$$\begin{aligned} L = {} & L_{gauge} + \bar{q}_L i\not{\partial} q_L + \bar{q}_R i\not{\partial} q_R + \frac{1}{2}\partial_\mu a \partial^\mu a \\ & - \frac{a}{v}(x m_u \bar{u} i\gamma_5 u + \frac{1}{x} m_d \bar{d} i\gamma_5 d + z m_e \bar{e} i\gamma_5 e) \\ & + \frac{a}{v} n_f (x + \frac{1}{x}) \frac{g_s^2}{32\pi^2} G^A_{\mu\nu} \tilde{G}^{A\mu\nu} \\ & + \frac{a}{v} n_f (\frac{4}{3} x + \frac{1}{3} x + z) \frac{g'^2}{16\pi^2} B_{\mu\nu} \tilde{B}^{\mu\nu} , \end{aligned} \tag{4.12}$$

where

$$B_{\mu\nu} = \cos\Theta_W F^\gamma_{\mu\nu} - \sin\Theta_W F^Z_{\mu\nu} ,$$

$$g' = \frac{e}{\cos\Theta_W} ,$$

and $F^{\gamma,Z}_{\mu\nu}$ are the abelian field strengths of photon and Z-boson. The lagrangian (4.12) contains all couplings of the axion to quarks, leptons, gluons, photon and Z-boson. The couplings to gauge bosons arise from the ABJ anomalies, the couplings to fermions are, as expected, pseudoscalar. All couplings scale like the inverse of the symmetry breaking vacuum expectation value v. In (4.12) n_f denotes the total number of quark-lepton generations.

The axion mass and its mixings with π° and $\tilde{\eta}$ (η') are most easily computed from the effective chiral meson lagrangian with approximate $U(2)_L \times U(2)_R \times U(1)_{PQ}$ symmetry. The nonlinear realization of the chiral symmetry on the meson fields

$$U = e^{\frac{i}{f_\pi}(\pi^a \tau_a + \tilde{\eta} 1)} \tag{4.13}$$

is defined through

$$U \to g_L U g_R^\dagger , \tag{4.14}$$

where $g_L(g_R)$ denote $U(2)_L$ $(U(2)_R)$ matrices. The invariant meson lagrangian reads

$$L = \frac{f_\pi^2}{4} \operatorname{tr}\left[(D_\mu U)^\dagger D^\mu U\right] + \frac{1}{2} \partial_\mu a \partial^\mu a + \Delta L_{cs} + \Delta L_{AN}, \tag{4.15}$$

where D_μ is the $SU(2)_W \times U(1)_Y$ gauge covariant derivative, and ΔL_{cs} and ΔL_{AN} denote the small explicit symmetry breakings due to current quark masses and the ABJ anomaly:

$$\Delta L_{cs} = \frac{1}{2} \kappa \operatorname{tr}\left[UM + M^\dagger U^\dagger\right]$$

$$+ \frac{i}{2} \kappa\, a \operatorname{tr}\left[UMX - X^\dagger M^\dagger U^\dagger\right], \tag{4.16a}$$

$$M = \begin{pmatrix} m_u & 0 \\ 0 & m_d \end{pmatrix}, \quad X = \begin{pmatrix} x & 0 \\ 0 & \frac{1}{x} \end{pmatrix},$$

and

$$\Delta L_{AN} = -\frac{1}{2} m_{AN}^2 \left(\frac{\tilde{\eta}}{f_\pi} + \frac{1}{2} n_f \left(x + \frac{1}{x}\right) a\right)^2 . \tag{4.16b}$$

The form of ΔL_{cs} follows by inspection from eqs. (4.12) and (4.14); ΔL_{AN} is obtained from (4.14) and the requirement that the effective lagrangian (4.15) changes under chiral U(1) transformations by the anomaly term (2.6).

Expanding the meson field U in powers of $\frac{1}{f_\pi}$, one easily finds

$$\Delta L_{cs} = (m_u+m_d)\kappa - \frac{(m_u-m_d)\kappa}{2f_\pi^2} \left(\vec{\pi}^2 + \tilde{\eta}^2 + 2 \frac{m_u-m_d}{m_u+m_d} \pi^\circ \tilde{\eta}\right) + \ldots$$

$$- \frac{(m_u-m_d)\kappa}{f_\pi} \frac{a}{v} \left(x \frac{m_u}{m_u+m_d} (\tilde{\eta} + \pi^\circ) + \frac{1}{x} \frac{m_d}{m_u+m_d} (\tilde{\eta} - \pi^\circ)\right) + \ldots \tag{4.17}$$

With $m_{AN}^2 \gg m_\pi^2$ and $v \gg f_\pi$, (4.16b) and (4.17) yield for pion and axion masses [16,17]:

$$m_\pi^2 = \frac{(m_u+m_d)\kappa}{f_\pi^2}, \tag{4.18}$$

$$m_a^2 = m_\pi^2 \left(\frac{f_\pi}{v}\right)^2 n_f^2 \left(x + \frac{1}{x}\right)^2 \frac{m_u m_d}{(m_u+m_d)^2} . \tag{4.19}$$

From (4.18) it is clear that the parameter κ plays the role of the chiral symmetry breaking condensate (2.2), $\kappa \sim \langle\bar{u}u\rangle_o \sim \langle\bar{d}d\rangle_o$. The axion mass formula (4.19) contains the factors (m_u+m_d) and $n_f^2(x + \frac{1}{x})^2$ reflecting the influence of the anomaly and chiral symmetry breaking on the axion mass, which we have qualitatively discussed in sect. 3 (cf. table I).

From (4.17) one also reads off the isotriplet and isosinglet axion mixings (cf. Fig. 3) which yield the transformation to physical states [16,17]:

$$\pi^\circ = \pi^\circ_{phys} + \xi_{a\pi}\, a_{phys}, \qquad (4.20a)$$

$$\tilde{\eta} = \tilde{\eta}_{phys} + \xi_{a\eta}\, a_{phys}, \qquad (4.20b)$$

$$a = a_{phys} - \xi_{a\pi}\, \pi_{phys} - \xi_{a\eta}\, \tilde{\eta}_{phys}, \qquad (4.20c)$$

$$\xi_{a\pi} = \frac{f_\pi}{v}\lambda_3, \quad \xi_{a\eta} = \frac{f_\pi}{v}\lambda_s, \qquad (4.20d)$$

$$\lambda_3 = \frac{1}{2}\left[(x - \frac{1}{x}) - n_f(x + \frac{1}{x})\frac{m_d-m_u}{m_d+m_a}\right], \qquad (4.20e)$$

$$\lambda_s = -\frac{1}{2}(n_f - 1)(x + \frac{1}{x}) \quad . \qquad (4.20f)$$

The axion couplings to heavy quarks of charge $\frac{2}{3}$ (U = c, t, ...) and charge $-\frac{1}{3}$ (D = s, b, ...), leptons (ℓ = e, μ, τ, ...) and photons is given by (4.12) ($x = \frac{v_2}{v_1}$, $z = x$ or $-\frac{1}{x}$, $n_f = 3 + \ldots$):

$$L_{af} = -x\frac{m_u}{v}\bar{u}i\gamma_5 u\, a - \frac{1}{x}\frac{m_D}{v}\bar{d}i\gamma_5 d\, a - z\frac{m_\ell}{v}\bar{\ell}i\gamma_5\ell\, a, \qquad (4.21)$$

$$L_{a\gamma} = \frac{1}{4}\frac{\xi\alpha}{v}F_{\mu\nu}\tilde{F}^{\mu\nu}a = -\frac{\xi\alpha}{v}\vec{E}\cdot\vec{B}\, a, \qquad (4.22)$$

$$\xi = \frac{n_f}{\pi}(\frac{4}{3}x + \frac{1}{3}x + z), \quad \alpha = \frac{e^2}{4\pi} \quad .$$

Fig. 3 Isotriplet and isosinglet axion mixings.

We note that the axion mass and all of its couplings scale like $\frac{1}{v}$. Numerically, one finds from (4.19):

$$m_a = 150 \text{ keV} \left[\frac{(\sqrt{2}G_F)^{-1/2}}{v} \frac{n_f}{3} \frac{1}{2} (x + \frac{1}{x}) \right] . \quad (4.23)$$

For $m_a < 2m_e$ the axion lifetime is determined by the two photon decay mode:

$$\tau(a \to 2\gamma) \cong \tau(\pi^\circ \to 2\gamma) (\frac{v}{f_\pi})^2 (\frac{m_\pi}{m_a})^3$$

$$\cong (\frac{100 \text{ keV}}{m_a})^5 \text{ sec.} \quad (4.24)$$

In the case $m_a > 2m_e$ the decay into an electron positron pair is dominant, and one obtains a much shorter lifetime:

$$\tau(a \to e^+e^-) = \frac{8\pi v}{m_e^2 (m_a^2 - 4m_e^2)^{1/2}} \frac{1}{z^2}$$

$$\cong 3 \cdot 10^{-9} \text{ sec} (\frac{1.7 \text{ MeV}}{m_a}) \frac{1}{z^2} . \quad (4.25)$$

This second case is of interest for "variant axion models" where due to a large value of x or $\frac{1}{x}$ the axion mass is substantially larger than the natural value of about 100 keV.

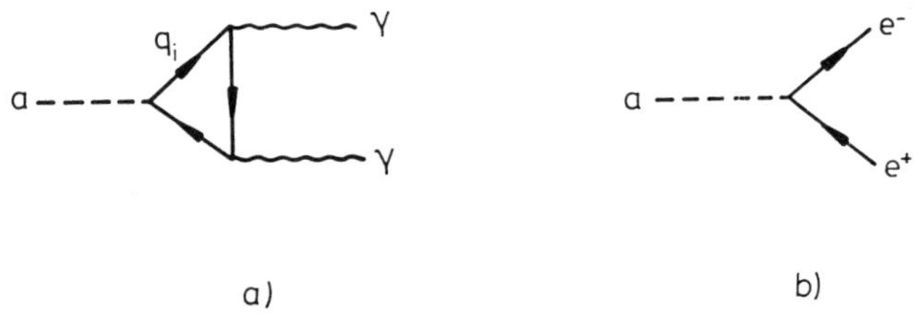

Fig. 4 Axion decays: a) $a \to 2\gamma$ (dominant for $m_a < 2m_e$), b) $a \to e^+e^-$ (dominant for $m_a > 2m_e$)

5. EXPERIMENTAL BOUNDS ON AXIONS

Since the invention of the Peccei-Quinn mechanism more than eight years ago axions have been extensively searched for experimentally in quarkonium decays, light meson decays, beam dump experiments and nuclear deexitations. Comprehensive reviews can be found in Refs. [18,19].

Table II. Processes providing bounds on axion couplings for $m_a < 2m_e$ and/or $m_a > 2m_e$.

	$m_a < 2m_e$	$m_a > 2m_e$
meson decays		
$J/\Psi,\ \Upsilon \to a\gamma$	*	-
$K^+ \to \pi^+$ nothing $(a, \bar{\nu}\nu, \ldots)$	*	-
$\pi^+ \to \underbrace{e^+ e^-}_{a} e^+ \nu_e$	-	*
beam dump experiments		
$p(e^-)N \to a\ X$, $a \to \gamma\gamma, e^+e^-$	*	*
nuclear deexcitations		
$N^* \to N\ a$, $a \to \gamma\gamma,\ e^+e^-$	*	*

In table II we have listed the processes which have provided intersting constraints on axion couplings in the two cases $m_a < 2m_e$ and $m_a > 2m_e$. In the following we will discuss only the most definitive bounds which suffice to rule out axions whose $U(1)_{PQ}$ symmetry breaking scale is the Fermi scale.

(i) Quarkonium decays

The decay of heavy quarkonium into axion and photon (cf. fig. 5) has the clean signal of a monochromatic photon whose energy equals half the quarkonium mass. The branching ratio of this decay relative to the decay into μ-pairs is given by [20]

$$\frac{\Gamma((Q\bar{Q}) \to a\gamma)}{\Gamma((Q\bar{Q}) \to \mu^+\mu^-)} = \frac{G_F m_Q^2}{\sqrt{2}\pi e_Q^2 \alpha_{EM}} \cdot \begin{cases} x^2 \quad , & e_Q = \frac{2}{3} \\ \frac{1}{x^2} \quad , & e_Q = -\frac{1}{3} \; . \end{cases} \tag{5.1}$$

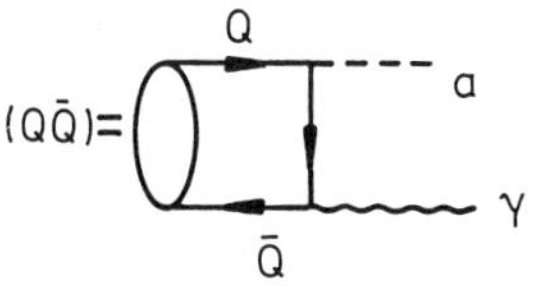

Fig. 5 Quarkonium decay into axion and photon

From the measured branching ratio into μ-pairs and eq. (5.1) one obtains the <u>theoretical</u> predictions

$$BR(J/\Psi \to a\gamma) = (4.9 \pm 0.8) \cdot 10^{-5}\, x^2 , \tag{5.2a}$$

$$BR(\Upsilon \to a\gamma) = (2.7 \pm 0.7) \cdot 10^{-4}\, \frac{1}{x^2} , \tag{5.2b}$$

which have to be compared with the <u>experimental</u> bounds [21]

$$BR(J/\Psi \to e\gamma) < 1.4 \cdot 10^{-5} \tag{5.3a}$$

$$BR(\Upsilon \to a\gamma) < 3 \cdot 10^{-4} \tag{5.3b}$$

(5.2) is a theoretically unambiguous prediction, and the comparison with (5.3) already rules out the standard Weinberg-Wilczek axion.

(ii) $K^+ \to \pi^+ a$

The theoretical prediction for the decay $K^+ \to \pi^+ a$ follows from the two pion decay of K^0 and the isosinglet axion mixing λ_s [16]:

$$BR(K^+ \to \pi^+ a) = 2\lambda_s^2 \left(\frac{f_\pi}{v}\right)^2 \frac{p_a}{p_\pi} BR(K^0 \to \pi^+\pi^-)$$

$$= 2.9 \cdot 10^{-5} \lambda_s^2 , \tag{5.4}$$

where p_a and p_π are axion and pion momenta in the final state. In the case $m_a < 2m_e$, where the axion escapes detection, the theoretical prediction (5.4) can be compared with the KEK limit of Asano et al. [22]:

$$BR(K^+ \to \pi^+ \text{ nothing}) < 2.7 \cdot 10^{-8} . \tag{5.5}$$

So far no bound exists for the decay chain $K^+ \to \pi^+ a(\to e^+e^-)$.

(iii) $\pi^+ \to e^+e^-e^+\nu_e$

This process provides a bound on the decay $\pi^+ \to ae^+\nu_e$ in the case $m_a > 2m_e$ where the axion can be observed through its decay into an electron-positron pair. One finds [16]

$$BR(\pi^+ \to ae^+\nu_e) = 3 \cdot 10^{-9} \lambda_3^2 . \tag{5.6}$$

At SIN Eichler et al. [23] have recently measured the branching ratio

$$BR(\pi^+ \to e^+e^-e^+\nu_e) = (3.4 \pm 0.5) \times 10^{-9}, \tag{5.7}$$

which agrees with standard model expectations and thereby provides the bound [23]

$$BR(\pi^+ \to ae^+\nu_e) < (1-2) \cdot 10^{-10} . \tag{5.8}$$

(iv) Nuclear deexcitation experiments

Interesting constraints on axion couplings also follow from nuclear transition $N^* \to N\, a$. The relative magnitude with respect to hadronic and electromagnetic transitions is given by the ratio of coupling constants:

Table III. Theoretical predictions and experimental bounds for $\Gamma(N^* \to N\, a)/\Gamma(N^* \to N\gamma)$

	Theory [16]	Experiment
$^{14}N(\Delta T = 1)$ Ref. [24]	$2 \cdot 10^{-5} \lambda_3^2$	$< 4 \cdot 10^{-4}$
$^{10}B(\Delta T = 0)$ Ref. [25]	$7.9 \cdot 10^{-4} \lambda_s^2$	$< 2.6 \cdot 10^{-3}$

$$\frac{g^2_{aNN}}{g^2_{\pi NN}} \sim (\frac{f_\pi}{v})^2 \sim 1.5 \cdot 10^{-7} , \qquad (5.9)$$

$$\frac{g^2_{aNN}}{e^2_{EM}} \sim 2.2 \cdot 10^{-4} . \qquad (5.10)$$

The ratio of the partial widths $\Gamma(N^* \to N\, a)/\Gamma(N^* \to N\gamma)$ can be rather reliably calculated. A comparison of theoretical predictions and experimental bounds is given in table III for two nuclear transitions.

The standard axion, where λ_s and λ_3 are O(1), is clearly ruled out from the K^+ and π^+ decays. Nuclear deexcitation experiments are important for variant axion models. Generally, bounds from beam dump experiments are not as stringent as the ones discussed in this section.

6. VARIANT AXION MODELS

Recent interest in axion models has partly been stimulated by heavy ion collision experiments at GSI where narrow peaks have been observed in e^+- and e^--energy spectra in coincidence. It is natural to consider the possibility that the observed e^+e^--pairs come from the decay of a new particle with a mass of 1.7 MeV given by the sum of electron and positron energies. Although the production mechanism of such a particle in heavy ion collisions has never been understood, it is an interesting question whether this hypothetical particle could be an axion.

Peccei, Wu and Yanagida [26] and Krauss and Wilczek [27] have indeed constructed "variant" axion model with an axion mass of 1.7 MeV. From

$$m_a = O(100\ \text{keV}) \cdot (x + \frac{1}{x}) \qquad (6.1)$$

one sees that either x or $\frac{1}{x}$ have to be rather large. The quarkonium bounds can be avoided by coupling c- and b-quark both to the same Higgs field, e.g. ϕ_1, and making $\frac{1}{x}$ small. From (6.1) one then finds $x \cong 70$ and hence a very short axion lifetime

$$\tau(a \to e^+e^-) \cong 6 \cdot 10^{-13}\ \text{see}. \qquad (6.2)$$

For such short-lived axions the decay $K^+ \to \pi^+$ nothing and beam dump experiments provide no constraints. One can not escape, however, the bounds from $\pi^+ \to e^+ e^- e^- \nu_e$ and nuclear deexcitation experiments. For the axion mixings with $\pi°$, η and η' one has two options [16]: $\lambda_s = 0$, $\lambda_3 \cong 26$ or $\lambda_s \cong -27$, $\lambda_3 = 0$. In general one finds the model independent relation [16] $(\lambda_3 - \lambda_s) \cong 25^2$. This is incompatible with the experimental bounds listed in the previous section. We conclude that axions are ruled out with a $U(1)_{PQ}$ symmetry breaking scale v_{PQ} of the order of Fermi scale $G_F^{-1/2}$.

7. SUPERSYMMETRY AND THE STRONG CP PROBLEM

Supersymmetric theories are intersting with respect to the strong CP problem for a number of reasons. First of all, the supersymmetric Higgs mechanism requires at least two Higgs doublet chiral superfields with $SU(2)_W \times U(1)_Y$ quantum numbers

$$H_1 \sim (2; \frac{1}{2}) , \quad H_2 \sim (2; -\frac{1}{2}) . \tag{7.1}$$

Hence an additional global $U(1)_{PQ}$ symmetry can occur already in the minimal supersymmetric standard model although, as we have seen in the last section, this symmetry has to be broken at a scale v_{PQ} larger than the Fermi scale $G_F^{-1/2}$. A special feature of supersymmetric theories is also that, due to non-renormalization theorems, the Θ-parameter at the Fermi scale is very small if, for some unknown reason, it vanishes at the unification or Planck mass [28].

The new particles introduced by supersymmetry can also contribute directly to the neutron electric dipole moment. The virtual exchange of scalar quarks and gluons yields, for $\mu_{\tilde{g}} \sim M_D \sim 100$ GeV and maximal phases [29] (cf. Fig. 6):

$$d_n \sim 4 \cdot 10^{-22} \text{ e cm}, \tag{7.2}$$

which exceeds the experimental bound (2.18) by three orders of magnitude! Fortunately, in a large class of supergravity models, the phases, and hence d_n, vanish at tree level.

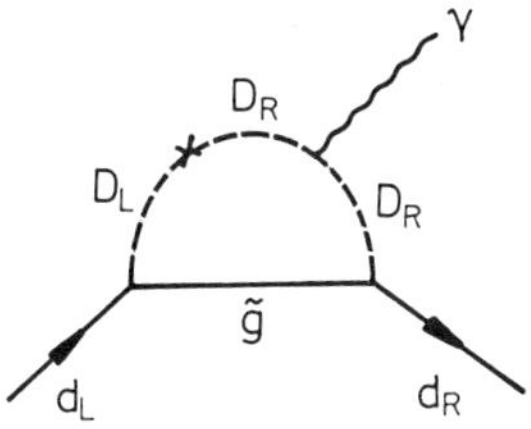

Fig. 6 Contribution to the neutron electric dipole moment from virutal gluinos and scalar quarks.

A particularly interesting aspect of supersymmetric theories is the chiral $U(1)_R$ invariance whose charge does not commute with the supersymmetry generator,

$$[R, Q_2] \neq 0 , \tag{7.3}$$

and which occurs in many supersymmetric models. Approximate R-invariance suppresses potentially large, radiatively induced operators which cause the transition $\mu \to e\gamma$ and B- and L- nonconservation. Hence one may expect [29], that in the effective low energy lagrangian R-invariance is only broken by the soft supersymmetry breaking terms:

$$\partial^{\mu} J^R_{\mu} = O(m_{3/2}) , \tag{7.4}$$

where $m_{3/2}$ is the gravitino mass. Because of the gluinos the R-current has a colour anomaly (cf. Fig. 7) and could hence serve as a PQ symmetry. Eq. (7.4) would then suggest that the axion is the pseudo-Goldstone boson of spontaneously broken R-invariance whose symmetry breaking scale is identical with the supersymmetry breaking scale $v_{SUSY} \sim (m_W m_{PL})^{1/2} \sim 10^{10}$ GeV.

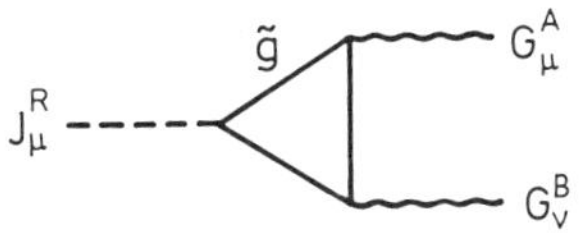

Fig. 7 Anomaly of the R-current

8. HOW TO FIND INVISIBLE AXIONS

Despite the so far fruitless experimental search for axions the Peccei-Quinn mechanism remains the most attractive solution of the strong CP problem. The experimental bounds discussed in section 5 then require that the $U(1)_{PQ}$ symmetry breaking scale is substantially larger than the Fermi-scale:

$$v_{PQ} >> G_F^{-1/2} \quad . \tag{8.1}$$

This means that the axion becomes (almost) massless ($m_a \sim \frac{1}{v_{PQ}}$), and that it (almost) decouples from ordinary matter ($g_a \sim \frac{1}{v_{PQ}}$). One may consider such "invisible" axions [30] to be an absurdity [31], but one may also regard them as a fascinating possibility, especially because recently various methods have been suggested by means of which invisible axions could become visible. In this section we will discuss these proposals. For recent theoretical work on invisible axions we refer the reader to the papers of ref. [32]. The interesting possibility of having truly massless Goldstone bosons ("arions"),with properties similar to those of axions, has been discussed by Anselm [33].

Constraints on the symmetry breaking scale v_{PQ} of invisible axions can be derived from astrophysical and cosmological considerations. A lower bound on v_{PQ} is obtained from the upper bound on the admisable energy loss of stars through axion emission, for instance via the Compton-type process $e\gamma \rightarrow ea$ (cf. Fig. 8). The recent analysis of Raffelt [34] yields

$$v_{PQ} > 10^7 \text{ GeV} \quad . \tag{8.2}$$

Fig. 8 Compton-type process for axion production

Less reliable is an upper bound inferred from cosmological arguments [35]

$$v_{PQ} < 10^{12} \text{ GeV}.$$

In the following we will normalize mass and Compton wave length of invisible axions to this scale:

$$m_a \cong 10^{-5} \text{ eV } \left(\frac{10^{12} \text{ GeV}}{v_{PQ}}\right), \tag{8.3a}$$

$$\nu_a = \frac{m_a}{h} \cong 2.4 \text{ GHz}, \tag{8.3b}$$

$$\lambda\!\!\!^{-}_a = \frac{\hbar}{m_a} \cong 2 \text{ cm}. \tag{8.3c}$$

If one accepts only the lower, astrophysical bound on v_{PQ}, one obtains from 10^7 GeV $< v_{PQ} < M_{PL}$ as possible range of axion masses

$$10^{-12} \text{ eV} < m_a < 1 \text{ eV} . \tag{8.4}$$

(i) Axion halo

It is conceivable that the dark halos of galaxies are made out of axions [35]. A method to search for axions of the Milky Way halo has been proposed by Sikivie [36] and further studied by Krauss et al. [37]. These axions, which have a number density [35]

$$\rho_a \sim \frac{10^{-24} \text{gr}}{m_a \ \text{cm}^3} \sim \frac{0.5 \text{ GeV}}{m_a \ \text{cm}^3} \sim \frac{5 \cdot 10^{13}}{\text{cm}^3} \left(\frac{v_{PQ}}{10^{12} \text{ GeV}}\right), \tag{8.5}$$

can be described by a classical field

$$a(t) \cong a_o \, e^{i\omega_o t} \tag{8.6}$$

with frequency $\omega_o \sim m_a$. The basic idea is to convert these axions via their electromagnetic coupling (cf. eq. (4.22)):

$$L = -\kappa \, \vec{E} \cdot \vec{B} \, a, \quad \kappa = \frac{\xi \alpha}{v_{PQ}} , \tag{8.7}$$

in an external magnetic field $\vec{B}_o$ into photons (cf. Fig. 9). The corresponding electric field is given by

$$\vec{\nabla}^2 \vec{E} - \frac{\partial^2}{\partial t^2} \vec{E} = \kappa \, \vec{B}_o \frac{\partial^2 a}{\partial t^2} , \tag{8.8}$$

and the generated electromagnetic energy is detected by means of a cavity with volume V, quality factor Q and modes $\vec{e}_j(\vec{x})$,

$$\frac{1}{V} \int_V d^3x\, \vec{e}_i(\vec{x})\, \vec{e}_j(\vec{x}) = \delta_{ij} \,. \tag{8.9}$$

The energy converted from axions of the halo into electromagnetic energy of mode i is [37]

$$U_i = \kappa^2 V \eta_i^2 \int \frac{d\omega}{2\pi} \frac{|a(\omega)|^2 \omega^4}{(\omega^2 - \omega_i^2)^2 + \frac{\omega^4}{Q^2}} \,, \tag{8.10}$$

where $a(\omega)$ is the Fourier transform of $a(t)$ and η_i the overlap of the external magnetic field with mode i:

$$\eta_i = \frac{1}{V} \int_V d^3x\, \vec{B}_o\, \vec{e}_i(\vec{x}) \,. \tag{8.11}$$

The frequency distribution $a(\omega)$ is expected to be very narrow [37],

$$Q_a = \frac{m_a}{\Delta\omega} \sim 9 \cdot 10^6 \,, \tag{8.12}$$

and in the zero-width approximation (8.6) one finds for $\omega_i = m_a$:

$$U_i = \kappa^2 V \eta_i^2 Q^2 \frac{\rho_a}{m_a} \,. \tag{8.13}$$

This corresponds to a detectable power $P \sim \frac{U_j \omega_o}{Q}$ $(\kappa \sim \frac{3\alpha}{v_{PQ}},\ \eta_i \sim B_o,\ Q \sim 10^6)$:

$$P \sim 8 \cdot 10^{-18}\ \text{Watt} \left(\frac{V}{5 \cdot 10^4\ \text{cm}^3}\right) \left(\frac{B_o}{8\ \text{Tesla}}\right)^2 \left(\frac{10^{12}\ \text{GeV}}{v_{PQ}}\right) \,. \tag{8.14}$$

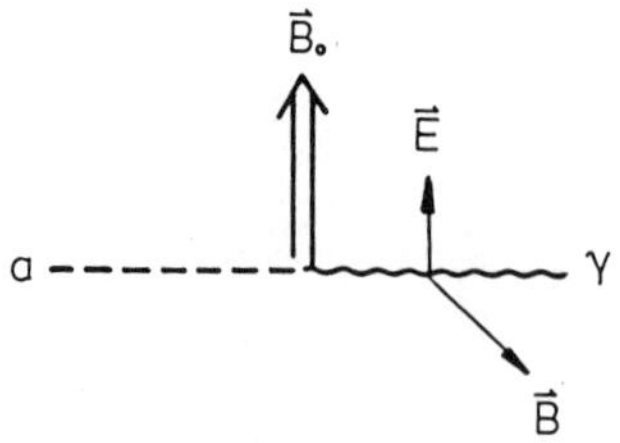

Fig. 9 Conversion of axions into photons in an external magnetic field

It appears that with a cavity of $5 \cdot 10^4$ cm^3, a magnetic field of 8 Tesla and a state of the art microwave detector it is possible to sean within two years a fequency interval corresponding to a range of the symmetry breaking scale [35]

$$10^{11} \text{ GeV} < v_{PQ} < 3 \cdot 10^{12} \text{ GeV} , \tag{8.15}$$

which lies inside the range allowed by astrophysics and cosmology.

(ii) Solar axions

If invisible axions exist they will be emitted from the sun with a rather broad energy spectrum peaking at about 1 keV. The axion flux expected for $\frac{v_{PQ}}{z} = 10^7$ GeV (cf. eq. (4.12)) is shown in Fig. 10. Similar to the photoelectric effect these axions have a rather large ionization cross section for certain elements ("axioelectric effect") which can be used to detect solar axions [38]. Recent results obtained with a germanium spectrometer with energy threshold of only 4 keV are shown in Fig. 11 together with theoretical expectations. Avignone et al. obtain from this figure the remarkable lower bound

$$\frac{v_{PQ}}{z} > 0.5 \cdot 10^7 \text{ GeV} , \tag{8.16}$$

which is almost as large as the astrophysical bound (8.2).

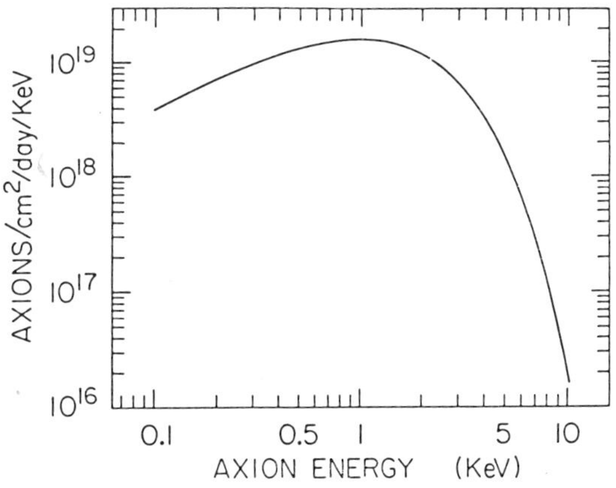

Fig. 10 Solar axion flux on earth for solar temperature T = 1 keV and bremsstrahlung production with $\frac{v_{PQ}}{z} = 10^7$ GeV. From ref. [38]

Axion Rate for Germanium

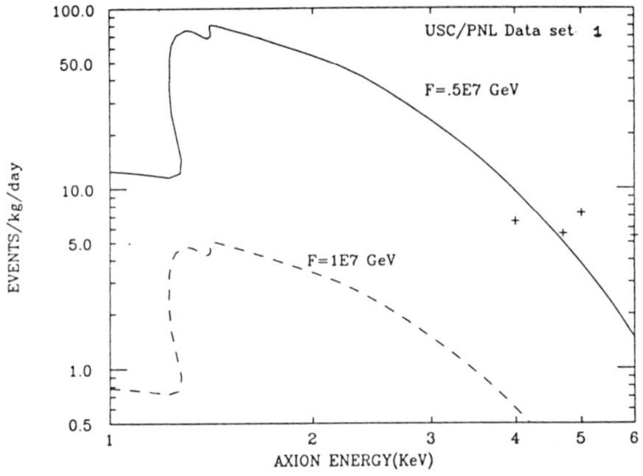

Fig. 11 Theoretical prediction for solar axion events per kg and day for Germanium with $\frac{v_{PQ}}{z} = 0.5 \cdot 10^7$ GeV (solid line) and $\frac{v_{PQ}}{z} = 10^7$ GeV (dashed line). The crosses denote recent experimental results of background count rates with $\omega \geqq 4$ keV. From ref. [38]

(iii) Long-range forces

Axions may also have CP violating scalar couplings to fermions which will then lead to additional nucleon-nucleon, nucleon-electron and electron-electron forces with range $\lambda_a = \frac{\hbar}{m_a}$. Moody and Wilczek have analyzed these "new macroscopic forces" [39], especially the monopole-dipole type nucleon-electron interaction. For 10^9 GeV < v_{PQ} < 10^{12} GeV all these forces are weaker than gravity with a range 0.002 cm < λ_a < 2 cm. Such deviations from gravity are the subject of the lectures given by John Bell at his school [40].

(iv) Laser experiments

An interesting method to search for invisible axions has recently been suggested by Maiani, Petronzio and Zavattini [41]. Due to the coupling (8.7) of axions to two photons the propagation of light in an external magnetic field is modified. One easily derives the coupled classical equations of motion

$$\Box \vec{E} - \kappa \frac{\partial^2 a}{\partial t^2} \vec{B} = 0 , \qquad (8.17a)$$

$$(\Box + m_a^2) a + \kappa \vec{E}\cdot\vec{B} = 0 , \tag{8.17b}$$

where, to first approximation, $\vec{B}$ can be replaced by the strong external field $\vec{B}_o$. Clearly, only the component of $\vec{E}$ parallel to $\vec{B}_o$ is modified through the interaction with the axion field whereas the orthogonal component of $\vec{E}$ propagates undisturbed (cf. Fig. 12).

Eqs. (8.17) admit plane wave solutions of the form ($\vec{B}_o = B_o \vec{e}_{\parallel}$, $\vec{E} = E_{\parallel} \vec{e}_{\parallel} + E_{\perp} \vec{e}_{\perp}$):

$$E_{\perp} = E_o \sin\alpha \, e^{-i(\omega t - \vec{k}\vec{x})} , \quad \omega^2 = \vec{k}^2 , \tag{8.18a}$$

$$E_{\parallel} = E_+ \, e^{-i(\omega_+ t - \vec{k}\vec{x})} + E_- \, e^{-i(\omega_- t - \vec{k}\vec{x})} , \tag{8.18b}$$

$$a = a_+ \, e^{-i(\omega_+ t - \vec{k}\vec{x})} + a_- \, e^{-i(\omega_- t - \vec{k}\vec{x})} , \tag{8.18c}$$

where the shifted frequencies $\omega_{\pm}$ are determined by

$$(\omega_{\pm}^2 - \omega^2 - \omega_m^2)(\omega_{\pm}^2 - \omega^2) - \omega_o^2 \, \omega_{\pm}^2 = 0 , \tag{8.19}$$

$$\omega_m = m_a , \quad \omega_o = \kappa B_o .$$

With initial conditions (at $\vec{x} = o$)

$$E_{\parallel}(0) = E_o \cos\alpha , \tag{8.19a}$$

$$a(0) = 0 , \tag{8.19b}$$

which correspond to linear polarization, one finds after propagation of length $L = ct$ $(|\omega_{\pm} - \omega| \ll \frac{2\pi}{L})$ [41]:

$$E_{\perp}(t) = E_o \sin\alpha \, e^{-i\omega t} , \tag{8.20a}$$

$$E_{\parallel}(\ ) = E_o \cos\alpha \, (1-\varepsilon) e^{-i(\omega t - \delta)} , \tag{8.20b}$$

where

$$\varepsilon(L) = \frac{1}{8} \kappa^2 \, B_o^2 \, L^2 , \tag{8.20c}$$

$$\delta(L) = \frac{\kappa^2 B_o^2 m_a^2 \, L^3}{48\omega} = 2 e . \tag{8.20d}$$

Eqs. (8.20) describe elliptically polarized light with ellipticity e. The major axis of the ellipse is rotated by $\delta\alpha = \frac{1}{2}\,\varepsilon(L)\,\sin2\alpha$ with respect to the axis of linear polarization.

Experimentally feasable values for the parameters B, L and ω are [41]

$$B_o = 10 \text{ Tesla}$$

$$L = 10^5 \text{ cm}$$

$$\omega = 2.4 \text{ eV} . \tag{8.21}$$

Furthermore measurements of $\delta(L)$ as small as

$$\delta(L) \sim 10^{-12} \text{ rad} \tag{8.22}$$

appear possible [41]. This has to be compared with the effect of light-light scattering induced by QED vacuum polarization as described by the Euler-Heisenberg effective lagrangian [41]:

$$\delta_{QED}(L) = \frac{\alpha^2}{30\pi}\,\frac{L\omega B_o^2}{m_e^4} \cong 5 \cdot 10^{-12} , \tag{8.25}$$

which is of the order of the experimental sensitivity of 10^{-12} rad.

For the angle $\delta(L)$ due to the interaction with the axion field one finds from eqs. (8.20d), (8.21) ($\kappa = \frac{\xi\alpha}{v_{PQ}}$) :

$$\delta(L) = 2 \cdot 10^{-21} \text{ rad } \left(\frac{\xi}{3}\right)^2 \left(\frac{10^{12}\text{ GeV}}{v_{PQ}}\right)^2 \left(\frac{m_a}{10^{-5}\text{ eV}}\right)^2 \left(\frac{2.4\text{ eV}}{\omega}\right) . \tag{8.26}$$

This result appears hopelessly tiny. We note, however, that the parameter ξ may be larger than 3 by an order of magnitude, and that the relation (8.3a) between axion mass and symmetry breaking scale is also model dependent. The experimentally accessible range in the m_a-v_{PQ} plane has been discussed more generally in ref. [41]. It is clear that laser experiments can probe a region which so far has not been explored in other laboratory experiments.

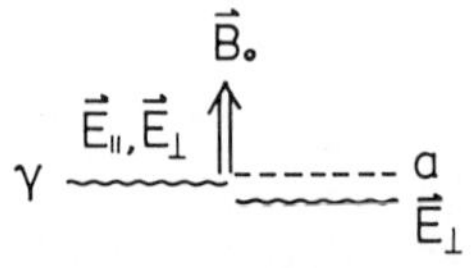

Fig. 12 Conversion of photons with parallel polarization into axions

9. CONCLUSIONS

In these lectures we have discussed some aspects of the strong CP problem. The solution of the U(1) problem in QCD through gauge field configurations with nontrivial topology and the associated Θ-vacuum gives rise to CP violation in strong interactions whose strength is governed by the new parameter Θ. This leads to a nonvanishing electric dipole moment d_n of the neutron proportional to $\overline{\Theta}$, the sum of Θ and the phase of the determinant of the quark mass matrix. The experimental upper bound on d_n implies an upper bound on $\overline{\Theta}$, $\overline{\Theta} < 10^{-9}$.

Within the Glashow-Weinberg-Salam theory it appears impossible to understand the smallness of $\overline{\Theta}$, especially the cancellation between the intrinsic QCD part Θ and the electroweak contribution from the quark mass matrix. The meaning of the parameter Θ in the standard model is tied to the absence of a global chiral symmetry with $SU(3)_c$ anomaly which would allow to "rotate Θ away". Hence the Peccei-Quinn mechanism, which rests on a spontaneously broken chiral $U(1)_{PQ}$ invariance in a minimal extension of the standard model, appears to be the natural solution of the strong CP problem.

The characteristic feature of the Peccei-Quinn mechanism is the appearance of a pseudo-Goldstone boson, the axion. However, a remarkable experimental effort has excluded the possibility that the $U(1)_{PQ}$ symmetry breaking scale v_{PQ} coincides with the scale $G_F^{-1/2}$ of electroweak symmetry breaking. Astrophysical and cosmological arguments limit the allowed range for the $U(1)_{PQ}$ symmetry breaking scale to 10^8 GeV $< v_{PQ} < 10^{12}$ GeV, corresponding to "invisible" axions of macroscopic Compton wavelength, which interact only very weakly with ordinary matter. It is intriguing that this mass range contains the preferred scale of supersymmetry breaking.

It is very remarkable that experiments are conceivable which could make such invisible axions visible. We have discussed Sikivie's axion haloscope, the search for axions from the sun, axion induced deviations from gravitational forces and laser experiments. At present the search for solar axions appears most promising, and a first experiment has already obtained a lower bound on v_{PQ} almost as large as the astrophysical bound. The other experiments are also interesting, even if they are sensitive only to smaller mass scales. For instance, the advantage of the laser experiment is that no outside source of axions is needed, because an axion beam is produced in the experiment!

The search for new interaction scales beyond the standard model of strong and electroweak interactions is an outstanding challenge in particle physics. It is likely that such new interaction scales give rise to very light, scalar or pseudoscalar particles which interact very weakly with ordinary matter. Such particles could be discovered in low energy precision experiments.

Acknowledgements

I would like to thank R.D. Peccei for helpful discussions during the preparation of these lectures, and the organizers of the school for providing a splendid atmosphere at Erice.

REFERENCES

[1] For recent reviews and references, see:
G. 't Hooft, Phys. Rep. 142 (1986) 357;
R.D. Peccei, Schladming Lectures 1986, preprint DESY 86-069 (1986)

[2] S.L. Adler, Phys. Rev. 177 (1969) 2426;
J.S. Bell and R. Jackiw, Nuovo Cim. 60A (1969) 47

[3] A.A. Belavin, A.M. Polyakov, A.S. Schwartz and Yu.S. Tyupkin, Phys. Lett. 59B (1975) 85

[4] G. 't Hooft, Phys. Rev. Lett. 37 (1976) 8; Phys. Rev. D14 (1976)3432;
R. Jackiw and C. Rebbi, Phys. Rev. Lett. 37 (1976) 172;
C.G. Callan, R.F. Dashen and D.J. Gross, Phys. Lett. 63B (1976) 334

[5] M. Göckeler, A.S. Kronfeld, M.L. Laursen, G. Schierholz and U.-J. Wiese, preprint DESY 86-107 (1986)

[6] R.D. Peccei and H.R. Quinn, Phys. Rev. Lett. 38 (1977) 1440;
Phys. Rev. D16 (1977) 1791

[7] S. Weinberg, Phys. Rev. Lett. 40 (1978) 223;
F. Wilczek, Phys. Rev. Lett. 40 (1978) 279

[8] For a recent review, see R.D. Peccei, in Proc. of the XXIII Int. Conf. on High Energy Physics, Berkeley, 1986

[9] S. Weinberg, Phys. Rev. D11 (1975) 3583

[10] E. Witten, Nucl. Phys. B156 (1979) 269

[11] G. Veneziano, Nucl. Phys. B159 (1979) 213

[12] V. Baluni, Phys. Rev. D19 (1979) 2227

[13] R. Crewther, P. di Vecchia, G. Veneziano and E. Witten, Phys. Lett. 88B (1979) 123; (E) 91B (1980) 487

[14] I.S. Altarev et al., Phys. Lett. 102B (1981) 13

[15] W. Buchmüller and D. Wyler, Nucl. Phys. B268 (1986) 621

[16] W.A. Bardeen, R.D. Peccei and T. Yanagida, Nucl. Phys. B279 (1987) 401

[17] W.A. Bardeen and S.-H.H. Tye, Phys. Lett. 74B (1978) 229

[18] A. Zehnder, in Proc. of the 1982 Gif-sur-Yvette summer school

[19] M. Davier, in Proc. of the XXIII Int. Conf. on High Energy Physics, Berkeley, 1986

[20] F. Wilczek, Phys. Rev. Lett 39 (1977) 1304

[21] For a review. see S. Yamada in Proc. of the Lepton Photon Symposium, Cornell University, eds. D.G. Cassel and D.L. Kreinick (1983)

[22] Y. Asano et al., Phys. Lett. 107B (1981) 159

[23] E. Eichler et al., Phys. Lett. 175B (1986) 101

[24] M.J. Savage et al., Phys. Rev. Lett. 57 (1986) 178

[25] A.L. Hallin et al., Phys. Rev. Lett. 57 (1986) 2105

[26] R.D. Peccei, T.T. Wu and T. Yanagida, Phys. Lett. 172B (1986) 435

[27] L.M. Krauss and F. Wilczek, Phys. Lett. 173B (1986) 189

[28] J. Ellis, S. Ferrara and D.V. Nanopoulos, Phys. Lett. 114B (1982) 231

[29] W. Buchmüller and D. Wyler, Phys. Lett. 121B (1983) 321

[30] J.E. Kim, Phys. Rev. Lett. 43 (1979) 103;
M.A. Shifman, A.I. Vainshtein and V.I. Zakharov, Nucl. Phys. B166 (1980) 493;
M. Dine, W. Fischler and M. Srednicki, Phys. Lett. 104B (1981) 199

[31] J. van der Bij and M. Veltman, Nucl. Phys. B231 (1984) 205

[32] D.B. Kaplan, Nucl. Phys. B260 (1985) 215;
M. Srednicki, Nucl. Phys. B260 (1985) 689;
H. Georgi, D.B. Kaplan and L. Randall, Phys. Lett. 169B (1986) 73

[33] A.A. Anselm, preprint CERN-TH 4349 (1986)

[34] G.G. Raffelt, Phys. Rev. D33 (1986) 897

[35] For a review and references, see: P. Sikivie, in Proceedings of the "Inner Space-Outer Space" Conference on Physics at the Interface of Astrophysics/Cosmology and Particle Physics, Batavia, 1984

[36] P. Sikivie, Phys. Rev. Lett. 51 (1983) 1415

[37] L. Krauss et al., Phys. Rev. Lett 55 (1985) 1797

[38] F.T. Avignone III et al., preprint SLAC-PUB-3872 (1986)

[39] J.E. Moody and F. Wilczek, Phys. Rev. D30 (1984) 130

[40] J.S. Bell, these proceedings

[41] L. Maiani, R. Petronzio and E. Zavattini, Phys. Lett. 175 (1986) 359

THE MUON ANOMALOUS g-VALUE*

Vernon W. Hughes

Gibbs Laboratory
Physics Department
Yale University
New Haven, Ct. 06520

I. INTRODUCTION

The muon g-2 value has played a central role in establishing that the muon obeys quantum electrodynamics (QED) and behaves like a heavy electron.[1,2] The experimental value for g-2 has been determined by three progressively more precise measurements at CERN,[3,4,5] the latest one[5] achieving a precision of 7.2 ppm. The theoretical value for g-2 has steadily become better known as higher order QED radiative contributions have been evaluated, and as knowledge of the virtual hadronic contributions to g-2 has been improved both by further measurements of the relevant quantity $R = \sigma(e^+e^- \to \text{hadrons})/\ \sigma(e^+e^- \to \mu^+\mu^-)$ and by calculations.[6] The theoretical value of g-2 is now known to 1.3 ppm, a factor of 6 better than the experimental value.[7] The present agreement of theory and experiment establishes that QED applies for the muon up to $Q^2 = 1000\ (\text{GeV}/c)^2$ and determines the hadronic contribution to the vacuum polarization to about 12%. Futhermore, one of the most sensitive limits on muon substructure ($\Lambda > 800$ GeV) is provided, as well as limits to various speculative modern theories.

The next stage in research on muon g-2, a measurement to 0.35 ppm at the AGS, is approved at Brookhaven and is now under development[7]. It will determine the contribution to g-2 of the newly discovered vector bosons W and Z, which mediate the weak interactions. This virtual radiative contribution is based on the renormalizability of the theory and is equivalent in the electroweak theory to the Lamb shift in QED.

Historically, the Lamb shift and the electron and muon g-2 values played a key role in the discovery and establishment of modern quantum electrodynamics. Precision measurements of hyperfine structure intervals and the Zeeman effect in hydrogen, muonium and positronium have provided important tests of QED bound state theory as well as values for the magnetic moments of fundamental particles. High energy experiments can explore the physics of W, Z, and other particles by direct production and are certainly the definitive way to identify new particles and new physics which are energetically accessible. On the other hand it is difficult in these experiments to study the properties of more massive particles that cannot be produced by the accelerators. In principle the existence of heavier particles can be detected through their effects on the behavior of lighter observed particles. Precise, sensitive experiments, such as the muon g-2 measurement, will give us a useful insight into the domain of physics which is not now accessible to high-energy experiments. In this sense low-energy, high-precision experiments play a role complementary to that of high energy experiments.

A high precision measurement of a fundamental quantity such as muon g-2 for which a precise value can be calculated from basic theory provides an important calibration point for modern particle theory. Not only is this valuable for the insight it provides about the very high energy regime beyond present accelerators, but also it may reveal new and deeper aspects of physics within the accessible energy regime, as was so dramatically illustrated for quantum electrodynamics by the Lamb shift in hydrogen and the anomalous g-value of the electron.

II. THEORY OF MUON g-2

The g-2 value for the electron has been measured to an accuracy of 4 parts per billion in an experiment in which a single electron is stored in a Penning trap at low temperature.[8] The experimental value is

$$a_e(\text{expt}) = 1\ 159\ 652\ 193\ (4) \times 10^{-12}\ (3.4\ \text{ppb}). \tag{1}$$

The theoretical value has been computed[9] through terms of order α^4:

$$a_e(\text{theor}) = 0.5\left(\frac{\alpha}{\pi}\right) - 0.328\ 478\ 966\left(\frac{\alpha}{\pi}\right)^2 + 1/1765\ (13)\left(\frac{\alpha}{\pi}\right)^3 - 0.8\ (2.5)\left(\frac{\alpha}{\pi}\right)^4. \tag{2}$$

Using the condensed matter value of α based on the ac-Josephson effect and the quantized Hall effect,[10]

$$\alpha^{-1} = 137.035\ 981\ 5\ (123)\ (0.090\ \text{ppm}), \tag{3}$$

we find

$$a_e(\text{theor}) = 1\ 159\ 652\ 263\ (113) \times 10^{-12}\ (0.092\ \text{ppm}) \tag{4}$$

in good agreement with a_e(expt). Alternatively, comparison of a_e(expt) and a_e(theor) yields the most precise value for the fine structure constant α:

$$\alpha^{-1} = 137.035\ 989(8)\ (27)\ (0.020\ \text{ppm}) \tag{5}$$

Only the electron and photon contribute significantly to a_e(theor); the contributions of virtual μ and τ leptons as well as virtual hadrons and $W^{\pm}$, Z particles are less than the present 4 ppb experimental error.

The theoretical value for the anomalous g-value of the muon, $a_\mu = (g_\mu - 2)/2$, in the standard theory[6,7] and the present experimental value[5] are given in Table 1. Here a_μ(QED) arises from the virtual radiative contributions of QED involving photons and charged leptons, a_μ(had) arises from virtual hadron contributions, and a_μ(weak) arises from the virtual radiative contributions involving the W, Z and Higgs particles. In contrast to the electron anomaly, which is dominated by the QED effect, the muon g-2 is much more sensitive to physics at smaller distances because of the larger mass scale.

The weak contribution[11] arises from single loop Feynman diagrams involving W,Z and ϕ as shown in Fig. 1 and is analogous to Lamb shift or g_e-2 radiative corrections in QED. The uncertainty arises from the quoted error in $\sin^2\theta_W$ and from the expected size of uncalculated higher order weak contributions. An essential feature of this calculation is the renormalizability of the electroweak theory, which has not yet been

TABLE 1. Theoretical and Experimental Values for a_μ.

$a_\mu(\text{theor}) = a_\mu(\text{QED}) + a_\mu(\text{had}) + a_\mu(\text{weak})$

$$a_\mu(\text{QED}) = C_1(\tfrac{\alpha}{\pi}) + C_2(\tfrac{\alpha}{\pi})^2 + C_3(\tfrac{\alpha}{\pi})^3 + C_4(\tfrac{\alpha}{\pi})^4 + a_\mu(\tau)$$

$C_1 = 0.5$; $C_2 = 0.765\ 857\ 577$;

$C_3 = 24.072\ 5\ (123)$; $C_4 = 137.96\ (250)$

$a_\mu(\tau) = 420 \times 10^{-12}$

$\alpha^{-1} = 137.035\ 989\ 5\ (61)\ (0.045\ \text{ppm})$ (1986 value)[28]

$a_\mu(\text{QED}) = 1\ 165\ 848\ 124\ (152)\ (52) \times 10^{-12}\ (0.14\ \text{ppm})$
↑ calculation ↑ α

$a_\mu(\text{had}) = 6940\ (142) \times 10^{-11}$, or (59.5 ± 1.2) ppm in $a_\mu(\text{theor})$

$a_\mu(\text{weak}) = 195\ (1) \times 10^{-11}$, or (1.7 ± 0.01) ppm in $a_\mu(\text{theor})$

$a_\mu(\text{theor}) = 116\ 591\ 947\ (143) \times 10^{-11}\ (1.3\ \text{ppm})$

$a_\mu(\text{expt}) = 11\ 659\ 230\ (84) \times 10^{-10}\ (7.2\ \text{ppm})$

$a_\mu(\text{expt}) - a_\mu(\text{theor}) = 35\ (85) \times 10^{-10}$

ELECTRON G-2 VALUE

$a_e(\text{expt}) = 1\ 159\ 652\ 193(4) \times 10^{-12}$

$a_e(\text{theor}) = 1\ 159\ 652\ 263(22)\ (104) \times 10^{-12}$
↑ calculation ↑ α

Using $\alpha^{-1} = 137.035\ 989\ 5(123)\ (0.090\ \text{ppm})$[10]

$a_e(\text{expt}) - a_e(\text{theor}) = -70(106) \times 10^{-12}$

$\alpha^{-1}\ (g-2) = 137.035\ 989\ 8\ (27)(0.020\ \text{ppm})$[9]

FUNDAMENTAL CONSTANT

$\mu_\mu/\mu_p = 3.183\ 345\ 47(47)(0.15\ \text{ppm})$ (1986 value)[28]

---

adequately tested by experiment. Radiative contributions are also tested from precise values of W and Z particle masses, but the present accuracies in mass measurements and also the projected accuracies from SLC, LEP and the $\bar{p}p$ colliders will not be sufficient to test single loop diagrams involving W and Z bosons.[7,12] Renormalization of a field theory is in general required to obtain finite predictions for the contributions of loop diagrams, which involve virtual particles of all momenta, and it is only through the study of loop diagrams or virtual radiative corrections that a test of the renormalization procedure of a theory can be made. No effects from loop diagrams involving virtual W and Z have yet been observed.

The hadronic vacuum polarization contribution is substantial and its uncertainty is the dominant error in $a_\mu(\text{theor})$. The strong interactions contribute to a_μ principally through the vacuum polarization diagram,

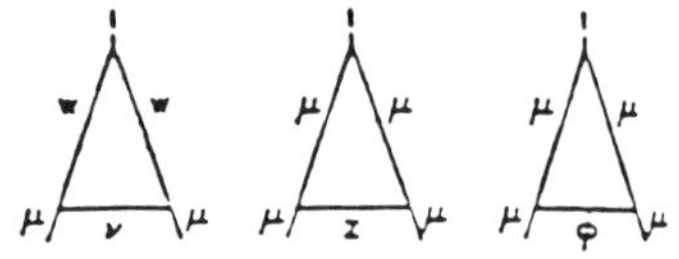

SINGLE LOOP DIAGRAMS CONTRIBUTING TO THE MUON g-FACTOR

$$\Delta a_\mu (W) = \frac{G_F \, m_\mu^2}{8\pi^2\sqrt{2}} \times \frac{10}{3} = +3.89 \times 10^{-9}$$

$$\Delta a_\mu (Z) = \frac{G_F \, m_\mu^2}{8\pi^2\sqrt{2}} \times \frac{1}{3}\left[(3-4\cos^2\theta_W)^2 - 5\right] = -1.94 \times 10^{-9}$$

$$\Delta a_\mu (\phi) = \frac{G_F \, m_\mu^2}{2\pi^2\sqrt{2}} \int_0^1 \frac{y^2\,(2-y)\,dy}{y^2 + (1-y)(m_\phi/m_\mu)^2}$$

$$= \frac{G_F \, m_\mu^2}{2\pi^2\sqrt{2}} \left(\frac{m_\mu}{m_\phi}\right)^2 \ln\left[\left(\frac{m_\phi}{m_\mu}\right)^2\right] \leq 0.01 \times 10^{-9}, \quad \text{IF } m_\phi \gg m_\mu$$

$$= \frac{3G_F \, m_\mu^2}{4\pi^2\sqrt{2}} \qquad m_\phi \ll m_\mu$$

$$\sin^2\theta_W = 0.226 \pm 0.004; \quad m_\phi > 7 \text{ GeV}$$

$$a_\mu(\text{WEAK}) = (1.95 \pm 0.01) \times 10^{-9}$$

Fig. 1. Weak interaction contribution to a_μ.

and this contribution can be expressed as shown in Fig. 2a. Since it is not yet possible to compute R(s) from QCD theory, measured values of R(s) from e^+e^- colliders are used to evaluate a_μ(had). A careful tabulation of measured values of R(s) and of their contributions to a_μ(had) has been made[13] and is summarized in the graph of Fig. 2b. Note that the principal contribution and error come from the energy region $\sqrt{s} \lesssim 1.2$ GeV and are associated with the ρ and ω resonances. Higher order hadronic contributions have been evaluated and found to be small.[13] Other evaluations of a_μ(had) have also been done recently.[14,15] The quantity a_μ(had) is now known a factor of about 6 better than when the CERN experiment was reported.[5] However, it will be necessary to determine a_μ(had) another factor of 3.5 better, that is to 0.6%, in order to utilize fully for deriving new physical information the projected accuracy of 0.35 ppm in a new measurement of a_μ. It now seems possible to obtain the required data for $\sqrt{s} \lesssim 1$ GeV by two independent approaches as indicated in Fig. 3. (1) With the e^+e^- collider at Novosibirsk,[14,16] and (2) with fixed target experiments at CERN[17] and at FNAL.

The primary purpose of the AGS experiment is to see whether the measurement of g-2 confirms the present theoretical picture of the standard model including the weak contribution. It is not at all a foregone conclusion, however, that the measurement simply confirms the electroweak effect, (Fig. 1), as described by the GSW theory (or any GUT theories which keep the electroweak sector intact). These may be nothing but the "low-energy limit" of a more fundamental theory which gives a completely different prediction for electroweak radiative contributions.

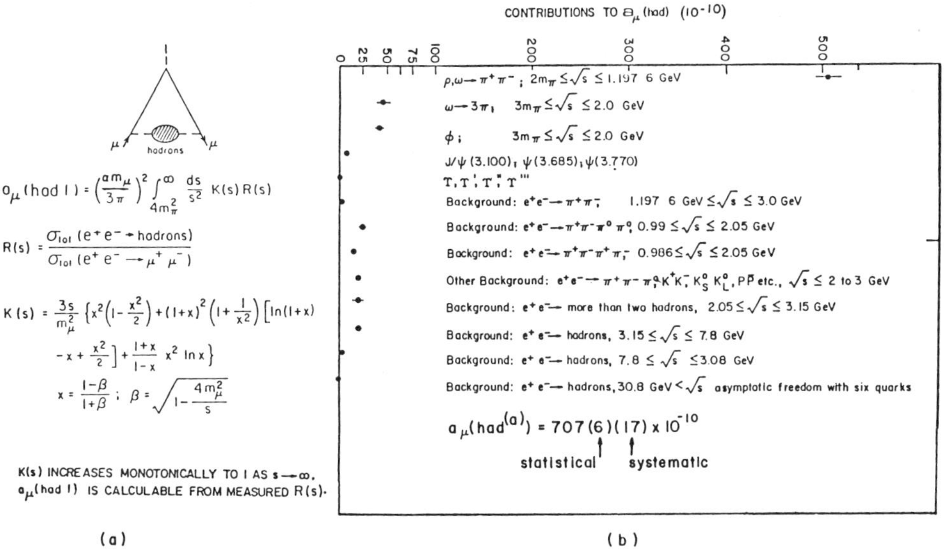

$a_\mu(\text{had}) = 703(19) \times 10^{-10}$ (60.3 ± 1.6) ppm

Fig. 2(a). Principal hadronic contribution to a_μ.

(b). Tabulation of specific hadronic contributions to a_μ.

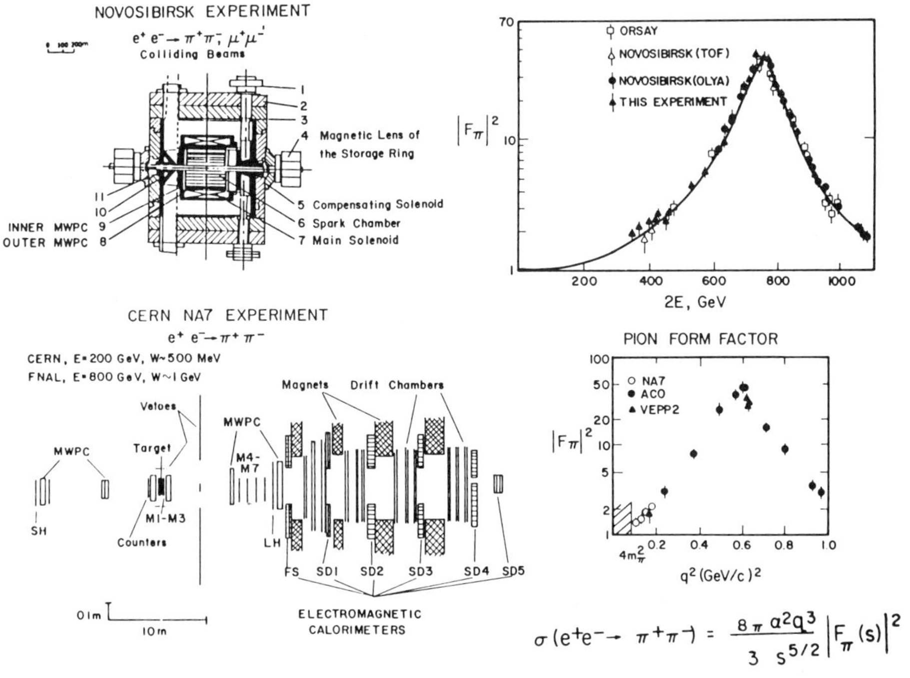

Fig. 3. Experiments to determine R at low $\sqrt{s}$.

The existence of additional heavier weak gauge bosons[18] would also affect g-2. Alternative electroweak models can have important contributions to a_μ(weak) from Higgs bosons.[19] Although the reported observation[20] of anomalous Z decays at CERN was not confirmed by more recent data[21], new physics could be lurking just above 100 GeV. Various attempts[22] were made to explain the originally reported events in terms of new interactions and new particles (excited leptons, for example), and it has been pointed out that the muon g-2 imposes a severe limitation on the choice of models.

Over a period of many years the muon g-2 value has provided a verygood constraint on theoretical models. Theories that attempt to generalize the standard model in one way or another all have their effect on the muon g-2 value. Such theories include left-right symmetric theories, theories with light Higgs scalars, technicolor theories, supersymmetric theories, Kaluza-Klein theories, and possibly others. Supersymmetric theories for example introduce many new particles with masses of the order of m_W, which would contribute to g-2. The present experimental accuracy eliminates significant mass ranges in some models[23] as indicated in Fig. 4. An improvement in sensitivity by a factor of 20 would provide a severe constraint on supersymmetry.

Another type of model building is motivated by the proliferation of the so-called elementary particles. Such models treat these particles as composites of several fundamental entities.[24] The present precision of muon g-2 imposes a lower limit of the order of 1 TeV on the mass of such constituent particles.[25] (In contrast the electron g-2 imposes a much weaker constraint on the constituent mass.) As the effect on a_μ of a massive constituent Λ is given by

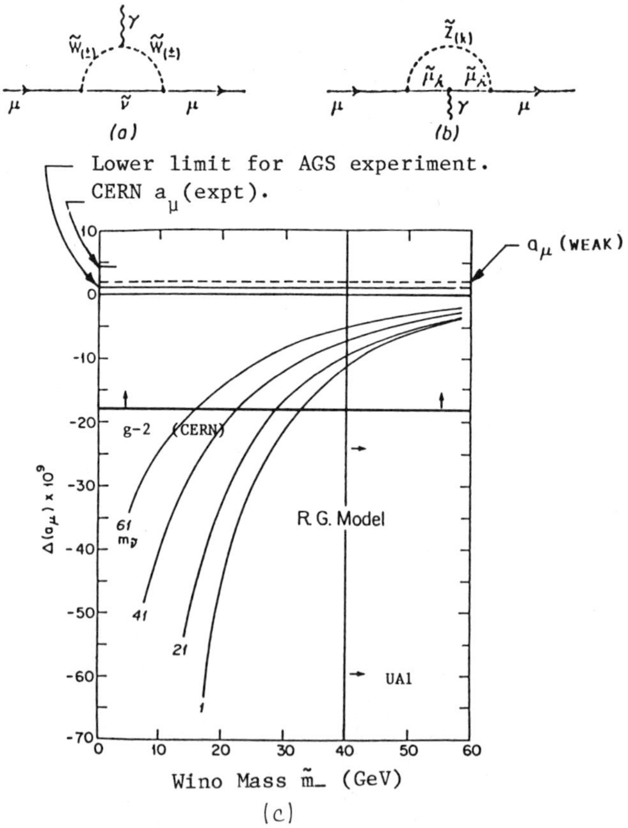

Fig. 4. Supersymmetry contribution to a_μ. a, b) One loop diagrams. c) Contribution in the renormalization group model. The zero is at a_μ(theor) = a_μ(QED) + a_μ(had). a_μ(weak) is the weak contribution.

$$\delta a_{\mu} \simeq m_{\mu}^{2} / \Lambda^{2}, \tag{6}$$

the more sensitive AGS experiment would test for substructure at the 4 to 5 TeV level.

Since the discovery of the muon it has been a mystery that the muon seems to be exactly identical with the electron except for its mass. After many years of investigation, this mystery has not yet been resolved. Instead in the standard theory it is simply accepted as a fact and used to rationalize the existence of the second and third generations of elementary particles. In this sense the mystery has merely deepened, and one of the goals of composite models is the attempt to solve this mystery. New experimental information treating this subject will be of fundamental importance to our understanding of the structure of matter, and in this context the muon g-2 has a crucial role to play.

III. THE EXPERIMENTAL DETERMINATION OF THE MUON g-2

The goal of the proposed new AGS experiment is a measurement of a_{μ} to 0.35 ppm, a factor of 20 improvement over the CERN experiment which achieved 7.2 ppm. After considering carefully a number of approaches to the measurement of muon g-2, the group proposing the new AGS experiment[7] chose a method similar to that used at CERN.[5] It will involve the storage of polarized muons with momentum of about 3 GeV/c in a 14 m diameter ring with a homogeneous magnetic field and with a quadrupole electric field to provide vertical focusing. Initially pions will be injected into the ring and polarized muons from pion decays will be captured. The g-2 precession is observed through the parity-violating correlation between the muon spin direction and the direction of emission of the decay electron. (Fig. 5).

The famous CERN experiment measured the g-2 precession frequency ω_a which is the difference between the spin precession frequency ω_s and the cyclotron frequency ω_c in a known magnetic field B:

$$\omega_{s} = \frac{eB}{mc\gamma} + \frac{e}{mc} aB \tag{7}$$

$$\omega_{c} = \frac{eB}{mc\gamma} \tag{8}$$

$$\omega_{a} = \omega_{s} - \omega_{c} = \frac{e}{mc} aB, \tag{9}$$

in which $a = (g-2)/2$ is the anomalous g-value of the muon and m is the muon rest mass.

The design of the CERN experiment and also of the proposed AGS experiment is based upon an important observation about the spin motion in a magnetic and electric field. The equation for the g-2 precession frequency is given by[26]

$$\omega_{a} = \frac{d\theta_{R}}{dt} = \frac{e}{mc}\left[aB + \left(a - \frac{1}{\gamma^{2}-1}\right)\left|\vec{\beta} \times \vec{E}\right|\right] \tag{10}$$

in which $\theta_R = (\vec{s}, \vec{\beta})$ is the angle between the muon spin ($\vec{s}$) direction in its rest frame and the muon velocity ($\vec{\beta}$) direction in the laboratory frame, and other quantities refer to the laboratory frame. If the bracketed factor $\left(a - \frac{1}{\gamma^{2}-1}\right)$ is zero, which is true when $\gamma = 29.3$ and $p_{\mu} =$ 3.094 GeV/c, then the g-2 precession frequency is determined by B alone and is independent of $\vec{E}$. This observation allows for a separated function

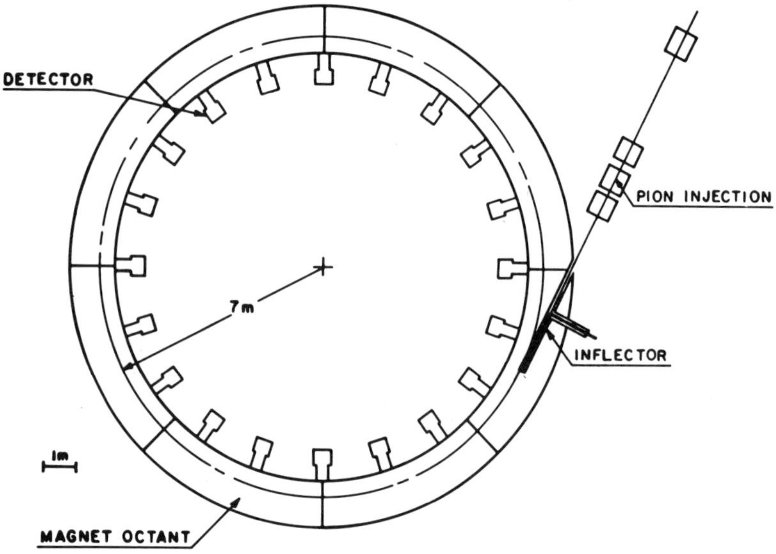

Fig. 5. General arrangement of AGS muon g-2 experiment.

muon storage ring in which a homogeneous magnetic field $\vec{B}$ determines the g-2 precession frequency, and an electric quadrupole field provides vertical weak-focusing for the muons.

Figures 5,6 and 7 show the overall arrangement, the muon storage ring, and the quadrupole electrodes. Polarized muons from pion decay in flight will be trapped in a large ring magnet with a uniform magnetic field B of 1.47 T, vertical focusing being provided by electric quadrupoles distributed more or less uniformly around the ring. The precession frequency ω_a of the spin relative to the momentum vector will be measured by observing the decay electrons emerging on the inside of the ring. As the angular distribution in the decay rotates with the muon spin, the counting rate for the high energy electrons (emitted forwards) will be modulated at the precession frequency. The superferric storage ring is designed to provide a field stable to 0.1 ppm and homogeneous to ~1 ppm over the muon storage area with cross sectional diameter of 9 cm. The effective mean field averaged over the muon orbits is to be determined to 0.1 ppm by NMR measurements. The electric quadrupole field (Fig. 7) will be pulsed with an electrode voltage of -40 kV. Fig. 8 shows the NMR trolley to be used to measure B in the muon storage region, and Fig. 7 indicates its movement. Fig. 9 shows an electron calorimeter based on a Pb-scintillator shower counter.

The g-2 precession frequency ω_a will be deduced from the decay electron counting data fitted to the formula

$$N = N_o \exp(-t/\tau) \left\{1 - A\cos(\omega_a t + \phi)\right\}, \tag{11}$$

in which τ = laboratory muon lifetime. The asymmetry parameter A is the product of the stored muon polarization and the asymmetry in μ-e decay which is a function of electron energy and the angular and energy spread of the electrons that are accepted by the detection system. The overall precision in time measurement by the electron detector must be about 30 ps in order to determine ω_a to <0.1 ppm. The statistical error in ω_a

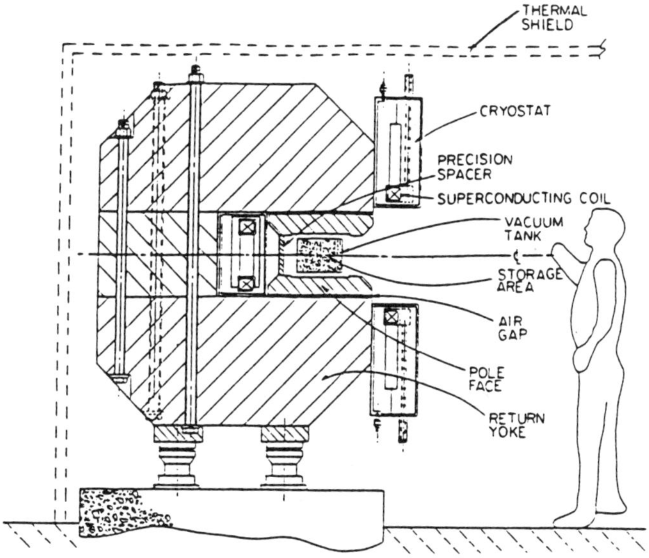

CROSS SECTION

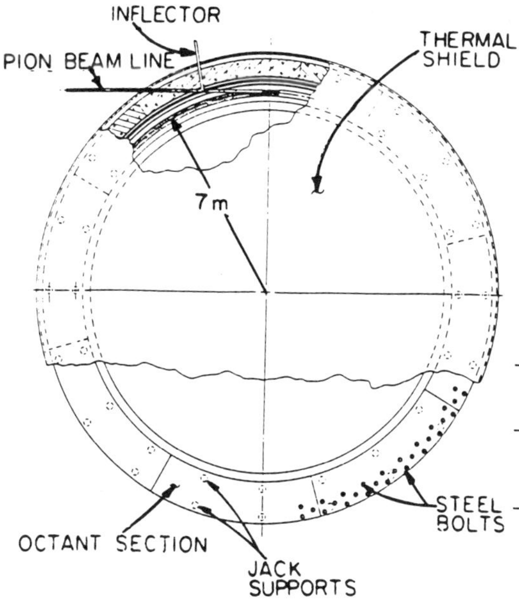

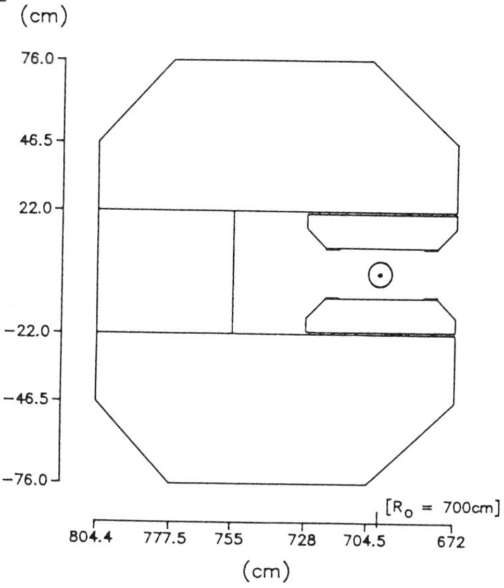

PLAN DRAWING

IRON CROSS SECTION

SHIMMING TECHNIQUES

Goal, Homogeneity to 1 ppm over muon storage volume.

1. Grinding and remachining of pole faces.
2. Iron shims on outside of yoke.
3. Iron shims in air gaps between pole pieces and yoke.
4. Circumferential current loop near pole faces.
5. Pole face windings.
6. Iron shims on pole faces.

Fig. 6. Superferric storage ring magnet.

TABLE 2. AGS g-2 EXPERIMENT

GENERAL APPROACH

Similar to CERN experiment.

A muon storage ring operating at the magic $\gamma = 29.3$ with $B \simeq 1.5$ T and electrostatic quadrupole focusing.

Pion injection.

IMPROVEMENTS

1. Primary Proton Beam Intensity [Injected π Beam] (with AGS Booster)
   x 100 [x 200]

2. Storage Ring Magnet: Field Homogeneity and Control
   Homogeneity ($\simeq$1 ppm) [CERN $\simeq$ 10-15 ppm]
   Control ($\simeq$ 0.1 ppm) [CERN $\simeq$ 0.5 ppm]
   Achieved by:
   (a) Superconducting coils
   (b) Larger magnet gap
   (c) Aximuthal symmetry in iron construction
   (d) Extensive shimming features
   (e) NMR feedback and control

3. Magnetic Field Measurement (NMR)
   Accuracy (0.1 ppm)
   Achieved by:
   (a) NMR trolley (movable within vacuum chamber)
   (b) Many (176) fixed NMR probes outside vacuum chamber
   (c) Insertable NMR probe for absolute calibration

4. Detector System
   Increased acceptance, data rate capacity, and time measurement accuracy.
   Achieved by:
   (a) Larger solid angle and thinner-walled vacuum vessel
   (b) Detector segmentation
   (c) Improved electronics
   (d) Improved digitrons

is given by

$$\frac{\Delta\omega_a}{\omega_a} = \frac{\sqrt{2}}{\omega_a A \tau N_e^{1/2}} \tag{12}$$

in which N_e = total number of detected electrons.

In addition to values of ω_a and B, determination of a_μ from Eq. (10) requires a value for the constant (e/mc). Since B is measured by a proton resonance frequency ω_p (corrected for molecular and bulk magnetic shielding), the constant needed is:[27,28]

$$\lambda = \mu_\mu/\mu_p, \tag{13}$$

and a_μ is then given by

$$a_\mu = R/(\lambda - R), \tag{14}$$

where $R = \omega_a/\bar{\omega}_p$.

Table 2 gives general features of the new AGS experiment compared to the CERN experiment[5] and Table 3 gives further parameters of the AGS experiment.

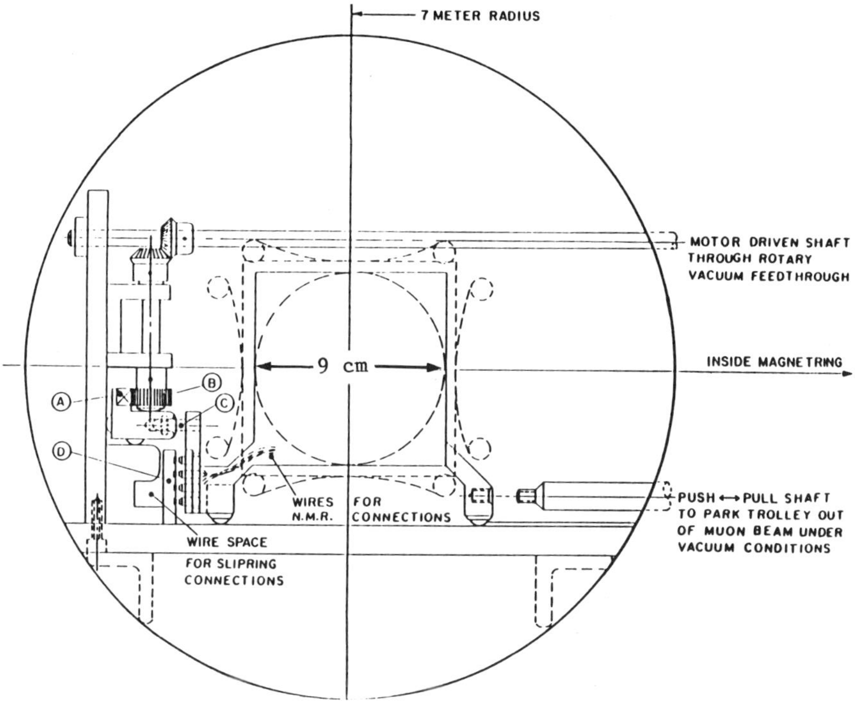

Fig. 7. Electrostatic quadrupoles and NMR trolley movement within vacuum vessel. The 9 cm muon storage region is indicated.

TABLE 3. Parameters of AGS Experiment

Magnet	
Orbit radius	7.0 m
Central magnetic field	1.47 T
Magnet gap	18 cm
Pole width	56 cm
Storage Systems	
Storage aperture diameter	9 cm
Vertical focusing by electric quadrupole field (pulsed 1 ms)	40 kV
Particle Injection	
Pulsed magnetic inflector	
π-μ decay	
Muon Motion in Storage Ring	
Gamma, γ	29.3
Momentum, p_μ	3.094 GeV/c
Lifetime, τ	64.4 μs
Orbital frequency, f_c	6.81 MHz
Orbital period, τ_c	147 ns
g-2 precession frequency, f_a	0.2327 MHz
g-2 period, τ_a	4.3 μs
Number electron shower detectors	80
Intensity data	
Proton per rf bunch with booster	4.2×10^{12}
Bunches injected per ring fill	2
Fills per AGS cycle	3
Pions injected per fill	1.07×10^{8}
Muons stored per fill	1.4×10^{4}
Electron counts per fill	2.8×10^{3}
Running Time	1500 hrs
Statistical error in a_μ	0.3 ppm
Systematic errors	0.2 ppm
Overall error	0.35 ppm

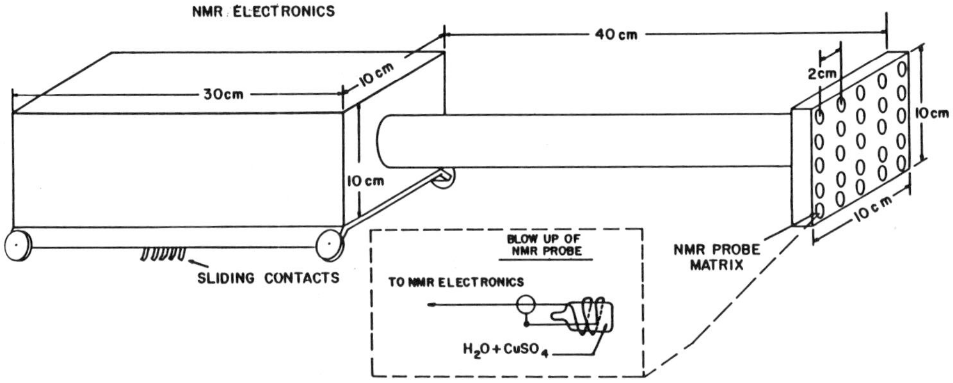

Fig. 8. Trolley and NMR probe matrix.

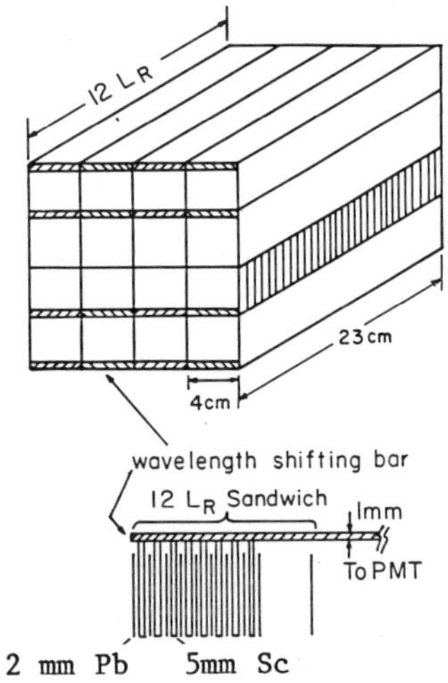

Fig. 9. Electron calorimeter.

The projected counting rates and errors are given in Table 4. The CERN experiment was dominated by the statistical error of 7.0 ppm, and the largest systematic error was an uncertainty of 1.5 ppm in $\bar{B}$. For the AGS experiment the primary proton beam average intensity will be about 100 times that available at CERN, so the statistical error in determining ω_a is projected to be 0.3 ppm. An improved storage ring magnet and NMR system should reduce the systematic error in $\bar{B}$ to about 0.1 ppm. Other systematic errors should be small enough to allow a determination of a_μ to 0.35 ppm.

This AGS experiment was approved at BNL in November, 1986. It is now being actively developed, with data-taking projected in less than 4 years.

TABLE 4. AGS Experiment Errors

COUNTING RATES AND STATISTICAL ERRORS

Storage aperture diameter	90 mm
Protons per rf bunch with booster	4.2×10^{12}
Bunches ejected per ring fill	2
Protons per fill	8.4×10^{12}
Pion $\Delta p/p$	±0.6%
Pions at inflector exit per fill(1.28×10^{7} per 10^{12} protons)	1.07×10^{8}
Muons stored per fill(at 134 ppm capture efficiency)	14×10^{3}
Electrons counted above 1.6 GeV per fill(20% of the decays)	2.8×10^{3}
Fills per AGS cycle(1.4 s)	3
Fraction of AGS protons used	50%
Fills per hour	7714
Electron counts per hour	22×10^{6}
Running time for 0.3 ppm (1 std. dev.)	1288 hours

SYSTEMATIC ERRORS

SOURCE	COMMENTS	ERROR (ppm)
Magnetic field	Includes absolute calibration of NMR probes and averaging over space, time, and muon distribution.	0.1
Electric field correction	0.7 ppm correction	0.03
Pitch correction	0.4 ppm correction	0.02
Particle losses		0.05
Timing error		0.01
	TOTAL (in quadrature)	0.12

REFERENCES

1. V.W. Hughes, and T. Kinoshita, in Muon Physics I, ed. by V.W. Hughes and C.S. Wu, (Academic Press, New York, 1977) p. 11.

2. V.W. Hughes and T. Kinoshita, Comments Nucl. Part. Phys. 14, 341 (1985).

3. G. Charpak et al., Nuovo Cim. 37, 1241 (1965).

4. J. Bailey et al., Il Nuovo Cimento 9A, 369 (1972).

5. J. Bailey et al., Nucl. Phys. B150, 1 (1979).

6. T. Kinoshita, B. Nizic and Y. Okamoto, Phys. Rev. Lett. 52, 717 (1984).

7. "A New Precision Measurement of the Muon g-2 Value at the Level of 0.35 ppm." , AGS Proposal 821, September, 1985; revised September, 1986, V.W. Hughes, spokesman. E. Hazen, C. Heisey, B. Kerosky, F. Krienen, E.K. McIntyre, D. Magaud, J.P. Miller, B.L. Roberts, D. Stassinopoulos, L.R. Sulak, W. Worstell - Boston University; H.N. Brown,, E.D. Courant, G.T. Danby, C.R. Gardner, J.W. Jackson, M. May A. Prodell, R. Shutt, P.A. Thompson - Brookhaven National Laboratory

M.S. Lubell - City College of New York; A.M. Sachs - Columbia University; T. Kinoshita - Cornell University; G. zu Putlitz - University of Heidelberg; W.P. Lysenko - Los Alamos National Laboratory; W.L. Williams - University of Michigan; J.J. Reidy - University of Mississippi; F. Combley - Sheffield University; K. Nagamine - University of Tokyo; S.K. Dhawan, A.A. Disco, F.J.M. Farley, V.W. Hughes, Y. Kuang, H. Venkataramania - Yale University.

8. P.B. Schwinberg, R.S. Van Dyck, Jr., and H.G. Dehmelt, Phys. Rev. Lett. 47, 1679 (1981); R.S. Van Dyck, Jr., P.B. Schwinberg and H.G. Dehmelt, in Atomic Physics 9, ed. by R.S. Van Dyck, Jr., and E.N. Fortson, (World Scientific Publ. Co., Singapore, 1984) p. 53.

9. T. Kinoshita and W.B. Lindquist, Phys. Rev. Lett. 47, 1573 (1981); Phys. Rev. D27, 853, 867, 877, 886 (1983); T. Kinoshita, "The Anomalous Magnetic Moment of the Electron and the QED Determination of the Fine Structure Constant." IEEE Trans. Instrum. Meas. IM-36, to be published June, 1987.

10. B.N. Taylor, J. Res. Natl. Bur. Stand. 90, 91 (1985).

11. R. Jackiw and S. Weinberg, Phys. Rev. D5, 2396 (1972); G. Altarelli, N. Cabibbo and L. Maiani, Phys. Lett. 40B, 415 (1972); I. Bars and M. Yoshimura, Phys. Rev. D6, 374 (1972); K. Fujikawa, B.W. Lee and A.I. Sanda, Phys. Rev. D6, 2923 (1972); W.A. Bardeen, R. Gastmans and B.E. Lautrup, Nucl. Phys. B46, 319 (1972).

12. W. Marciano, "$\sin^2\theta_W$ and Radiative Corrections" in XXIII Int. Conf. on High Energy Physics, Berkeley, July, 1986. To be published.

13. T. Kinoshita, B. Nizic and Y. Okamoto, Phys. Rev. D31, 2108 (1985).

14. L.M. Barkov et al., Nucl. Phys. B256, 365 (1985).

15. J.A. Casas et al., Phys. Rev. D32, 736 (1985).

16. L.M. Barkov et al., Novosibirsk Institute of Nuclear Physics Preprint 85-118 CMD-2 Detector for VEPP-2M.

17. S.R. Amendolia et al., Phys. Lett. 138B, 454 (1984); 146B, 116 (1984).

18. R.W. Robinett and J.L. Rosner, Phys. Rev. D25, 3036 (1982).

19. S.R. Moore, K. Whisnant, and B.-L. Young, Phys. Rev. D31, 105 (1985).

20. G. Arnison et al. (UA1 Collaboration), Phys. Lett. 126B, 398 (1983); P. Bagnaia et al. (UA2 Collaboration), Phys. Lett. 129B, 130 (1983).

21. C. Rubbia; P. Langacker, 1985 Intl. Symp. on Lepton and Photon Interactions at High Energy, ed. by M. Konuma and K. Takahashi (Nissha Printing Co., Kyoto, Japan, 1986) p. 242; 186.

22. R.D. Peccei, Phys. Lett. 136B, 121 (1984); N. Cabibbo, L. Maiani, and Y. Srivastava, Phys. Lett. 139B, 359 (1984); M.J. Duncan and M. Veltman, Phys. Lett. 139B, 310 (1984); L. Bergstrom, Phys. Lett. 139B, 102 (1984); F.M. Renard, Phys. Lett. 139B, 449 (1984); M. Leurer, H. Harari, and R. Barbieri, Phys. Lett. 141B, 455 (1984); M. Suzuki, Phys. Lett. 143B, 237 (1984); S.D. Drell and S.J. Parke, Phys. Rev. Lett. 53, 1993 (1984).

23. P. Nath in Supersymmetry and Supergravity/Non-Perturbative QCD, ed. by P. Roy and V. Singh, (Springer-Verlag,Heidelberg, 1984) p. 113; T.-C. Yuan et al., Zeit. fur Physik C - Particles and Fields 26, 407 (1984); R. Arnowitt and P. Nath, Intersections Between Particle and Nuclear Physics, ed. by D.F. Geesaman, (AIP, New York, 1986) p. 582.

24. L. Lyons, Prog. Part. Nucl. Phys. 10, 227 (1983); M.E. Peshkin, in Proc. of Intl. Symp. on Lepton and Photon Interactions at High Energies, Bonn, 1981, ed. by W. Pfeil (Bonn, 1981) p. 880; E.J. Eichten, K.D. Lane, and M.E. Peshkin, Phys. Rev. Lett. 50, 811 (1983); M. Suzuki, Phys. Lett. 153B, 289 (1985).

25. M.E. Peshkin, 1985 Intl. Symp. on Lepton and Photon Interactions at High Energy, ed. by M. Konuma and K. Takahashi (Nissha Printing Co. Kyoto, Japan 1985) p.714

26. V. Bargmann, L. Michel and V.A. Telegdi, Phys. Rev. Lett. 2, 435 (1959); F.J.M. Farley, Cargese Lectures in Physics, (Gordon and Breach, 1968) Vol. 2, p. 55; J. Bailey and E. Picasso, Progr. in Nucl. Physics 12, 43 (1970).

27. F.G. Mariam et al., Phys. Rev. Lett. 49, 993 (1982).

28. E.R. Cohen and B.N. Taylor, "1986 Adjustment of the Fundamental Physical Constants, A Report of the CODATA Task Group of Fundamental Constants" CODATA Bull. 63, (Pergamon Press, Oxford, December 1986).

*Research supported in part by DOE under contract DE-AC02-ER03075 and DE-FG02-84ER40243.

MUONIUM*

Vernon W. Hughes

Gibbs Laboratory
Physics Department
Yale University
New Haven, CT 06520

I. INTRODUCTION

Muonium ($M\equiv\mu^+e^-$) was discovered[1] in 1960 through observation of its characteristic Larmor precession in a magnetic field. Since then research on the fundamental properties of M has been actively pursued[2,3], as has also the study of muonium collisions in gases, muonium chemistry and muonium in solids.[4]

The principal reason that muonium continues to be important to fundamental physics is that it is the simplest atom composed of two different leptons. The muon retains a central role as one of the elementary particles in the modern standard theory, but we still have no understanding as to "why the muon weights" and in all respects behaves simply as a heavy electron. Muonium is an ideal system for determining the properties of the muon, for testing modern quantum electrodynamics, and for searching for effects of weak, strong, or unknown interactions in the electron-muon bound state. Basically muonium is a much simpler atom than hydrogen because the proton is a hadron and, unlike a lepton, has a structure that is determined by the strong interactions.

II. GROUND STATE HYPERFINE STRUCTURE AND ZEEMAN EFFECT

After the discovery of muonium measurements of its energy levels could be undertaken by microwave magnetic resonance spectroscopy utilizing the facts that the incident μ^+ are polarized so that polarized muonium is formed and that the decay positrons have an asymmetric angular distribution with respect to the muon spin direction.[5] The Breit-Rabi energy level diagram for the ground state of muonium is shown in Fig. 1. With the aim of determining the hyperfine structure interval $\Delta\nu$ and the muon magnetic moment μ_μ transitions at both low and high magnetic fields have been measured as indicated. Starting in 1962 a series of increasingly accurate measurements were undertaken[2,3] by both the Yale-Heidelberg and Chicago groups. The latest experiment at the Los Alamos Meson Physics Facility (LAMPF) was a strong field measurement (Fig. 2).[6] Typical resonance curves are shown in Fig. 3. The history of measurements of $\Delta\nu$ is shown in Fig. 4.

The experimental results for $\Delta\nu$ and μ_μ and the current theoretical value for $\Delta\nu$ are given in Table I. The radiative and recoil corrections to the leading Fermi value for $\Delta\nu$ have been computed to high order.[7] The principal error of 1.4 kHz or 0.31 ppm is due mostly to uncertainty in the value of the constant μ_μ/μ_p appearing in the Fermi term E_F. (The condensed matter value[8] of α is accurate to 0.09 ppm.) The error of 0.2 kHz is an estimate of numerical uncertainties in the calculated terms in ε_{QED}, and the 1.0 kHz error is an estimate of the size of uncalculated higher order radiative and recoil terms. A small hadronic vacuum polarization contribution of 0.22(4) kHz is included in the factor with 18.18 in the last term of δ'_μ. The experimental value for $\Delta\nu$ is known to 36 ppb. The experimental and theoretical values agree well within the theoretical error of 0.39 ppm, and this agreement constitutes one of the important, sensitive tests of quantum electrodynamics and of the behaviour of the muon as a heavy electron.

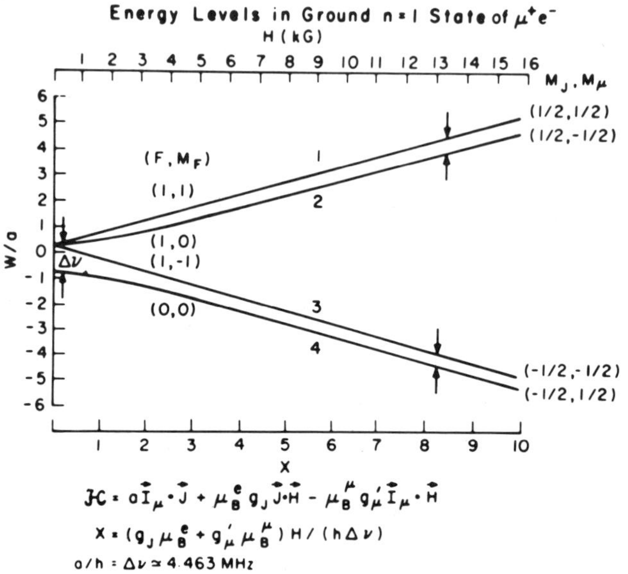

Fig. 1. Breit-Rabi energy level diagram for the ground state of muonium.

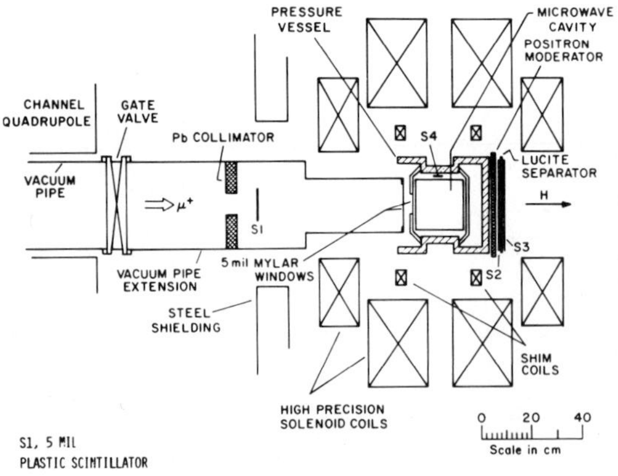

Fig. 2. Experiment at LAMPF in which the latest precision measurement of the hyperfine structure interval $\Delta\nu$ in muonium was made.

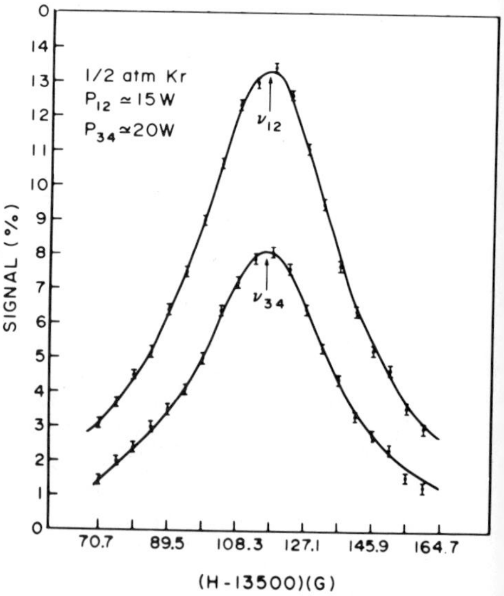

Fig. 3. Fitted resonance lines from the experiment shown in Fig. 2.

Improvement by a factor of about 10 in the sensitivity of the comparison of theory and experiment for $\Delta\nu$ appears possible at this time.[9] With the use of a chopped muon beam now available at Los Alamos, line narrowing techniques can be employed.[10] Use of a higher magnetic field value will improve the accuracy in determining μ_μ/μ_p. Finally, the intensity and quality of the muon beam has been improved since the last measurement. Considering all these factors, a measurement of $\Delta\nu$ to about 5 ppb and of μ_μ/μ_p to 30 ppb appears possible. Theoretical computation to about 0.1 kHz or 20 ppb appears realistic. We might remark that weak interaction contributions to $\Delta\nu$ are predicted[11] at the level of 20 ppb.

Table I. Theoretical value of muonium $\Delta\nu$ and comparison with experiment.

$$\Delta\nu_{th} = \left[\tfrac{16}{3}\alpha^2 c R_\infty(\mu_\mu/\mu_B^e)\right]\left[m_\mu/(m_e+m_\mu)\right]^3\left[1+\epsilon_{QED}\right]$$

$$\Delta\nu_{th} = E_F(1+\epsilon_{QED})$$

$$\epsilon_{QED} = \tfrac{3}{2}\alpha^2 + a_e + \epsilon_1 + \epsilon_2 + \epsilon_3 - \delta'_\mu$$

$$a_e = (g_e-2)/2;\ \epsilon_1 = \alpha^2(\ln 2 - 5/2)$$

$$\epsilon_2 = -\tfrac{8\alpha^3}{3\pi}\ln\alpha\left(\ln\alpha - \ln 4 + \tfrac{281}{480}\right);\ \epsilon_3 = \tfrac{\alpha^3}{\pi}\,(15.38\pm0.29)$$

$$\delta'_\mu = \Big\{\tfrac{3\alpha}{\pi}\tfrac{m_R}{m_\mu-m_e}\ln m_\mu/m_e + \alpha^2\tfrac{m_R}{m_\mu+m_e}\left[2\ln\alpha + 8\ln 2 - 3\tfrac{11}{18}\right]$$
$$+(\alpha/\pi)^2 m_e/m_\mu \times\left[2\ln^2(m_\mu/m_e) - \tfrac{13}{12}\ln(m_\mu/m_e) - 18.18(63)\right]\Big\}\frac{1}{1+a_\mu}$$

where $m_R = m_e m_\mu/(m_e+m_\mu)$

$R_\infty = 1.097\,373\,152\,1\ (11)\times10^5\text{cm}^{-1}$ (0.001 ppm)

$c = 2.997\,924\,580\times10^{10}$ cm/sec

$\alpha^{-1} = 137.035\,981\ (12)$ (0.09 ppm)

$a_e = 1\,159\,652\,193\ (4)\times10^{-12}$ (3.4 ppb)

$\mu_\mu/\mu_B^e = (\mu_\mu/\mu_p)(\mu_p/\mu_B^e);\ \mu_p/\mu_B^e = 1.521\,032\,209\ (16)$ (0.01 ppm)

$m_\mu/m_e = 206.768\,259\ (62)$ (0.3 ppm)

$\mu_\mu/\mu_p = 3.183\,345\,47\ (95)$ (0.3 ppm)

$[\mu_\mu/\mu_p = 3.183\,346\,1\,(11)$ (0.36 ppm)] from muonium Zeeman effect

$a_\mu = (g_\mu - 2)/2 = 0.001\,165\,923\,0(84)$ (7.2 ppm)

$E_F = 4\,459\,033.4\ (1.5)$ kHz

$\Delta\nu_{th} = 4\,463\,303.6(1.5)(0.2)(1.0)$ kHz (0.4 ppm)

$\Delta\nu_{exp} = 4\,463\,302.88(0.16)$ kHz (0.036 ppm)

$\Delta\nu_{th} - \Delta\nu_{exp} = (+0.7\pm1.8)$ kHz

Determination of α : $\alpha^{-1} = 137.035\,988\ (20)$ (0.15 ppm), μ^+e^-

137.035 994 (5) (0.038 ppm), $g_e - 2$

137.035 981 (12) (0.09 ppm), condensed matter

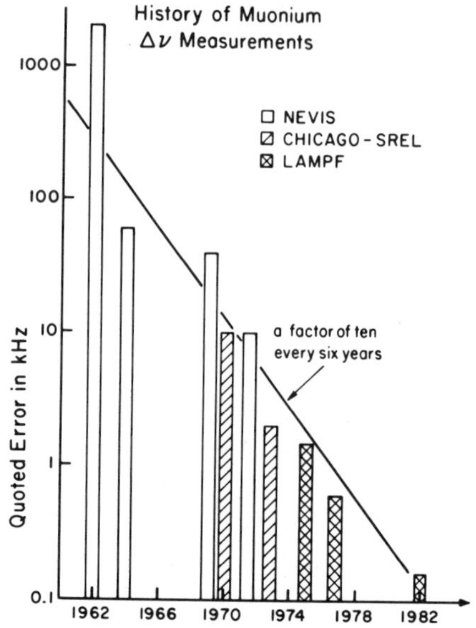

Fig. 4. History of measurements of Δν.

III. LAMB SHIFT IN MUONIUM

The experiments on ground state muonium have involved muonium in a gas. Two important new experiments have recently become possible with the development of a muonium beam in vacuum. One of these is the measurement of the Lamb shift in muonium, and the other is a sensitive search for the spontaneous conversion of muonium (μ^+e^-) to antimuonium (μ^-e^+). For the Lamb shift measurement the 2S metastable state is readily quenched in an atomic collision, and for the $M \to \bar{M}$ conversion the conversion rate is greatly reduced by atomic collisions; hence both of these experiments require muonium in vacuum.

A beam foil method is used to produce muonium in vacuum.[12] A low energy μ^+ beam is passed through a thin foil in vacuum and μ^+e^- emerges from the downstream side of the foil due to electron capture by μ^+ from the foil. Figure 5 shows the expected equilibrium fractions of the different charge states μ^+,μ^+e^- and $\mu^+e^-e^-$, based on experimental data for protons scaled to muons using the Born approximation result that muon and proton charge capture cross sections will be the same at equal velocities.

An experiment at Los Alamos,[12] indicated in Fig. 6, used an incident surface μ^+ beam of about 30 MeV/c which was degraded in a moderator and windows and then passed through a foil. The downstream beam was analyzed for its μ^+e^- component by requiring that it be undeflected by a magnetic field and that it stop in a target and then yield decay positrons with the characteristic Michel spectrum for μ^+ decay. Electron counts versus energy as observed with the NaI detector are shown in Fig. 7. For the vacuum case above 30 MeV the spectrum agrees well with the predicted Michel spectrum folded with the energy resolution of the NaI detector. Below 30 MeV the spectrum is dominated by background e^+ in the beam and associated bremsstrahlung. With the experimental arrangement shown the ratio of M downstream of the foil to incident μ^+ was about $1/(3 \times 10^3)$. With even a low pressure of He in the system, the muonium beam is ionized in collisions and does not reach the beam stop.

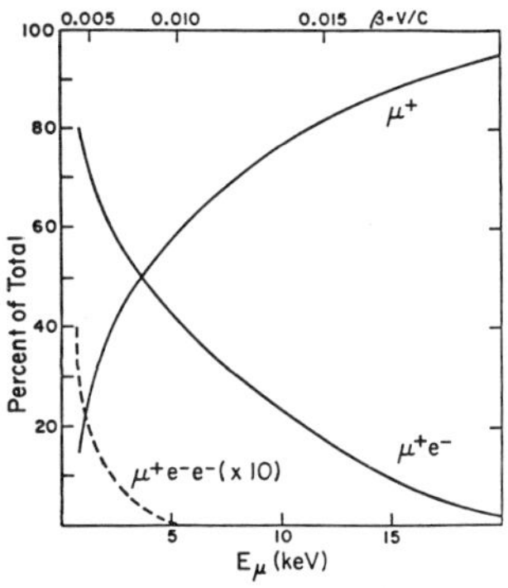

Fig. 5. Expected charge-state distribution for muons emerging from foil targets.

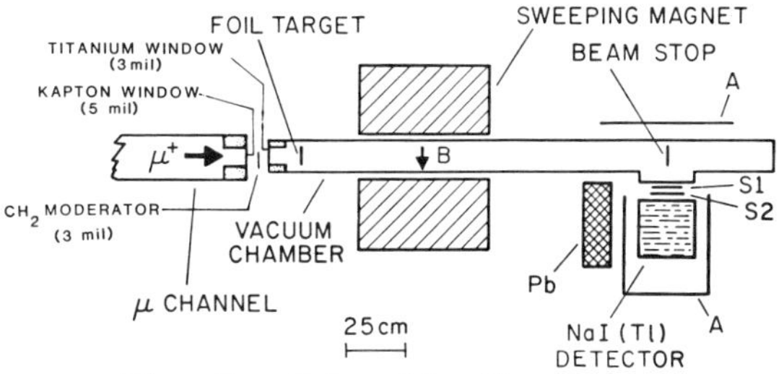

Fig. 6. Schematic diagram of experimental apparatus to observe muonium in vacuum.

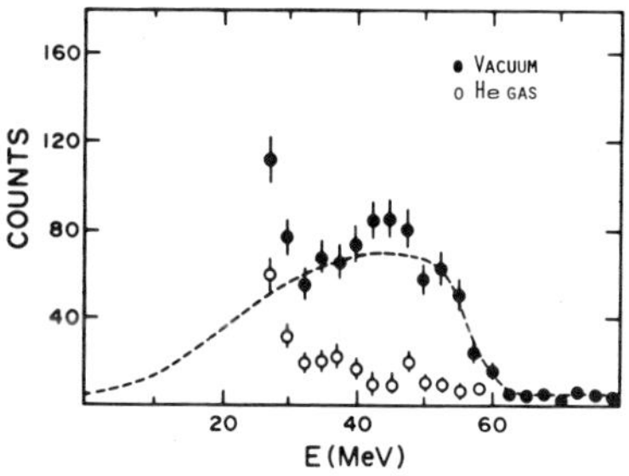

Fig. 7. Measured NaI spectrum indicating muonium formation.

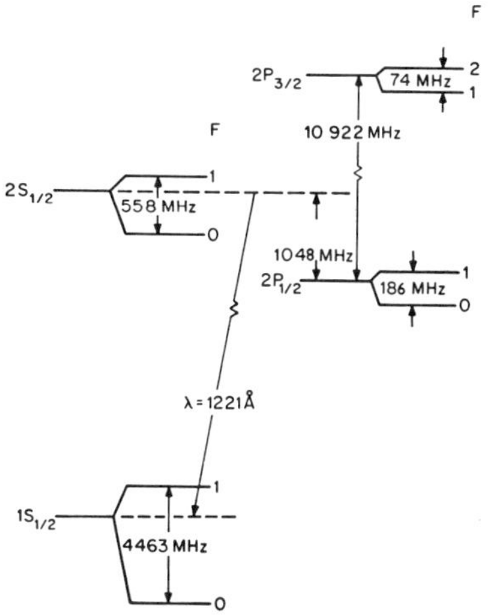

Fig. 8. Energy level diagram of the n=1 and n=2 states of muonium.

Having a muonium beam in vacuum a measurement of the Lamb shift in muonium could be attempted. Such a measurement would provide a further test of the behaviour of the muon as a heavy electron and, if done with sufficient precision, would provide an ideal test of the Lamb shift-more ideal than in H where proton structure contributes importantly to the Lamb shift and to its theoretical uncertainty. The energy level diagram for the n=1 and n=2 states of muonium is shown in Fig. 8. The separation $\mathcal{S}_M$ = 1048 MHz between the $2S_{1/2}$ and $2P_{1/2}$ levels in the absence of hfs interaction is the predicted Lamb shift in M.

The method of the experiment[13] was to form a beam of M(2S) atoms by the beam foil method, induce a microwave transition from a $2S_{1/2}$ state to a $2P_{1/2}$ state, and observe the resulting 2P→1S Lyman-α photons. The experimental arrangement is shown in Fig. 9. A separated low momentum "subsurface" muon beam[14] of 10 MeV/c was used and resulted in a production efficiency for $M(2S)/\mu^+$ of about $4/10^3$. Following the rf coaxial line, there is a region with an applied static electric field viewed by the UV phototubes. Finally a microchannel plate is located to detect M atoms. Our first experiment sought to establish the existence of a M(2S) beam by the method of static electric field quenching. A static E field of adequate strength (~600 V/cm) will mix in 2P state with the 2S state and lead to the 2P→1S Lyman-α photon emission. A signal event is a delayed triple coincidence which involves first a count of an incident μ^+ in the plastic scintillator, then detection of a Lyman-α photon, and finally detection of a M(1S) atom by the microchannel plate. The observed signal induced by the static E field is shown in Fig. 10. The shape corresponds roughly to the expected range of kinetic energies of the M(2S) atoms; the event rate is only about 5 per hr.

Observation of the Lamb shift transition was made by applying a microwave field with the static E field still on, so that inducing a microwave transition results in a decrease in the event rate. The observed signal for the transitions 2S(F=1) → 2P(F=1) and 2S(F=1) → 2P(F=0) is shown in Fig. 11. A theoretical line shape involving two overlapping Lorentzian lines is fitted to the observed points assuming the theoretically predicted values of the hfs intervals. The most complete analysis of the data which utilizes the timing information for each event gives $\mathcal{S}_M$ = (1054 ± 22) MHz. A similar experiment was also done at TRIUMF.[15]

The theoretical value for the muonium Lamb shift is given in Table II. It differs from the Lamb shift in H by the absence of a proton structure term and by the relatively greater importance of recoil terms. Our experimental value agrees well within its limited accuracy with the theoretical value. We believe that substantial improvement in the accuracy of measurement of $\mathcal{S}_M$ can and will be made.

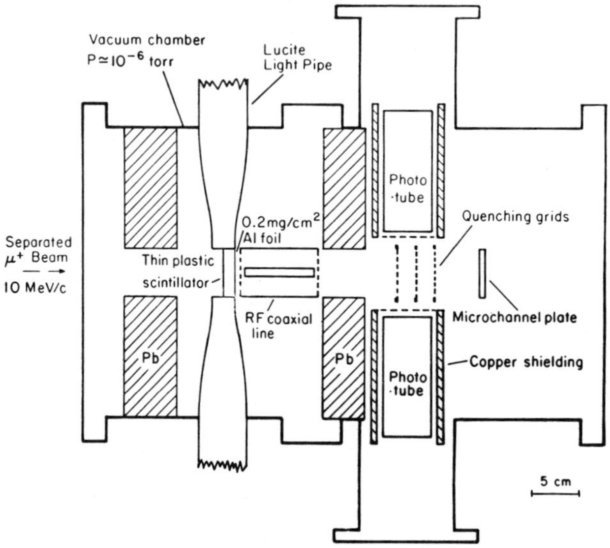

Fig. 9. Experimental apparatus used in muonium Lamb shift measurement.

Table II. Theoretical and experimental values for the muonium Lamb shift.

CORRECTION	ORDER (mc^2)	VALUE (MHz)
Self energy	$\alpha(Z\alpha)^4[\ln(Z\alpha)^{-2},1,Z\alpha,\ldots]$	1 085.812
Vacuum polarization	$\alpha(Z\alpha)^4[1,Z\alpha,\ldots]$	-26.897
Fourth order	$\alpha^2(Z\alpha)^4$	0.101
Reduced mass	$\alpha(Z\alpha)^4 m/M_\mu[\ln(Z\alpha)^{-2},1]$	-14.626
Relativistic recoil	$(Z\alpha)^5 m/M_\mu[\ln(Z\alpha)^{-2},1]$	+3.188
Total		1 047.578(300)MHz

The estimated uncertainty in the theoretical value is due to uncalculated terms of higher order in m/M_μ and in $\alpha m/M_\mu$, that is the terms (m/M_μ)(reduced mass term) and α (reduced mass term). An estimate of the size of these terms is 0.3 MHz.

Experimental Value

$\mathcal{S}_{expt}$ = 1054 ± 22 MHz LAMPF

1077^{+12}_{-15} MHz TRIUMPF

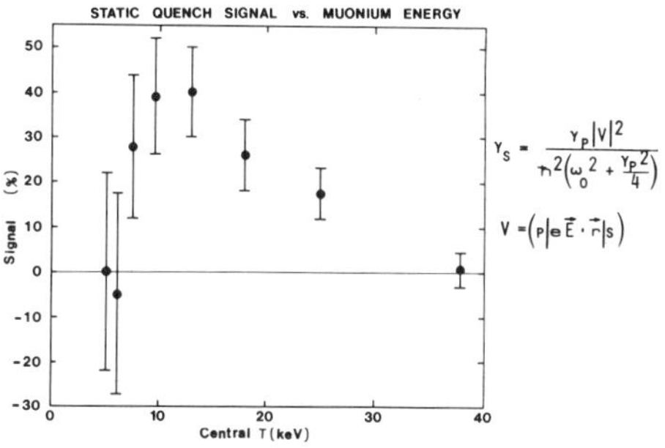

Fig. 10. Observation of the static electric field quenching of muonium in the 2S state in the experiment indicated in Fig. 9.

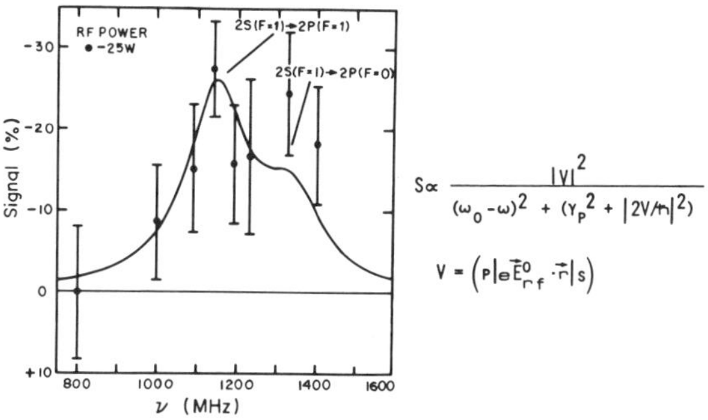

Fig. 11. Observation of the rf induced 2S→2P transitions which determine the muonium Lamb shift in the experiment indicated in Fig. 9.

IV. MUONIC HELIUM ATOM

The muonic helium atom which consists of an α particle, a negative muon, and an electron, is an interesting atom in which μ^- and the μ^-,e^- interaction can be studied.[16] As indicated in Fig. 12a this atom can be considered to consist of a pseudo-nucleus $(^4He\mu^-)^+$ and an electron, where the pseudo-nucleus is a heavy charge +1 system with the magnetic moment of the μ^-. Hence this atom is a muonium-like system and has the ground state energy level system shown in Fig. 12b. The atom can be formed by stopping μ^- in He gas with a small admixture of Xe. The μ^- is captured by a He atom and cascades to the 1S state of $(^4He\mu^-)^+$. Then an electron is captured from a Xe donor atom to form $^4He\mu^-e^-$. The hyperfine structure interval Δν and the Zeeman effect have been measured in microwave magnetic resonance experiments similar to those for muonium.[17] Typical resonance curves are shown in Fig. 12c. The theoretical and experimental values for Δν are also given and are seen to be in good agreement within the relatively large theoretical error associated with knowledge of the Schroedinger wavefunction.[18] The hfs interval of the muonic ^{3}He atom $(^3He\mu^-e^-)$ has also been measured recently.[19]

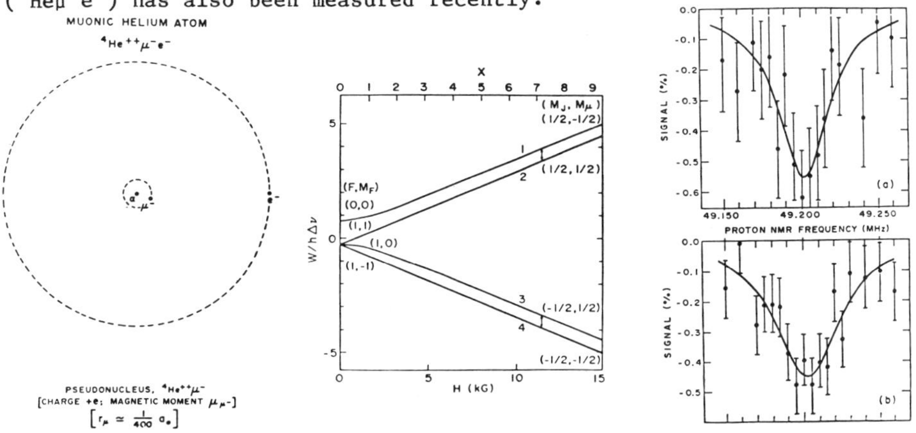

(a) (b) (c)

$$\Delta\nu_{exp} = 4\ 465.004(29)\ \text{MHz}\ (6.5\ \text{ppm})$$
$$\Delta\nu_{th} = 4\ 465.0(0.3)\ \text{MHz}$$

Fig. 12. The muonic helium atom. (a) Bohr picture, (b) Breit-Rabi energy level diagram, (c) resonance curves for strong field transitions. For curve (a) helium pressure was 15 atm. and for (b), 5 atm.

V. MUONIUM → ANTIMUONIUM CONVERSION

The muon and the electron may be considered to belong to two different generations of leptons, which thus far appear to remain separate because of the independent conservation laws of muon number and of electron number. Any connection between the muon and the electron, such as a process which would violate muon number conservation, would be an important clue to the relationship between the two generations.[20] Speculative modern theories which seek a more unified theory of particles and their interactions predict muon number violating processes.[21] As yet no such rare decay process has been observed,[22] and with our present knowledge theory has little useful predictive power.

The conversion of muonium (μ^+e^-) to its antiatom antimuonium (μ^-e^+) would be an example of a muon number violating process[2], and like neutrinoless double beta decay would involve $\Delta L_e=2$. The M-$\bar{M}$ system also bears some relation to the K°-$\bar{K}^\circ$ system, since the neutral atoms M and $\bar{M}$ are degenerate in the absence of an interaction which couples them. In Table III a four-Fermion Hamiltonian term coupling M and $\bar{M}$ is postulated, and the probability that M formed at time t=0 will decay from the $\bar{M}$ mode is given. Present experimental limits[3] for the coupling constant G are indicated and are still much larger than the Fermi constant G_F.

The first search[23] for the M→$\bar{M}$ conversion was made in an experiment at the Columbia Nevis sychrocyclotron which involved a search for muonic atom x-rays which would accompany M→$\bar{M}$ conversion. If muonium is formed initially in a gas (e.g., Ar) by electron capture by μ^+ and if an $\bar{M}$ component of the wavefunction develops due to G, then μ^-Ar can be formed and its x-rays will be emitted. Figure 13a indicates the physical processes involved, and Fig. 13b shows the experimental arrangement. A

Table III. Muonium-antimuonium conversion including present experimental limits.

$\mu^+e^- \rightarrow \mu^-e^+$

Muon number, L_μ, +1(-1) for $\mu^-(\mu^+),\nu_\mu(\bar{\nu}_\mu)$, 0 for other particles.

Violates additive conservation law for muon number, ΣL_μ = constant.

Allowed by multiplicative conservation law, $(-1)^{\Sigma L_\mu}$ = constant.

$$\mathcal{H}_{M\rightarrow\bar{M}} = \frac{G}{\sqrt{2}}\, \bar{\psi}_\mu\, \gamma_\lambda(1+\gamma_5)\, \psi_e \bar{\psi}_\mu \gamma^\lambda(1+\gamma_5)\, \psi_e + \text{H.C.}$$

$$P(\bar{M}) = \int_0^\infty \gamma\, e^{-\gamma t} \left|\langle\bar{M}|\psi(t)\rangle\right|^2 dt = 2.5 \times 10^{-5} \left(\frac{G}{G_F}\right)^2$$

Probability of decay of M from $\bar{M}$ state reduced by collisions in a gas by factor of (number of collisions during muon lifetime)$^{-1}$.

$G < 5800\ G_F$ (M→$\bar{M}$ in gas; Nevis: 1968)[23]

$G < 600\ G_F$ ($e^-e^- \rightarrow \mu^-\mu^-$; HEPL; 1969)[24]

$G < 20\ G_F$ (M→$\bar{M}$ in powers; TRIUMF; 1982,1986)[25,32]

signal event is defined by an incoming μ^+, no electron detected, and a γ ray of energy 643 keV, which is the 2P-1S μ^-Ar transition energy. Fig. 13c shows the observed spectrum. The absence of a signal at the energy of 643 keV can be interpreted as providing the limit $G < 5800\ G_F$ at a 90% confidence level. We note that the sensitivity of the measurement is greatly reduced by collisions of $\bar{M}$ in the Ar gas due to the removal of the degeneracy of M and $\bar{M}$ in a collision.

A much more sensitive experiment of this type has been done recently at LAMPF (Fig. 14).[26] With a muonium beam in vacuum the development of an $\bar{M}$ component proceeds without interruption due to atomic collisions. The large crystal box NaI detector system used at Los Alamos[27] to search for the rare decay $\mu^+ \to e^+\gamma$ was employed to search for coincident high Z muonic atom x-rays. Data-taking has been completed and data analysis is in progress.

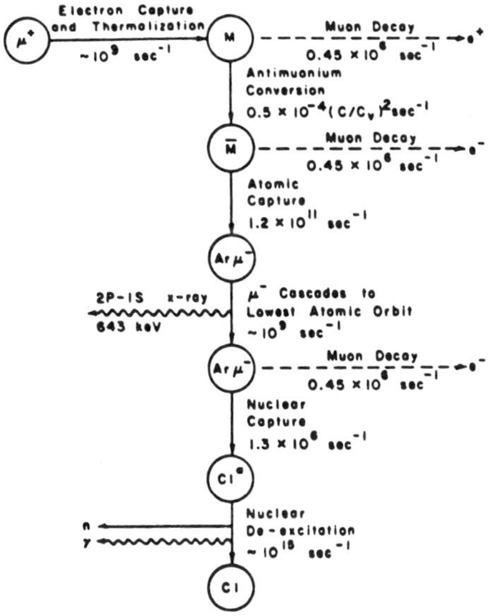

Physical processes for μ^+ in argon.
(a)

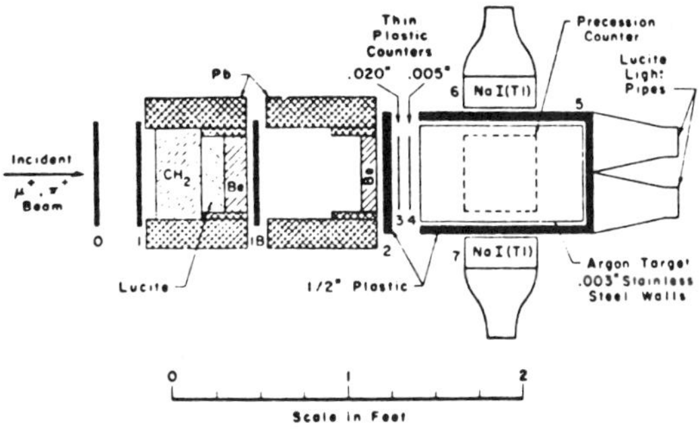

Experimental arrangement
(b)

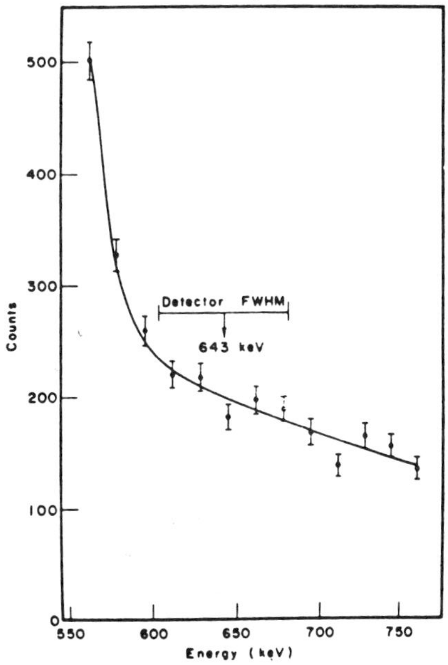

γ ray energy spectrum.
(c)

Fig. 13. Experiment at Nevis (Columbia) in which a search for muonium→antimuonium conversion was made (Ref. 23).

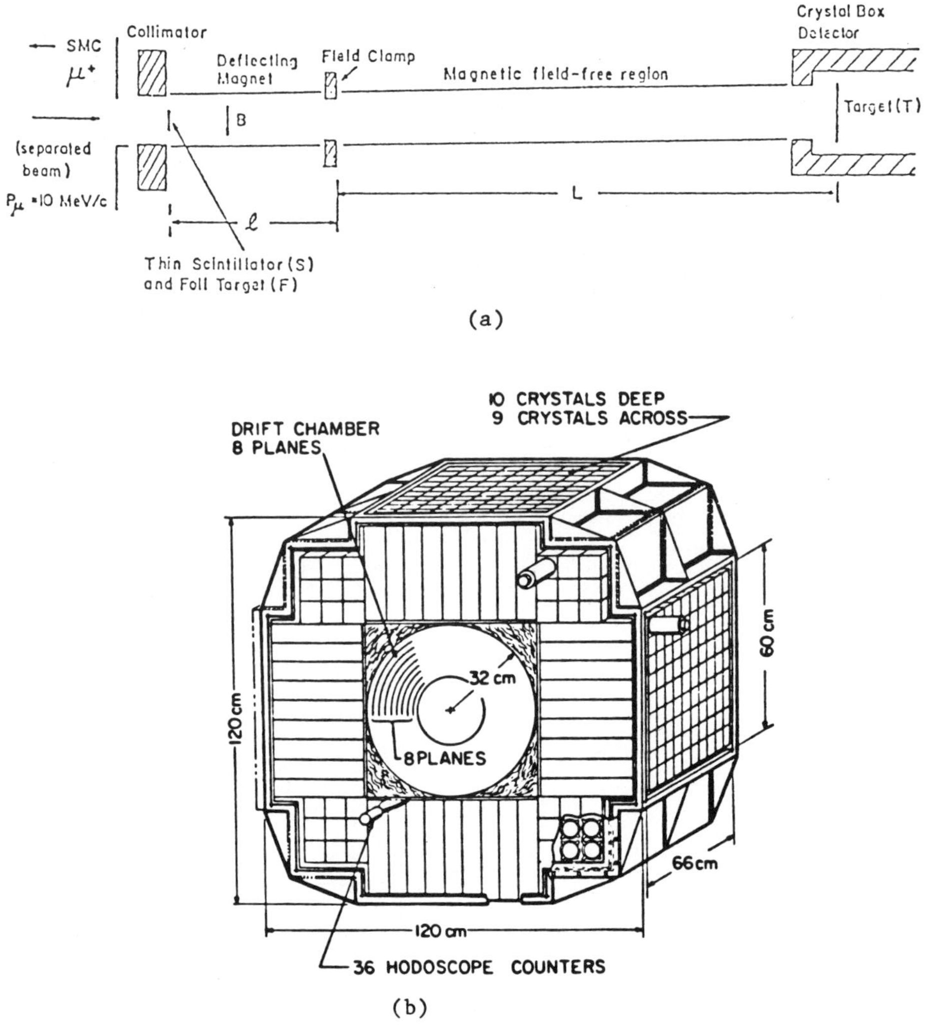

Fig. 14. An experiment done at LAMPF to search for $M \rightarrow \bar{M}$ conversion. (a) Experimental setup, (b) Crystal box detector.

VI. NEGATIVE MUONIUM ION

Recently the negative muonium ion has been produced in an experiment at Los Alamos.[28] The method used is to pass a low momentum μ^+ beam through a thin foil. Double electron capture occurs as the muon emerges from the downstream side of the foil and a small fraction of the emerging beam is in the M^- state. The expected charge state distribution for the emerging muons as a function of their kinetic energy is shown in Fig. 5 and is based on proton data scaled under the Born approximation to muons. The curve predicts a branching ratio of M^- to M of about 10^{-2} with M^- principally in the energy range below 2 keV.

The experimental arrangement is shown in Fig. 15. A separated subsurface muon beam of 10 MeV/c is incident on the production target. The M^- produced from the target are accelerated through a potential of about 20 kV, are selected in charge and momentum by a bending magnet, and are transported through a solenoid to the detector. The detector consists of a microchannel plate to detect the incoming M^-, scintillation counters to observe energetic e^+ from μ^+ decay, and also a NaI detector to measure the energies of the e^+. Time-of-flight of the M^- from the production foil to the microchannel plate is measured (Fig. 16). The scintillation data provide a measurement of the M^- lifetime and the NaI data measure the Michel energy spectrum of the e^+ decays-both confirming that the particle counted in the microchannel plate contains a muon.

From all these data we conclude that M^- is formed and the ratio of outgoing M^- to incident μ^+ on the foil is about 0.5×10^{-4}, which is consistent with the predictions from Fig. 5. Spectroscopy or collision experiments with M^- can now be considered.

VII. FUTURE

As mentioned in section II, for the ground state of muonium improvement in the accuracy of determining $\Delta\nu$ and μ_μ/μ_p by factors of 5 and 10, respectively,seem possible. These improvements will allow a still more sensitive test of QED and of the relativistic two body equation. A more accurate value of m_μ/m_e will also result. The improved determination of μ_μ/μ_p is needed for the more precise measurement of muon g-2 now being undertaken at Brookhaven.[29]

Large improvement in the measurement of the Lamb shift will no doubt take place, particularly ultimately with the use of muon beams with higher phase space density.[30]

The $M \rightarrow \bar{M}$ problem is an important one for flavor-changing reactions and the sensitivity of the search will surely also be extended.

The important development of thermal muonium beams in vacuum,[31,32] opens the possibility of laser spectroscopy of the excited states of M, e.g. the $1S \rightarrow 2S$ transition, as well as more sensitive experiments searching for the $M \rightarrow \bar{M}$ transition.

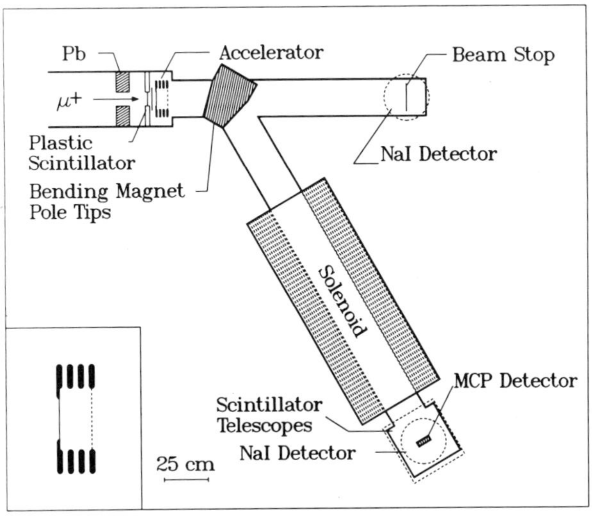

Fig 15. Experimental apparatus used to produce and detect muonium negative ion.

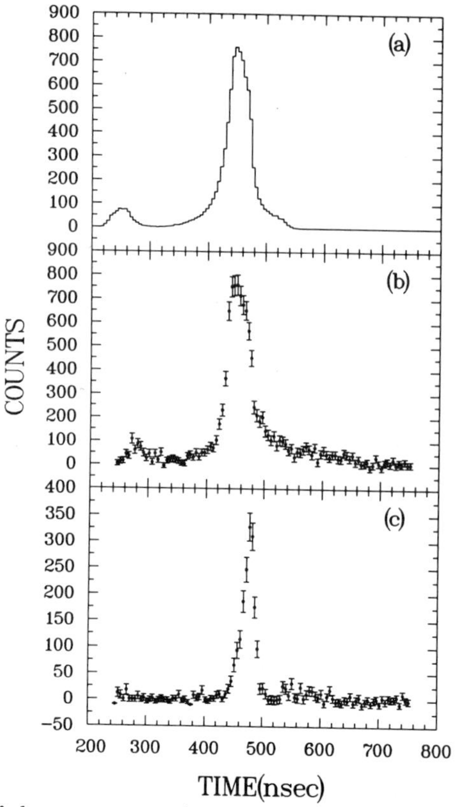

Fig. 16. Time-of-flight spectra: (a) Monte Carlo simulation for muons, (b) measured data for positive muons, (c) measured data for negative muonium ions with -17 kV acceleration voltage. Each bin is 5 ns.

*Research supported in part by DOE under contract DE-AC02-76ER03075.

REFERENCES

1. V.W. Hughes, et al., Phys. Rev. Lett. 5, 63 (1960); Phys. Rev. A1, 595 (1970).

2. V.W. Hughes and T. Kinoshita, in Muon Physics I, ed. by V.W. Hughes and C.S. Wu, (Academic Press, New York, 1977), p. 11.

3. V.W. Hughes and G. zu Putlitz, Comments in Nucl. Part. Phys. 12, 259 (1984); V.W. Hughes, Ann. Phys. Fr. 10, 955 (1985).

4. J.H. Brewer et al., in Muon Physics, ed. by V.W. Hughes and C.S. Wu, (Academic Press, New York, 1975), Vol. III p. 3; J.H. Brewer and K.M. Crowe, Ann. Rev. Nucl. Part. Sci. 28, 239 (1978).

5. V.W. Hughes, Ann. Rev. Nucl. Sci. 16, 445 (1966).

6. F.G. Mariam, et al., Phys. Rev. Lett. 49, 993 (1982).

7. J.R. Sapirstein, et al., Phys. Rev. Lett. 51, 982 (1983); J.R. Sapirstein, Phys. Rev. Lett. 51, 985 (1983); T. Kinoshita and J. Sapirstein, in Atomic Physics 9, ed. by R.S. Van Dyck, Jr. and E.N. Fortson, (World Scientific, Singapore, 1985) p. 38; E.A. Terray and D.R. Yennie, Phys. Rev. Lett. 48, 1803 (1982); J. Sapirstein, in Intersections Between Particle and Nuclear Physics, ed. by D.F. Geesaman, (AIP, N.Y., 1986) p. 567.

8. B.N. Taylor, J. Res. Natl. Bur. Stand. 90, 91 (1985).

9. "Ultrahigh Precision Measurements on Muonium Ground State: Hyperfine Structure and Muon Magnetic Moment" LAMPF Proposal, November 1986, V.W. Hughes, G zu Putlitz, P.A. Souder, Spokesmen. F.G. Mariam, K.P. Arnold, H.J. Mundinger, M. Eckhause, P. Guss, J. Kane, S. Dhawan, S. Kettell, Y. Kuang, B. Matthias, B. Ni and R. Schaefer.

10. V.W. Hughes, in Quantum Electronics, ed. C.H. Townes (Columbia Univ. Press. N.Y., 1960), p. 582; D.E. Casperson, et al., Phys. Lett. 59B, 397 (1975); D. Favart et al., Phys. Rev. Lett. 27, 1336 (1971); Phys. Rev. A8, 1195 (1973).

11. M.A.B. Bég and G. Feinberg, Phys. Rev. Lett. 33, 606 (1974); 35, 130 (1975).

12. P.R. Bolton, et al., Phys. Rev. Lett. 20, 1441 (1981).

13. A. Badertscher, et al., Phys. Rev. Lett. 52, 914 (1984); A. Badertscher, et al., in Atomic Physics 9, ed. by R.S. Van Dyck, Jr. and E.N. Fortson, (World Scientific, Singapore, 1985) p. 83.

14. A. Badertscher, et al., Nucl. Inst. Meth. A238, 200 (1985).

15. C.J. Oram, et al., Phys. Rev. Lett. 52, 910 (1984); in Atomic Physics 9, ed. by R.S. Van Dyck, Jr. and E.N. Fortson , (World Scientific, Singapore, 1985) p. 75.

16. P.A. Souder, et al., Phys. Rev. Lett. 34, 1417 (1975); P.A. Souder, et al., Phys. Rev. A22, 33 (1980).

17. H. Orth, et al., Phys. Rev. Lett. 45, 1483 (1980); C.J. Gardner, et al., Phys. Rev. Lett. 48, 1168 (1982).

18. K.N. Huang and V.W. Hughes, Phys. Rev. A26, 2330 (1982); S.D. Lakdawala and P.J. Mohr, Phys. Rev. A22, 1572 (1980); A29, 1047 (1984).

19. "Neutral Muonium Helium (3)." K.-P. Arnold et al, Abstract X'th Int. Conf. on Particles and Nuclei, Heidelberg (1984); To be submitted to Z. Physik.

20. G. Feinberg and L.M. Lederman, Ann. Rev. Nucl. Sci. 13, 431 (1963); T.D. Lee and C.S. Wu, Ann. Rev. Nucl. Sci. 15, 381 (1965); E.D. Commins and P.H. Bucksbaum, Weak Interactions of Leptons and Quarks, (Cambridge Univ. Press, Cambridge, 1983); L.B. Okun, Leptons and Quarks, (North Holland, Amsterdam, 1982); Muon Physics II, ed. by V.W. Hughes and C.S. Wu, (Academic Press, New York, 1975).

21. H. Primakoff and S.P. Rosen, Ann. Rev. Nucl. Part. Sci. 31, 145 (1981); A. Halprin, Phys. Rev. Lett. 48, 1313 (1982); R.N. Mohapatra, in Quarks, Leptons, and Beyond, ed. H. Fritzsch, (Plenum Press, N.Y., 1985), p. 219.

22. H.K. Walter, in Particles and Nuclei, ed. by B. Povh and G. zu Putlitz, (North-Holland, Amsterdam, 1985), p. 409c.

23. J.J. Amato, et al., Phys. Rev. Lett. 21, 1709 (1968).

24. W.C. Barber et al., Phys. Rev. Lett. 22, 902 (1969).

25. G.M. Marshall et al., Phys. Rev. D25, 1174 (1982).

26. LAMPF Proposal 985, "Search for Muonium to Antimuonium Spontaneous Conversion," V.W. Hughes, J.R. Kane, C. Hoffman. (November, 1985).

27. R.D. Bolton et al, Phys. Rev. Lett. 53, 1415 (1984).

28. K.P. Arnold et al., Bull. Am. Phys. Soc. 31, 35 (1986); Y. Kuang et al., "First Observation of the Negative Muonium Ion Produced by Electron Capture in a Beam Foil Experiment" To be published in Rapid Comm. Phys. Rev. A, 1986; D.R. Harshman et al., Phys. Rev. Lett. 56, 2850 (1986).

29. "A New Precision Measurement of the Muon g-2 Value at the Level of 0.35 ppm.", AGS Proposal 821, September 1985; revised September, 1986. V.W. Hughes, spokesman. Boston University; Brookhaven National Laboratory; City College of New York; Columbia University; Cornell University; University of Heidelberg; Los Alamos National Laboratory; University of Michigan; University of Mississippi; Sheffield University; University of Tokyo; Yale University;
V.W. Hughes, "The Muon Anomalous g-Value", this volume.

30. C. Petitjean, Proceedings of the Workshop on Fundamental Muon Physics: Atoms, Nuclei, and Particles, (Los Alamos, LA-10714C 1986) p. 138; D. Taqqu, ibid. p. 141; L. Simons, ibid. p. 147.

31. A.P. Mills, Jr. et al., Phys. Rev. Lett. 56, 1463 (1986).

32. A. Olin, in Intersections Between Particle and Nuclear Physics, ed. by D.F. Geesaman, (AIP, N.Y., 1986) p. 587; G.A. Beer, et al., Phys. Rev. Lett. 57, 671 (1986).

THE IDENTIFICATION OF $\bar{p}p$ K X-RAYS AT LEAR

C. A. Baker[5], C. J. Batty[5], J. D. Davies[1], C. W. E. Van Eijk[3], R. Ferreira Marques[2], R. W. Hollander[3], D. Langerveld[3], E. W. A. Lingeman[4], J. Lowe[1], J. Moir[5], J. M. Nelson[1], W. J. C. Okx[3], G. J. Pyle[1], S. Sakamoto[5], A. Selvarajah[1], G. T. A. Squier[1], R. E. Welsh[6], R. G. Winter[6], and A. Zoutendijk[3]

[1]Birmingham, [2]Coimbra, [3]Delft, [4]NIKHEF-K Amsterdam, [5]Rutherford Appleton Laboratory, [6]William and Mary

The main aim of PS 174 was to measure the spectrum of L and K X-rays coming from $\bar{p}$ stopping in gaseous hydrogen. Comparing $\bar{p}p$ and electronic hydrogen atoms, atomic energies are raised by m_p/m_e but radii lowered, bringing the innermost orbits within the range of the strong interaction. QED gives L and K X-ray energies in the intervals 1.7-3 and 8-12 kev. Conventional nuclear potential models would shift (ΔE_{1s}) and broaden (Γ_{1s}) the 1s level, and hence the K X-ray energies, by ½ - 1 kev and broaden the 2p level (Γ_{2p}) to 20 - 60 mev.[1] Γ_{2p} is determined from

$$\frac{\text{total L yield}}{K_\alpha\text{ yield}} = \frac{\Sigma Y_L}{Y_{K_\alpha}} = \frac{\Gamma_{2p}}{\Gamma_{2p}(\text{radiation})}.$$

That Γ_{2p} (rad.) = 0.4 mev shows the very strong p-wave absorption. Shifts and widths outside these values are explained by exotica such as baryonium.

The main problems were low X-ray yields, obtaining a well-calibrated X-ray detector of resolution ≃ ½ kev and good efficiency 2 - 20 kev, line background, continuum background and, above all, reliably identifying the K X-rays.

The Apparatus - (figure 1)

The 300 mm^2 Si(Li) detector had an in-beam resolution of 300 ev FWHM at 6.4 kev and good efficiency uniform to above 20 kev. The low yield results from p-wave annihilation and Stark mixing where the intense electrical fields experienced during inter-atomic collisions shifts high n, high ℓ states to high n, low ℓ from whence annihilation rapidly occurs. The low momentum and with $\frac{\Delta p}{p} \sim 0.2\%$ of LEAR beams permits gas targets having decreased atomic scattering. Line background was reduced by using a large Aℓ flask of high purity whose 1 mm walls screened against external X-rays. The Si(Li) was collimated to see only $\bar{p}$ in the target gas or Aℓ.

The continuum background was dominated by Compton scattering of low energy γ-rays in the Si(Li) crystal. The γ-rays come from the decay of annihilation π^0s to 2γ which then shower. Hence the system was kept to very low mass; a NaI(Tℓ) anti-Compton shield around the crystal reduced the continuum by x(2 - 3) in accordance with that calculated from the

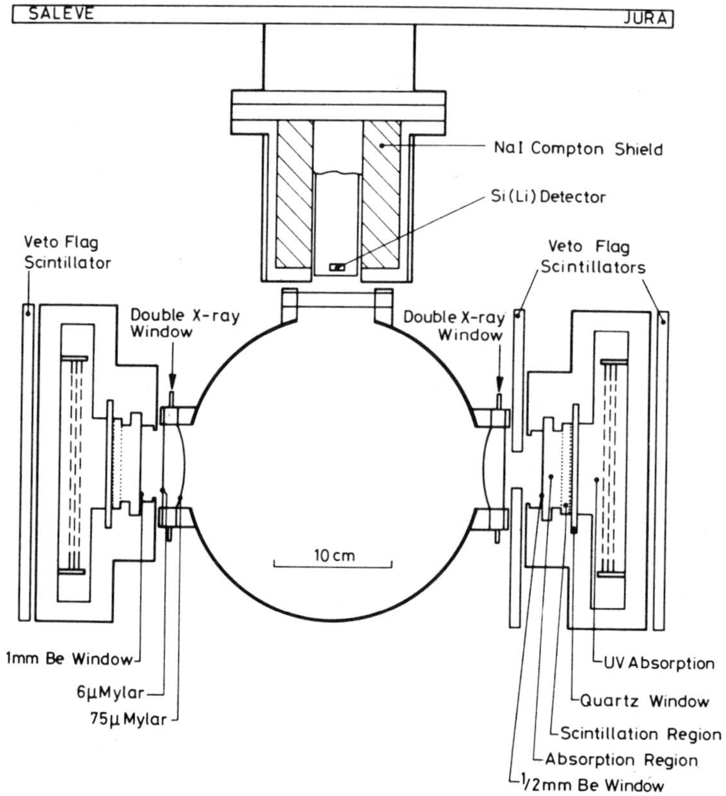

Figure 1 Apparatus - Vertical section normal to Beam

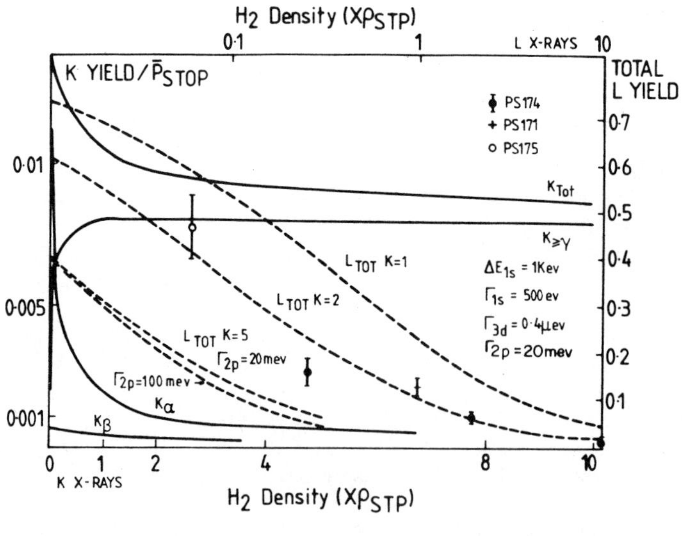

Figure 2 Total $\bar{p}$-p L X-ray yields as a function of H_2 density (1985 data). Predicted $\bar{p}$-p K X-ray yields for Stark mixing parameter, K = 2

γ-ray spectrum measured at the Si(Li) position. The ungated/gated ratio was 25% higher because the NaI itself contributed to the continuum.

Identification

Three separated K X-ray peaks are possible, K_α, K_β and $\Sigma K_{\gamma,\delta...}$; the K_γ, K_δ ... lines appear as one peak because the expected broadening is greater than their (decreasing) line separation.

As for our successive visits to LEAR the lowest momentum was reduced $300 \to 200 \to 105$ MeV/c then we could reduce the gas density $10 \to 2 \to \frac{1}{4}\,\rho_{STP}$ keeping the same stopping distribution and fraction (high at 80%). Cooling the gas $300 \to 30$°K gave the higher densities and reducing the pressure of room temperature gas gave $\frac{1}{4}\rho_{STP}$. For each density, equal numbers of $\bar{p}$ were stopped in hydrogen and deuterium with frequent interchanging; $\bar{p}$ He spectrum for calibration were obtained several times daily. Gas changing was quick, easy and reliable.

Over this density range our cascade calculation, agreeing with that published[2], predict only a small increase in total K yield, K_β always low and the strength transferring $\Sigma K_{\geq\gamma}$ to K_α at low densities. Figure 2 shows predictions for the density behaviour of K X-ray yields (Stark mixing parameter, K = 2) and total L X-ray yields (1985 data).
Equality of K_α and $\Sigma K_{\geq\gamma}$ yields is expected at about $\frac{1}{4}\rho_{STP}$ since K = 1 moves this point only to $0.4\rho_{STP}$.

Figures 3a and b show our analysed data at 10 and 2 ρ_{STP}. The peak is identified as $\Sigma K_{\geq\gamma}$.

In 1986 we took data at 105 MeV/c, 1 and $\frac{1}{4}\rho_{STP}$ with X3 the number of $\bar{p}$ into the completely new apparatus of figure 1. The double windows prevented water vapour diffusing through the mylar. Preliminary SiLi spectra are shown in figures 3c and d and clearly show the expected change in spectral distribution. Each peak already has $>5\sigma$ significance and the continuum background will be halved by known further analysis.

Presently SiLi detectors are limited by small area but high Compton scattering because, in order to reduce junction capacitance, their thickness is x 10 greater than that required to absorb the X-rays. At > 1 kev the resolution of proportional drift counters is too large and their argon filling significantly loses efficiency for the higher energy X-rays.

The Xe filling of the Gas Scintillation Proportional Counters of figure 1 maintains this efficiency. X-rays are absorbed in the first cell and the electrons drifted into the scintillation region. Here the electron field is sufficiently intense to enable the electrons to excite but not ionize the Xe atoms. U.V. photons pass the quartz window to be absorbed by the TMAE doping of the MWPC detector. That amplification is by excitation and photon emission, not ionization, reduces the spread in pulse heights and improves the resolution.

GSPCs had not previously been used in the fierce backgrounds from hadron beams, especially from $\bar{p}p$ annihilations. Considerable precautions were taken against charged particles, viz the scintillator vetos. By having the Be window at HT then the 2nd grid could be at earth and hence lie on the quartz window; this removed the need for a 3rd grid so reducing the lateral extent of the MWPC required to catch the UV photons. The potential on the 1st grid was increased so that there was a small amount of prescintillation light from the absorption region. Figure 4 shows the stylized pulses obtained from an X-ray and from a charged

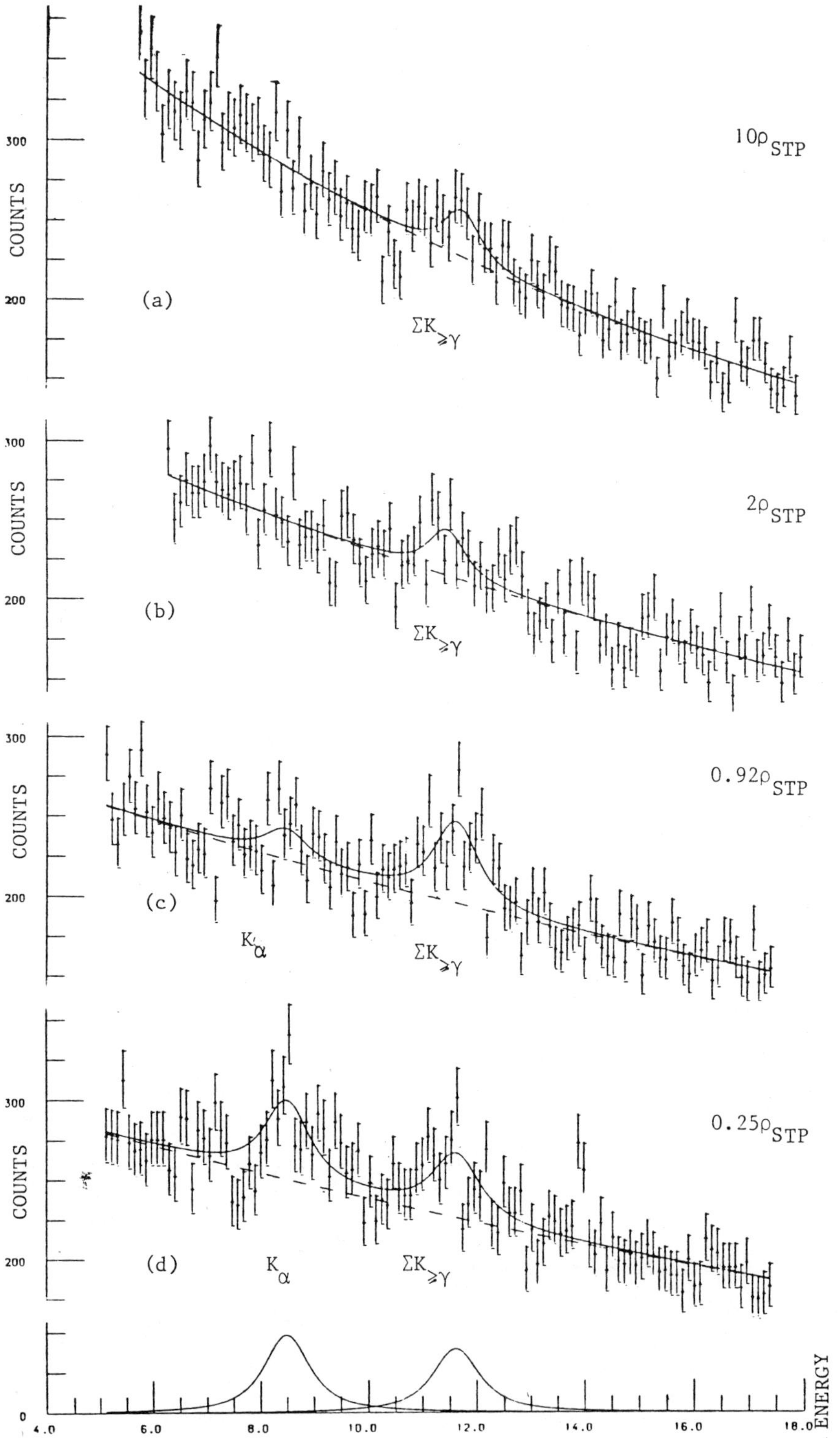

Figure 3 $\bar{p}$-p X-ray spectra at H_2 densities (a) $10\rho_{STP}$ (b) $2\rho_{STP}$ (c) $0.92\rho_{STP}$ (d) $0.25\rho_{STP}$.

particle. Setting $t_1 - t_0 > t_{min.}$ and tight limits on $t_2 - t_1$, $t_3 - t_2$, $\frac{Q}{V}$ derived from frequent $\bar{p}$ He and source calibrations enabled a 300:1 charge-particle rejection. In comparison with the SiLi looking at $\bar{p}$ He spectra, the GSPC resolution is x 1.7, area x 10 and signal: noise x 6.

Further verification of $\bar{p}p$ K X-rays having been seen comes from the $\bar{p}d$ spectrum. In hydrogen and deuterium these X-rays have the same negligible absorption and $\bar{p}$ the same stopping distribution if the densities are the same. Also $\bar{p}d$ energies are increased by x 1.33 because of the greater deuteron mass. Continuum background from hydrogen and deuterium are very similar and we found no evidence in the $\bar{p}d$ spectra for the $\bar{p}p$ K X-ray structure. Every hydrogen spectrum had a corresponding deuterium one and the results were null for all SiLi and GSPC measurements. Figures 5a and b show the $\bar{p}p - \bar{p}d$ spectral differences for the raw on-line data at ¼ and $1\rho_{STP}$ from one GSPC. This measure is independent of figure 3 but shows the same spectrum change with density.

Spectral regions around K lines are used to search for contaminant lines. Some air was found. So a spectrum from an air target with the same $\bar{p}$ stopping distribution was taken to give the intensity of sister lines in the K X-ray regions; their presence does not affect the results.

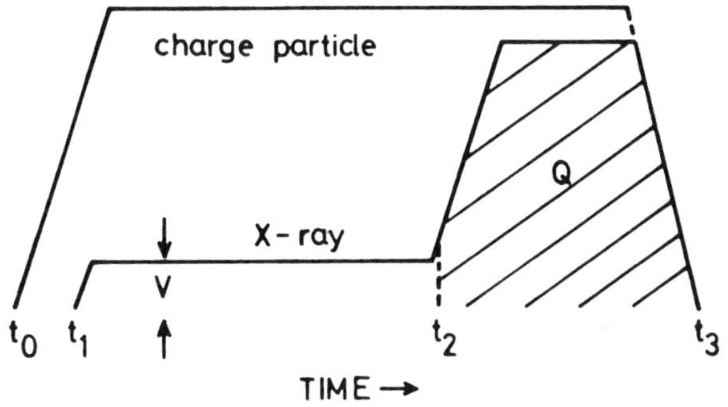

Figure 4 Stylised pulses from a charge particle and from an X-ray in a GSPC

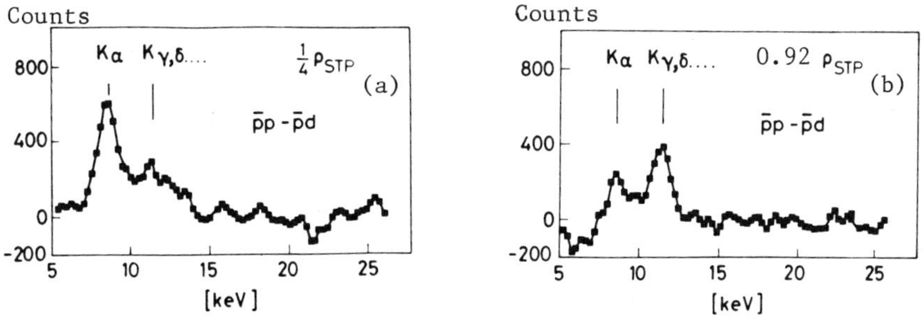

Figure 5 Raw data from one GSPC. $\bar{p}p - \bar{p}d$ spectral difference at (a) $\frac{1}{4}\rho_{STP}$ (b) $0.92\rho_{STP}$

Results

p(Mev/c)	ρ_{STP}	ΔE_{1s}(keV)	Γ_{1s}(keV)		
300	10	-0.65 ± 0.14	0.77 ± 0.46	SiLi	Published[3]
200	2	-1.00 ± 0.25	1.10 ± 0.80	SiLi	Published[3]
105	0.92	-0.91 ± 0.08	0.96 ± 0.26	SiLi	Preliminary
105	¼	-0.90 ± 0.08	0.94 ± 0.29	SiLi	Preliminary
105	¼ and 0.92	-0.8 ± 0.2	$\leq$ 0.95	GSPC	raw data

References

1. J.M. Richard and M.E. Sainio, Phys. Lett. 110B (1982) 349.
2. E. Borie and M. Leon, Phys. Rev. A21 (1980) 1460.
   E. Borie, Physics at LEAR , U. Gastaldi and R. Klapisch ed., Plenum (1984) 185.
3. T.P. Gorringe et al., Phys. Lett 162B (1985) 71.

LOW AND HIGH RESOLUTION PROTONIUM SPECTROSCOPY AT LEAR IN ACOL TIME: TOOLS FOR GLUEBALL, HYBRID AND LIGHT MESON SPECTROSCOPY AND FOR MEASURING THE $N\bar{N}$ STRONG INTERACTION DEPENDENCE ON ANGULAR MOMENTUM

Ugo Gastaldi

CERN
CH-1211 Geneva 23
Switzerland

INTRODUCTION

The $p\bar{p}$ atom is a simple system that permits to study $N\bar{N}$ strong interactions at threshold in depth and to exploit $N\bar{N}$ annihilations for light and exotic mesons spectroscopy. The study of $N\bar{N}$ interactions in depth at threshold requires to measure individually all the annihilation channels, to measure the global effects of all open annihilation channels (that reflect in the hadronic width of protonium atomic levels) and to measure the global effects of virtual elastic and charge exchange and of all open virtual annihilation channels (that reflect in the hadronic shift of the atomic levels). Ideally these studies should be done for all J^{PC} states of protonium with L=0 and 1 in order to get the complete picture of the dynamics of the strong interaction and of its dependence on the quantum numbers of the initial state. The search of new mesonic resonances in $p\bar{p}$ annihilation requires to measure exclusive annihilation channels and the intermediate resonant states. This search and the study of the dynamics of $N\bar{N}$ annihilations at threshold are one a byproduct of the other. Experimentally two lines of research are open with different requirements from the X-ray detectors used for protonium spectroscopy.

RESEARCH WITH LOW RESOLUTION PROTONIUM SPECTROSCOPY

$N\bar{N}$ interactions

One line of research is the study of the annihilation channels of $p\bar{p}$ atoms from the S and P initial states populated by the atomic cascade. This study will permit to learn experimentally the dependence of the dynamics of $p\bar{p}$ annihilation from the angular momentum of the initial state. It requires complex and complete detectors able to identify and measure precisely the pions, gammas and kaons that may be present in the final state so that ideally all the exclusive final states and the possible resonances in the intermediate states can be identified and measured. Protonium spectroscopy helps in this work as the L X-rays of the radiative transitions from the nD levels to the 2P can be used to select clean samples of P-wave annihilation events. The 2P level decays dominantly via annihilation (with probability >98%[1]) and the L X-rays populating the 2P level can be detected efficiently with a dedicated large acceptance detector[2] that works inside a magnetic field and complements the final state detector for triggering, tracking and identifying charged particles of the final state[3]. The contamination of S-wave annihilations in this sample is below 10% in a H_2 gas target surrounded by a XDC/SPC[2,3] detector, as the non L X-ray background in the L X-ray energy region (1.7-3.1 keV) is mostly due to internal bremstrahlung from the charged particles emitted in the annihilation process, and the measured yield of these bremstrahlung X-rays is about 10% of the L X-rays yield[4]. The programme to study P and S-wave $p\bar{p}$ annihilation events at rest at LEAR was suggested in 1978[5]. It has been started by the ASTERIX experiment[6] that has used the first operational X-ray drift chamber[7] and will be extended by the OBELIX[8] and Crystal Barrel[9] experiments in ACOL time. For the purpose of this programme the energy resolution of the X-ray detector is not critical provided the detector can select L X-rays (1.7 to 3.1 keV) against M (0.6 to 1.4 keV) and K (8.9 to 12.1 keV) X-rays (see fig. 1). It is instead critical the demand that the detection efficiency for L- X-rays be good (~50%) and that the physical background and detector noise levels be well below the L X-ray signal.

Exotic and light meson searches

In the same experimental context, when the physics interest is focused on subnuclear phenomena, the L X-ray emitted by protonium offers a new tool to search for new resonances (particularly exotic mesons containing gluons as constituents), including broad ones.

The general idea[10÷13] is to study and compare in production experiments with the same detector identical decay channels (e.g. $\pi^+\pi^-$ or K^+K^-) of possible resonant states present in exclusive final states of the same type (e.g. final states with three pions or final states with a K^+K^- pair plus one pion) produced from $N\bar{N}$ initial states of different discrete quantum numbers (spin, isospin, angular momentum) but of equal energy. Phase space, detector and software efficiencies, acceptances and biases and detector plus operation instabilities can be factorized away in this comparison if data are taken simultaneously for different quantum numbers of the initial state. Differences in the spectra of identical decay channels may indicate new structures (besides giving informations about dynamical effects). In practice it is not always possible to switch one discrete quantum number of the initial state but it may be feasible to collect simultaneously two data sets that have initial states with different distributions of that quantum number. Changing the spin of the initial state necessitates a polarized $\bar{p}$ beam or a dense polarized pure hydrogen target. These facilities are not available at present, but work is in progress to explore the realizability of recent suggestions to develop polarized $\bar{p}$ beams[14,16] and dense polarized H_2 targets[15,16]. We hoped to be able to distinguish and select singlet and triplet 1S initial states of protonium if they would have been split and separated enough in energy by the spin dependent strong interaction[17]. It turns out that, apart for $p\bar{p}$ annihilation with only gammas present in the final state, the physical X-ray background due to internal bremstrahlung is larger that the K line signal, which is broad and hard to appreciate in the X-ray spectrum. No evidence in favour nor against the ground state splitting is available, the K_α and K_β lines are not resolved and it is not clear whether this is an instrumental problem, or due to a large broadening of the S states or due to strong interaction shift of the singlet and triplet with opposite signs so that approaching or even crossing occurs between K_α and K_β lines feeding different spin states. The isospin composition of the initial state can be changed selecting $p\bar{p}$ (I=0 and 1) and $\bar{p}n$ (I=1) annihilations with $\bar{p}$ stopped in a D_2 target and measuring the recoil neutron direction and the recoil proton momentum for a complete kinematical reconstruction of the events. As already mentioned in the previous paragraph the L X-rays of protonium permit to select P-wave annihilations (occuring from the 2P atomic levels) with an S-wave contamination inferior to 10%. For the purpose of observing differences in the spectra of candidate decay channels of new mesons, data sets with L X-rays in coincidence can be compared with data sets where no request is made about X-rays. This second data set has a markedly different S-wave annihilation fraction (the precise value of this fraction is still under evaluation, the ball-park for the S-wave annihilation fraction is in a NTP H_2 gas target between 20 and 50%[18÷20].

RESEARCH WITH HIGH RESOLUTION PROTONIUM SPECTROSCOPY

$N\bar{N}$ interactions

The second line of research offered by protonium to study $p\bar{p}$ strong interactions relies critically on the energy resolution of the detector that measures the X-rays emitted in the atomic cascade of the $p\bar{p}$ atom. The strong interaction modifies the Coulomb wave function of S and P orbitals and causes a shift ΔE and a broadening Γ^A of the levels of protonium. ΔE and Γ^A depend on the principal quantum number n, on the angular momentum and on the spin of each level. The shift ΔE of a level is associated, as already mentioned, to the virtual elastic and charge exchange channels and to the sum of all virtual annihilation channels open to the $p\bar{p}$ system in that level. The broadening Γ^A of a level is due to the sum of all the real annihilation channels open to that level. Fig. 2 illustrates the scheme of protonium levels. Table 1 summarizes the contributions to the binding energy and to the broadening of the lower levels of protonium. The electromagnetic and relativistic effects inserted in the table have been calculated by Borie[21]. The strong interaction contributions are the values by Richard and Sainio[22]. Their potential model calculation gives results compatible with other approaches (see for example refs. 23 and 24), however the numerical values are valuable to illustrate the expected orders of magnitude of the strong interaction effects and to compare with electromagnetic effects, but they must not be used for more, as the annihilation term used in the potential is assumed as independent on the quantum numbers of the initial states (this is experimentally not the case in real annihilation already for individual exclusive annihilation channels). Theoretical work is necessary to create both the tools to calculate the individual channels of real annihilation (branching ratios of exclusive final states and of intermediate resonant states, angular distributions and correlations) and the global effects of real and virtual annihilation on the protonium levels (broadening and shift). Motivations for this work may be of several types: (i) it is the first time that the dependence on angular momentum can be investigated directly, (ii) $p\bar{p}$ annihilation is a good example of "hadronization" and the hadronization phenomenology of this process will be eventually explored at LEAR in great detail (initial state quantum number dependence, final state flavour dependence) and with statistics limited essentially by the analysis time of the experimental data, (iii) there is the possibility and the time for "previsions" and there are experimental data available from bubble chambers[25] and coming shortly from pre-ACOL LEAR experiments for exercising theoretical and computation models.

Table 1. Contributions to the binding energy of the lower levels of the $p\bar{p}$ atom

State		Coulomb	Vac. Pol $\alpha Z\alpha$	Vac. Pol $\alpha^2 Z\alpha$	Rel. Corr.	Spin orbit + Spin-Spin	Strong	
J^{PC}	$n^{2S+1}L_J$	(eV)	(meV)	(meV)	(meV)	(meV)	ΔE (meV)	$\Gamma/2$ (meV)
0^{-+}	1 $^1S_0$	12491					$-540\ 10^3$	$510\ 10^3$
1^{--}	1 $^3S_1$	12491					$-760\ 10^3$	$450\ 10^3$
0^{-+}	2 $^1S_0$	3123					$-68\ 10^3$	$66\ 10^3$
1^{--}	2 $^3S_1$	3123					$-96\ 10^3$	$59\ 10^3$
1^{+-}	2 $^1P_1$	3123	2037	16	27	0	+ 26	13
0^{++}	2 $^3P_0$	3123	2037	16	27	-249	+ 74	56
1^{++}	2 $^3P_1$	3123	2037	16	27	- 16	- 36	10
2^{++}	2 $^3P_2$	3123	2037	16	27	60	+ 5	15
0^{-+}	3 $^1S_0$	1388					$-20\ 10^3$	$20\ 10^3$
1^{--}	3 $^3S_1$	1388					$-29\ 10^3$	$18\ 10^3$
1^{+-}	3 $^1P_1$	1388	538	4	11	0	+ 9	4.4
0^{++}	3 $^3P_0$	1388	538	4	11	- 74	+ 26	20
1^{++}	3 $^3P_1$	1388	538	4	11	- 5	- 13	3.6
2^{++}	3 $^3P_2$	1388	538	4	11	18	+ 1.2	5.4
2^{-+}	3 $^1D_2$	1388	152	1	4	0	$< 10^{-3}$	$<10^{-3}$
1^{--}	3 $^3D_1$	1388	152	1	4	- 16	"	"
2^{--}	3 $^3D_2$	1388	152	1	4	- 1	"	"
3^{--}	3 $^3D_3$	1388	152	1	4	7	"	"

$\Gamma^{RAD}(2P) = 0.37$ meV $\quad \Gamma^{A}_{BSL}(2P) \cong 30$ meV $\quad \Delta E_{vp}(nS) \simeq \frac{1}{n^3}\ 40$ eV

$\Gamma^{RAD}(3D) = 0.038$ meV $\quad \Gamma^{A}_{BSL}(3D) \cong 10^{-4}$ meV

From table 1 it appears that (i) an energy resolution of better than 1 keV for X-ray energies of about 10 keV is necessary to study conveniently the strong interaction effects on the ground state of protonium, (ii) an energy resolution of better than 20 meV (20 millielectron volts) for X-ray energies of about 2 keV is necessary to study strong interaction effects on the 4 sublevels of the 2P level.

Pre-ACOL LEAR experiments linked or dedicated to protonium spectroscopy have used different types of detectors with characteristics summarized in table 2.

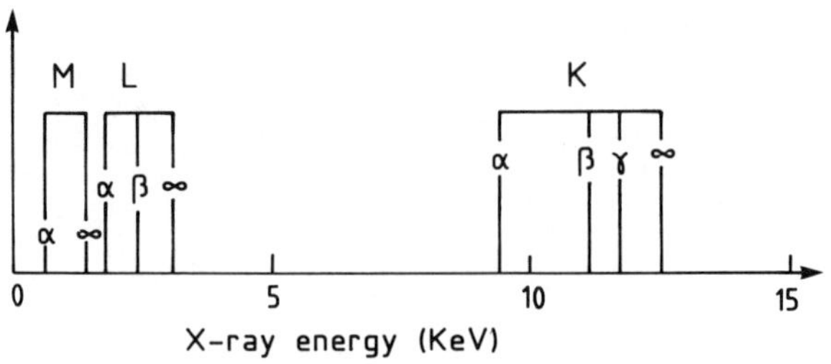

Fig. 1. Energy windows for K, L and M protonium X-rays. There is no overlap of the three windows

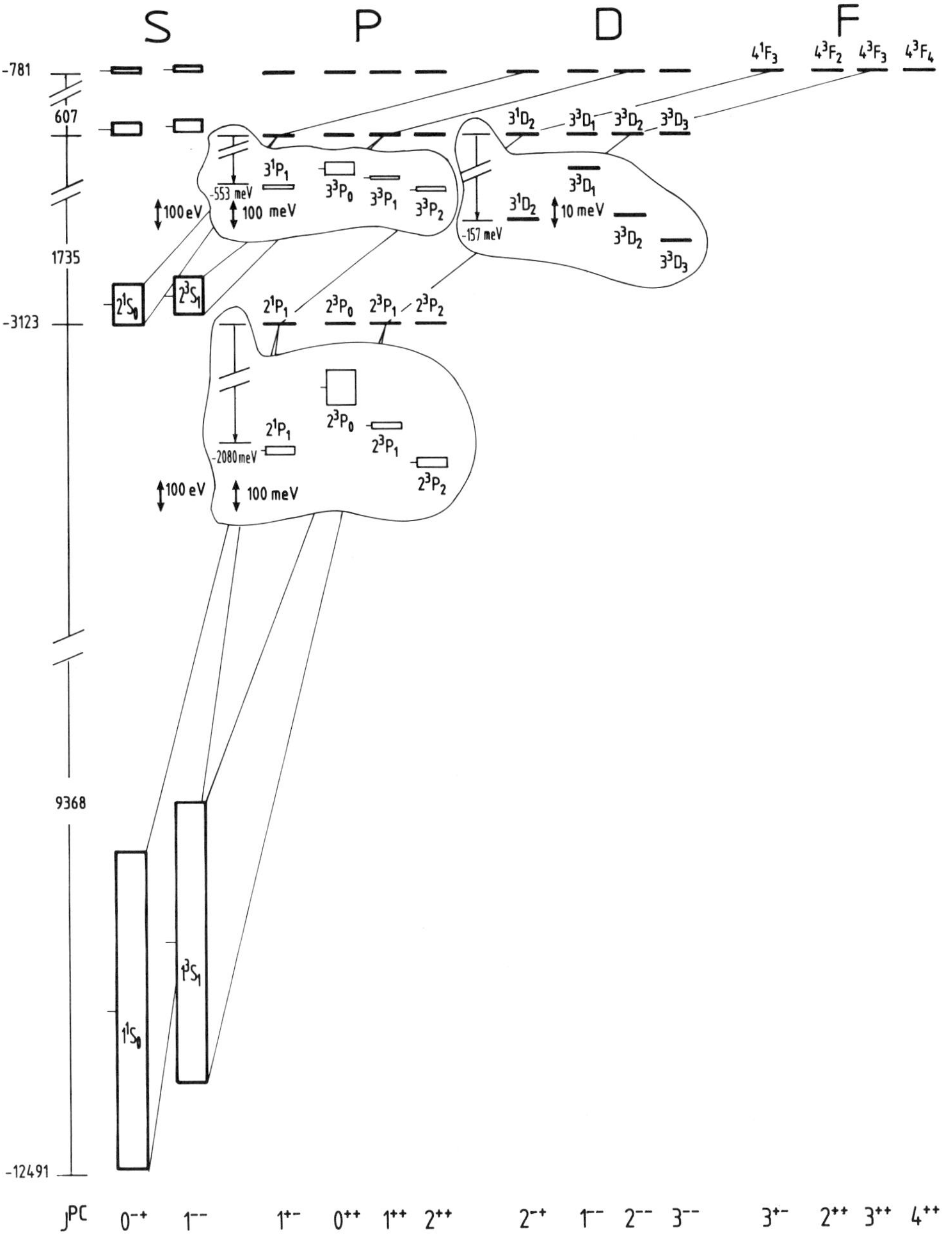

Fig. 2. Level scheme of the low levels of protonium, where X-ray emission spectroscopy is feasible

Table 2. Characteristics of the X-ray detectors used at LEAR before ACOL for protonium spectroscopy

Experiment	ASTERIX (PS171)	PS174		PS175
Detector	XCD/SPC	SiLi	GSPC	SiLi
References	2,3,6,7,26	29	27,28,29	30
Geometry of active volume	(a)	(b)	(c)	(d)
Surface of entrance window (cm^2)	4000	3	39	3
Angular acceptance ($\Omega/4\pi$)	~ 90%	~0.1%	~ 1%	~0.1%
Thickness of active volume (cm)	7	≈0.1	2	≈0.1
Entrance windows (material-thickness)	- - mylar:6μ	mylar:81μ air:⩾2cm detector window	mylar:81μ air:⩾2cm Be:1000μ	- - detector window
Granularity (nr of cells of lattice in active volume	⩾ 10^6	1	1	1
Spacial resolution inside active volume	$\sigma(r)$~200μm $\sigma(z)$~ 2cm $\sigma(\phi)$= 4°	-	-	-
Energy threshold (e) (keV)	⩽ 1	>3.5	>3.5	>1
Energy resolution FWHM(eV) at 1.7 keV at 5.9 keV at 6.4 keV	 700 (f) 1400 (g)	 300	 530 eV 	 260 (f)
Special capabilities	(h) (k) (l)	-	-	-

(a) torus t_i=8cm t_e=15cm l=90cm filled with Ar/Ethane 50/50 NTP
(b) disk
(c) disk r=3.5cm l=2cm filled with Xe NTP
(d) disk
(e) transmission of entrance window is >10% above threshold
(f) measured on $p\bar{p}$ L_α line
(g) measured on ^{57}Co 6.4 keV line
(h) gives position of X-ray line source
(k) vertex detector for $N\bar{N}$ annihilations
(l) can determine energy of X-ray line also by mean free path measurement

Obviously with those detectors (that have been used in the conventional calorimetric approach to measure the X-ray energies) only the spectroscopy of the ground state could be attacked because of the moderate energy resolution. The predictions for the sign and the order of magnitude of $\Delta E(1S)$ have been confirmed by two experiments[1,31] but, as already mentioned, separate measurements are not available on $\Delta E\ (1^3S_1)$ and $\Delta E\ (1^1S_0)$ and only coarse upper limits can be given on $\Gamma^A(1S)$. The upper limits on Γ^A are coarse and model dependent as the singlet and triplet lines are unresolved and it is very hard to extract the four independent quantities $\Delta E(1^3S_1)$, $\Delta E(1^1S_0)$, $\Gamma^A(1^3S_1)$, $\Gamma^A(1^1S_0)$ from one single bump in the energy spectrum which is in addition sitting on top of a large background and that might contain unresolved both singlet (triplet) K_α and triplet (singlet) K_β transitions.

In the following of this paper we focus onto a new approach to protonium spectroscopy which will be possible only at LEAR and that aims at very high resolution, so to permit to study in emission spectroscopy the 2P level by measuring the strong interaction effects onto the 2P singlet sublevel and on the three 2P triplet sublevels. The relative energy resolution will be after some efforts of about 10^{-4} independently of the X-ray energy. The energy resolution will then be about 200 meV in the 2 keV region, thus permitting also -as a bonus- to study accurately the 2S level: the 200 meV resolution at 2 keV has to be confronted with the expected shifts and broadening of the 2^3S_1 and 2^1S_0 sublevels of about 80'000 meV. The 1 eV energy resolution in the 10 keV energy region has to be compared to shifts of the 1^1S_0 and 1^3S_1 sublevels of about 600 eV and to the best todays energy resolution of ≈300 eV in the K lines energy region.

HIGH RESOLUTION EMISSION SPECTROSCOPY OF PROTONIUM

Fig. 3 summarizes the basic ideas of the approach. Protonium atoms are produced in flight in LEAR by having $\bar{p}$ and H^- coasting beams corotating at the velocity required by the transition to be studied. Resonant spectroscopy on the H^0 beam produced together with the $p\bar{p}$ atomic beam offers an absolute calibration of the $p\bar{p}$ beam velocity and complements momentum measurements in LEAR. A vertex detector aligned to the $p\bar{p}$ atoms beam line measures $p\bar{p}$ annihilation vertices and annihilation channels and provides timing and trigger for the X-ray detectors. X-ray Drift Chambers measure the X-rays emitted by the $p\bar{p}$ atoms that annihilate in the region surrounded by the vertex detector. The thin windows at the detector entrances or the active gas itself of the detectors are used as differential X-ray absorbers. X-rays emitted by the $p\bar{p}$ atom are Doppler shifted and their energy in the laboratory is given by

$$E_{p\bar{p}} = E_{LAB} \frac{1-\beta\cos\theta}{\sqrt{1-\beta^2}} \tag{1}$$

For a given velocity $v=\beta c$ of the $p\bar{p}$ atoms, the X-rays can traverse the window or penetrate the detector gas depending from their emission angle Θ in the laboratory frame. For each detected X-ray the angle Θ is determined by the $p\bar{p}$ beam direction, the annihilation vertex and the X-ray absorption point. Below or above the critical angle Θ^* the X-rays of a monochromatic line appear or disappear in a detector and the measurement of Θ^*, together with v and the absorption energy E_k of the element used as differential absorber, permits to measure the energy $E_{p\bar{p}}$ of the transition under investigation:

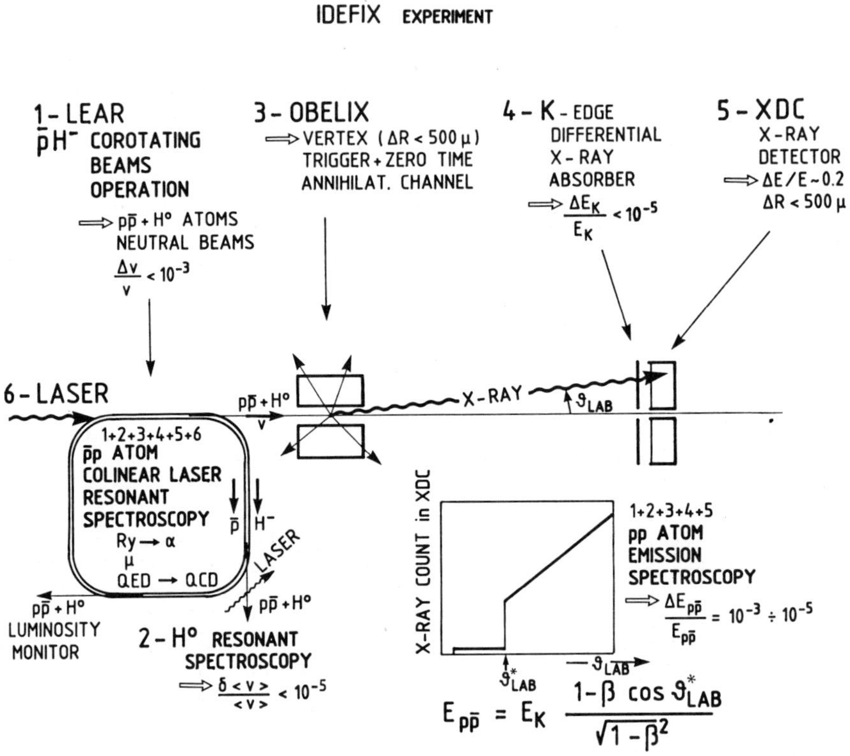

Fig. 3. Overall scheme for spectroscopy of protonium in flight

$$E_{p\bar{p}} = E_k \frac{1-\beta\cos\theta^*}{\sqrt{1-\beta^2}} \qquad (2)$$

The Doppler swing in energy possible at LEAR is up to a factor 2 for forward emitted X-rays and down by a factor 1/2 for backward emitted X-rays. This will permit to study with commercially available differential absorbers the protonium L lines and K lines. The factors limiting the resolution are the velocity spread $\Delta\beta$ of the $\bar{p}$ beam, the error on Θ^* and the error on E_k according to the relation[32]:

$$\left[\frac{\Delta E_{p\bar{p}}}{E_{p\bar{p}}}\right]^2 = \left[\frac{\Delta E_K}{E_K}\right]^2 + \left[\frac{\beta-\cos\Theta^*}{1-\beta\cos\Theta^*}\,\frac{1}{1-\beta^2}\right]^2 \Delta\beta^2 + \left[\frac{\beta\sin\Theta^*}{1-\beta\cos\Theta^*}\right]^2 \Delta\Theta^{*2} \qquad (3)$$

The relative energy resolution is mainly controlled by the momentum spread Δp of the $\bar{p}$ beam circulating in LEAR and by the angular resolution and it is practically independent from the energy as $\Delta E_K/E_K \leqslant 10^{-5}$ with convenient K-edge absorbers in all the 1÷20 keV energy window.

The IDEFIX approach to protonium spectroscopy was considered first in 1976[33] with respect to the advantage of boosting the X-ray energies in the lab with beams of $p\bar{p}$ atoms. It was made realistic in 1977 with the ideas of exploiting differential absorbers in $p\bar{p}$ emission spectroscopy[34] and of producing $p\bar{p}$ atoms with $\bar{p}$ and H^- co-rotating in the same storage ring[35,5]. After the letter of intent proposing protonium spectroscopy in flight at LEAR[36] and the proposal of the X-ray Drift Chamber[2], it was realized[37] (i) the possibility of using a XDC (with a geometry different from the one that was at that time in construction for ASTERIX[17]) to measure with large acceptance and precisely the absorption point of each detected $p\bar{p}$ X-ray and (ii) that -in conjunction with a good measurement of the $p\bar{p}$ annihilation vertex- this offers a precise measurement of Θ^* [and reduces $\Delta\Theta^*$ in (3)] together with an increase of the angular acceptance. Later on it was realized[38] the possibility of measuring accurately β of the $p\bar{p}$ atoms beam by resonant spectrosocpy of the H^0 atoms produced together with and more copiously than the $p\bar{p}$ atoms in the $\bar{p}H^-$ corotating beams scheme. This measurement is complementary to LEAR machine measurements of the momentum of the circulating beam, it might be more precise[39] and permits to compare directly the spectroscopies of the $p\bar{p}$ and pe^- systems[38].

Doppler-tuned beam-foil X-ray spectrometry[32] is used with non relativistic beams at several linear and circular ion accelerators[40]. The main new features at LEAR will be large granularity position-sensitive X-ray detectors and beams in the relativistic regime. Resonant spectroscopy on relativistic H^0 beams has been performed in the context of an experimental programme of study of the H^- ion at LAMPF at Los Alamos[41÷43]. The typical relative energy resolution is around 10^{-3} and it is dominated by the momentum spread of the H^- bunches accelerated by the linac. At LEAR the H^- beam will be stored and cooled and an improvement by one order of magnitude is well feasible.

The way we envisage at present the detection system is sketched in fig. 4, where it is not shown for economy the apparatus with axial symmetry that surrounds the ~1m long central part of the biconical beam pipe sector and that measures the annihilation channel and vertex.

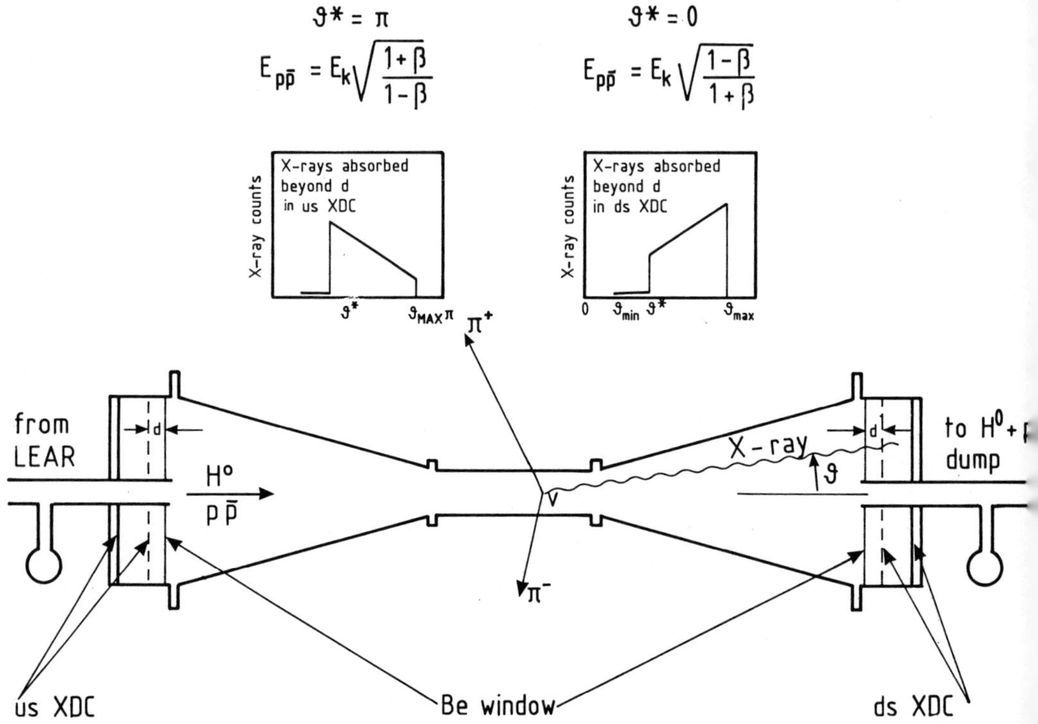

Fig. 4. Set-up for X-ray detection in IDEFIX

Planar XDC's working at low pressure (to reduce the mechanical stress on the Be windows) are flanged upstream and downstream of the biconical beam pipe. The distance of each detector from the centre of the useful annihilation vertex region is ≈3m, the maximum accepted opening angle between the X-ray direction and the $p\bar{p}$ beam direction is 15° in both directions. The planar XDC should have a diameter of ~1.5m to cover the usable solid angle at a distance of ~3m from the annihilation point. The XDC on the left receives $p\bar{p}$ X-rays with energies reduced by the doppler shift by a factor down to $\sqrt{(1-\beta/1+\beta)}$. The XDC on the right receives $p\bar{p}$ X-rays with energies doppler enhanced by a factor up to $\sqrt{(1+\beta/1-\beta)}$. For a given protonium transition a convenient β is chosen for the operation of LEAR in the $\bar{p}H^-$ mode so that $\Theta^*<15°$ or $\Theta^*>165°$ in one of the two XDC detectors. The upstream detector is used if no convenient differential absorber is available with $E_k>E_{p\bar{p}}$, the other way around for the downstream XDC. As differential absorber the most elegant option is to use the noble gas of the drift chamber (Ne: 866.2÷867.6 edge, Ar: 3203 eV edge, Xe: 4782 eV edge) for the L X-rays, while foils can be comfortably used in the 10 keV energy region for work on the K lines. If a foil is used the situation is illustrated in fig. 3. In the forward XDC if the counter gas is used as differential absorber, all X-rays that cross the Be window are absorbed in the detector very near to the Be entrance window for X-rays with $\Theta<\Theta^*$. The complementary situation will occur in the downstream XDC. The two XDC will have the same physical background for X-rays not associated to the moving $p\bar{p}$ atom. The essence of the measurements will be to plot as a function of Θ the X-ray signals associated to L or K transitions (distinguishable by pulse amplitude measurement) that will have been absorbed inside one XDC active volume beyond a plane (defined by drift time) parallel to the entrance window and chosen at a z position convenient for optimizing the signal/-background ratio. A sharp step in the Θ plot will indicate Θ^*.

In order to get a feeling of the orders of magnitude for rates and resolution consider the figures of table 3.

Table 3: Scenario of IDEFIX data taking

Beam flux	$B = 10^4\ p\bar{p}\ sec^{-1}$
Length of beam instrumented with vertex detector	$L = 1$ m
Average weighted distance between annihilation vertex & X-ray absorption point	$D = 3$ m
Angular acceptance of one X-ray detector	$\Theta^{max} \leqslant 15°$ $0 \leqslant \Phi \leqslant 2\pi$ $\Omega/4\pi \simeq 3.4\%$
Accuracy in vertex reconstruction (a)	$\sigma_v(r) \leqslant 500\mu$
Accuracy in absorption point reconstruction (b)	$\sigma_x(r) \leqslant 500\mu$
Population of 2P level (c)	$P \geqslant 90\%$
Fraction of $p\bar{p}$ atoms that annihilate in the instrumented region	$F \simeq 1\%$

(a) conservative value, but sensible in view of the large diameter of the beam pipe (~30 cm)
(b) conservative. 70μ accuracy is obtainable by equipping the amplification and read out part of the XDC with a parallel plate and a two dimensional read out[44] and a comparable resolution would be attainable with a plane of amplification wires and measuring the induced pulses on circular printed strips coaxial to the $p\bar{p}$ beam
(c) from calculation of the protonium cascade in vacuum[45]

The signal X is given by

$$X = B \cdot F \cdot P \cdot \frac{1}{4\pi} \int_{\Theta^*}^{\Theta^{max}} 2\pi \sin\Theta \, d\Theta \qquad (4)$$

with the figures of table 3 we get

$$X = 45 \cdot (\cos\Theta^* - \cos\Theta^{max}) \ sec^{-1} \leqslant 1.5 \ sec^{-1}.$$

Concerning resolution, in the Θ range $\Theta \leqslant 15°$ the term multiplying $\Delta\beta$ in (3) is always comprised between 0.8 and 1.6 for all values of β possible with the $\bar{p}H^-$ operation. The term multiplying $\Delta\Theta$ varies with Θ between 0 and 0.03 at β=0.1, between 0 and 0.1 at β=0.3 and between 0 and 0.3 at β=0.6. The typical angular resolution is $\Delta\Theta \cong 1mm/3m = 3 \cdot 10^{-4}$. We have also

$$\Delta\beta = \beta(\Delta p/p) \qquad (5)$$

where Δp is the momentum dispersion in LEAR of the $\bar{p}$ beam. Already in the first years of operation we had typically $\Delta p/p \sim 10^{-3}$ and in dedicated $\bar{p}H^-$ operation the $\bar{p}$ beam could be cooled down to 10^{-4}. From these numbers it emerges that resolutions in the 10^{-4} region are attainable at β around 0.6 and resolutions in the 10^{-5} region are in scope at $\beta \sim 0.1$, where however the operation mode of the corotating beams is more complex. Notice that with 10^{-5} relative resolution the energy resolution in the 2 keV region becomes 20 eV and the study of all the 4 sublevels of the 2P level becomes feasible. To appreciate the Θ^* bin in which the jump in count rate occurs, at least 4 counts are necessary in total absence of background. Since the solid angle coverage ω of one angular bin in Θ of the detector is

$$\omega = \tfrac{1}{2}(\Delta\Theta)\sin\Theta \simeq 1.5 \cdot 10^{-4} \sin\Theta \simeq 1.5 \cdot 10^{-4}\,\Theta \qquad (6)$$

there is interest to choose β so that Θ^* occurs in the region $\Theta \geqslant 10°$ to have good count rates and a clear experimental situation not limited by statistics.

HIGH RESOLUTION RESONANT SPECTROSCOPY OF PROTONIUM

Collinear laser spectroscopy has also been considered. Monitoring of the resonant frequency could be done -for transitions between levels with different principal quantum number n- by observing the $p\bar{p}$ atom dissociation occuring only in the higher n level in the motional electric field experienced by the $p\bar{p}$ atom when traversing a static magnetic field of intensity conveniently chosen. This is a technique exploited with success in the study of H^- spectroscopy[43]. Another way to monitor the resonant frequency would be by observing variations -at resonance- in the $p\bar{p}$ atomic cascade when a transition is induced between levels with the same n (e.g. depression of the L_α line intensity when inducing transitions from 3D to 3P sublevels). For this type of measurements the instrumentation necessary for emission spectroscopy would be readily available and usable. By these means direct measurements of the $p\bar{p}$ Rydberg and of the antiproton magnetic moment with resolutions at the 10^{-4} level would be feasible. Considering that these measurements are likely to be done -if they are possible- in a second phase of the IDEFIX experiment, resolutions as good as 10^{-5} can be envisaged with the electron cooling of the $\bar{p}$ beam that quite likely will be fully operational at LEAR by that time. Some thinking has been put on resonant spectroscopy of protonium some time ago[5,34], but more conception activity and conceptual design work are necessary to establish its feasibility and select convenient measurements.

THE $\bar{p}H^-$ CO-ROTATING BEAMS SCHEME

The scheme proposed in 1977 has been studied from the machine point of view in the LEAR design report[46], it has been summarized in ref. 37 and reviewed by Möhl[47] at the Tignes LEAR Workshop.

Operation scheme

Schematically an operational sequence could be:

- Inject 10^8-10^9 $\bar{p}$ at 600 MeV/c
- Cool and decelerate to 300 MeV/c (H^- linac momentum)
- Cool again
- Bunch over part of the circumference of LEAR
- Inject H^- at 300 MeV/c onto the free part of the circumference
- Merge $\bar{p}$ and H^- beams by adiabatic debunching
- Accelerate or decelerate to the momentum providing a convenient β for the transitions to be measured in protonium ($\sim$100 MeV/c$\leqslant$ p$\leqslant$700 MeV/c)
- Data taking as long as a convenient number (flux) of $p\bar{p}$ atoms is produced. The overall $p\bar{p}$ atoms production could be initially up to 10^6/sec[47]. The coast that is consumed faster is the H^- one as the H^0 production is 10^2-10^3 higher than protonium one. The spill length is determined hence by the H^- lifetime.
- Cooling, when possible, in parallel to data taking
- Back to H^- injection momentum, bunching, H^- injection, adiabatic debunching, back to data taking momentum every 10 to 100 sec.
- Reinjection of fresh $\bar{p}$ from AA+ACOL every 300 to 3000 sec.

Luminosity and cross sections

The luminosity integrated over the full LEAR circumference for both $\bar{p}H^-$ and H^-H^- interactions is[5,47]

$$L = \frac{N_1\, N_2\, \Delta v}{C\, S} \cdot \frac{1}{\gamma^2} \qquad (7)$$

$N_1 \sim 10^8 \div 3\cdot10^9$ is the number of $\bar{p}$ for $\bar{p}H^-$ interactions or the number of H^- for H^-H^- interactions

$N_2 \sim 10^8 \div 6\cdot10^9$ is the number of H^- present in the ring

$C = 7850$ cm is the LEAR circumference

Δv is the r.m.s. velocity spread within and between the two beams in the co-rotating reference system

S is the effective average beam cross section

γ^2 is the relativistic factor which takes into account the target shortening in the $\bar{p}$ c.m. and the time conversion factor.

The difference between the average longitudinal velocities of the $\bar{p}$ and H^- beams is zero if electron cooling equalizes the two beam velocities in a region where $\alpha_p = 0$ and is negligible ($\delta v_L \leqslant 2\cdot10^{-4}\beta c$) if an RF system imposes the same revolution frequency to the two beams. The magnitude of the longitudinal and of the two transverse components of Δv is then determined by the momentum dispersion of the two beams and by the amplitude of the betatron oscillations. With the notations of the design report[46]

$$\Delta v_L = (\Delta p/p)\beta c,\quad \Delta v_H \cong \beta c\sqrt{(\varepsilon_H/\pi\beta_H)} \text{ and } \Delta v_V \cong \beta c\sqrt{(\varepsilon_V/\pi\beta_V)} \qquad (8)$$

β_H and β_V change along the circumference of LEAR and depend on the selected working point. In the regions where β_H and β_V are higher the luminosity is smaller, but the $p\bar{p}$ atom production rate may be higher because of the dependence of the formation cross section on Δv.

Due to the mass difference between $\bar{p}$ and H^- the two beams will be in general slightly displaced radially with respect to each other. With the working point foreseen in the design report[46], this displacement Δr is of the order of 3 mm in the regions where the momentum compaction function α_p is different from zero [$\alpha_p = (\Delta p/p)^{-1}\Delta r$ is the radial dispersion of the orbit position around the nominal orbit; α_p changes along the LEAR circumference and is determined by the focusing characteristics of the storage ring]. This occurs both if a radiofrequency system imposes the same revolution frequency to the two types of particles and if electron cooling equalizes their velocities. Quantitatively, if a RF system or a stochastic cooling system imposes the same revolution frequency we have[46)]

$$\Delta r = \alpha_p \gamma_t^2 \, 1/(\gamma_t^2 - \gamma^2)(\Delta m_0/m_0) \qquad (9)$$

and

$$\delta v_L = 1/(\gamma_t^2 - \gamma^2)(\Delta m_0/m_0) \qquad (10)$$

where $\Delta m_0/m_0 \simeq 10^{-3}$ is the ratio of the mass difference between the $\bar{p}$ and the H^- over the $\bar{p}$ mass and γ_t is the imaginary transition energy of the machine. If the average velocities of the $\bar{p}$ and H^- beams are equalized by the action of electron cooling, then

$$\Delta r = \alpha_p(\Delta m_0/m_0) \qquad (9')$$
$$\delta v_L = 0. \qquad (10')$$

The transverse size of each beam is given by

$$S = \pi \, a_H \, a_V \qquad (11)$$

where

$$a_H{}^2 = \beta_H \varepsilon_H/\pi + [\alpha_p \, (\Delta p/p)]^2 \text{ and } a_V^2 = \beta_V \varepsilon_V/\pi \qquad (12)$$

and changes along the machine circumference according to the variation of β_H and β_V. With the emittances and momentum spread achievable without electron cooling $\varepsilon_H \sim \varepsilon_V > 10\ \pi$ mm mrad, $\Delta p/p \sim 10^{-3}$ the horizontal size a_H of the $\bar{p}$ and H^- beams is of about 1 cm, the radial displacement of $\sim$ 3 mm due to the mass difference is negligible, and the two beams overlap all around the machine. When it will be possible to have beams extremely cold ($\varepsilon_H \sim \varepsilon_V \sim \pi$ mm mrad, $\Delta p/p < 5\cdot10^{-4}$) the $\bar{p}$ and H^- beam sizes a_H and a_V will shrink to ~2mm and the two beams will superpose only in the regions where $\alpha_p \simeq 0$.

The H^- and $\bar{p}$ beams form $p\bar{p}$ atoms by the Auger capture reaction $\bar{p} + H^- \rightarrow p\bar{p}_{atom} + 2\,e^-$. The cross section $\sigma_A(\bar{p}H^-)$ is several orders of magnitude higher than the one of the radiative capture process with parallel going p and $\bar{p}$ beams that was initially envisaged for doppler tuned $p\bar{p}$ atom spectroscopy[33,50)]. The idea to have co-rotating $\bar{p}$ and H^- beams[35)] came as an extrapolation of the suggestion by the VAPP-NAP group of Novosibirsk to produce $p\bar{p}$ atoms with a H^0 beam superposed to a circulating $\bar{p}$ beam[51)]. The $\bar{p}H^-$ scheme combines the advantage of the high $\bar{p}H^0$ cross section to form protonium with the unique advantages to have in storage the particle containing the proton: recirculation (gain factor ~10^6), adjustable velocity via the ring RF, cooling, relativist regime accessible. In 1977 we had only the order of magnitude of the cross section[35)] that has been calculated in 1979 by Bracci et al[52]. It turns out[52] that the $\bar{p}H^-$ Auger capture cross section $\sigma_A(\bar{p}H^-)$ is zero below the threshold of single ionization of the H^- (the binding energy of the peripheral electron is 0.75 eV) and goes abruptly to zero at $\bar{p}$ and H^- relative velocities $\Delta v > 2.4 \cdot 10^{-4}c$. $\sigma_A(\bar{p}H^-)$ is 10^8 times larger than the radiative capture cross section $\sigma_R(p\bar{p})$, it is roughly inversely proportional to Δv and $\sigma_A(\bar{p}H^-, \Delta v = 10^{-4}c) \simeq 4 \cdot 10^{-16} cm^2$. Fig. 5 from ref. 47 plots as a function of Δv $\sigma_A(\bar{p}H^-)$ and all the other cross sections important for the $\bar{p}H^-$ operation.

$p\bar{p}$ atomic beam

At 300 MeV/c using performance parameters already experienced in LEAR[53] ($\varepsilon_H \sim \varepsilon_V \sim 10\pi$ mm mrad, $\Delta p/p \sim 10^{-3}$) one expects a total production rate of about 10^5 $p\bar{p}$ atoms sec^{-1} with 10^9 H^- and 10^9 $\bar{p}$ stored in the ring. Four beams of about 10^4 $p\bar{p}$ s^{-1} each would emerge in the prolongation of the straight sections. The divergence and size at the origin of these $p\bar{p}$ beams will be roughly that of the $\bar{p}$ beam in the straight sections of the machine ($S \sim 1cm^2$, $\Delta\Theta \sim 1$mrad). A factor 100 in production rate can be gained by additional cooling ($\varepsilon_H \sim \varepsilon_V \sim 1\pi$ mm mrad, $\Delta p/p \sim 3 \cdot 10^{-4}$). A special working point with α_p very little in the first straight section of LEAR and unconstrained elsewhere would maximize the $p\bar{p}$ flux in the beamline that points to the LEAR experimental area where the experiments on protonium can be performed like on the usual extracted beams without direct interference with the machine.

The time structure of the $p\bar{p}$ beam is essentially controlled by the lifetime of the H^- beam stored in LEAR, that is 100 to 1000 times shorter than the lifetime of the $\bar{p}$ coast in data taking conditions. In practice the $\bar{p}$ beam lifetime wil be reduced by losses during the complex gymnastics associated to H^- refillings. The $p\bar{p}$ beam is "contaminated" by a H^0 beam with 100 to 1000 times larger intensity. The maximum β allowed to the $p\bar{p}$ beam is $\beta \simeq 0.7$. Above this value electromagnetic stripping prevents H^- storage in LEAR, as it will be discussed later on.

H^- and H^0 beams

The most critical point in the $\bar{p}H^-$ scheme is the lifetime of the H^- coast. This aspect has been object of early investigations[54], theoretical[55,56] and, recently, of experimental study[57÷59]. H^- ions can be dissociated by several processes:

a) motional electric stripping in the bending magnets[54,47]
b) gas stripping with the residual gas of LEAR[54,47]
c) beam beam stripping between the co-rotating $\bar{p}$ and H^- beams[47]
d) intrabeam stripping[47,55,56]
e) radiation stripping[59]

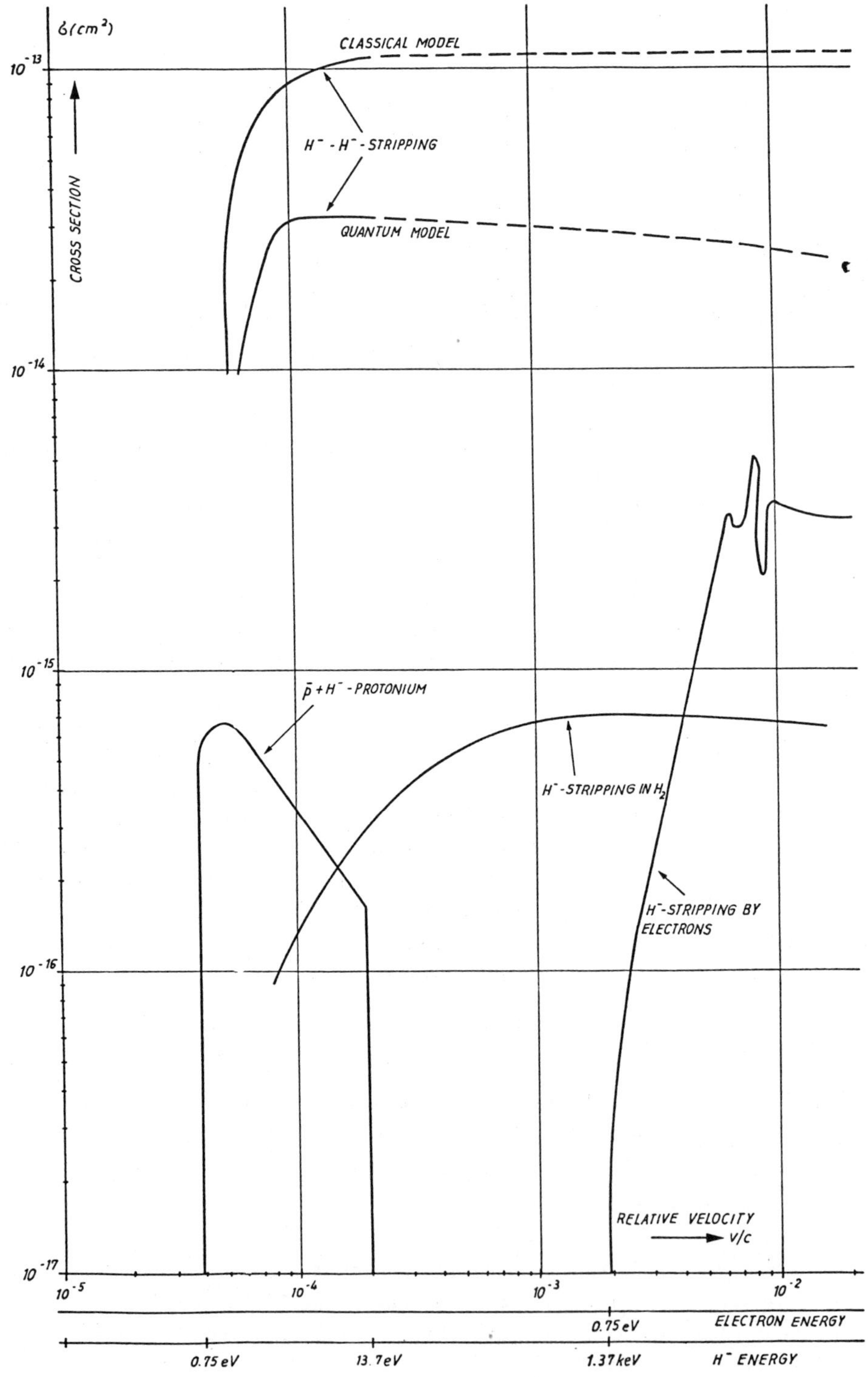

Fig. 5. Cross sections relevant to the $\bar{p}H^-$ co-rotating beams operation

f) electron beam stripping in collisions with cooling electrons[47) (in case electron cooling is used).

The cross section for electric stripping (induced in bends by the external electric field experienced by the H^- in its rest frame) depends very strongly on β and limits H^- operation to momenta below 0.7 GeV/c. Below this value electric stripping will be completely negligible, above this value it will inevitably destroy the beam. Quantitatively speaking the motional electric field experienced in the c.m. system of a H^- in the bends of LEAR is[54)

$$E = 9\beta^2\gamma^2 R^{-1} \text{ MV cm}^{-1} = 9(\gamma^2-1)R^{-1} \text{ MVcm}^{-1} \qquad (13)$$

where R is the bending radius measured in metres. The dependence of the H^- lifetime from a motional electric field is given by

$$\tau_E(\text{sec}) \cong \frac{8\cdot 10^{-14} \text{ MV cm}^{-1}}{E} \exp\left(\frac{42.56 \text{ MV cm}^{-1}}{E}\right) \qquad (14)$$

Fig. 9 in ref. 54 plots τ_E versus E and fig. 5 in ref. 47 plots τ_E versus p(MeV/c) in LEAR. For $p \leqslant 700$ MeV/c $\tau_E > 10^3$ sec, for $p > 730$ MeV/c $\tau_E < 1$ sec.

The gas stripping lifetime depends on the gas composition and on the H^- momentum. For pure H_2 and pure N_2 gases, for example, one has

$$\tau_{N_2} \sim \frac{700}{10^{12}p(\text{torr})} \beta\gamma^2 \text{ and } \tau_{H_2} \sim \frac{5000}{10^{12}p(\text{torr})} \beta\gamma^2 \qquad (15)$$

Extremely low pressures (p in the region $10^{-12} \div 10^{-11}$ torr, at the limit of the technical feasibility in accelerators and storage rings) are then required to have lifetimes in excess of 10 sec, which is the condition to have reasonable time to execute a H^- cycle in the $\bar{p}H^-$ co-rotating beam operation. At the time of the project of the machine the LEAR vacuum was designed with ISR standards ($\sim 10^{-12}$ torr N_2 equivalent) mainly having in mind the $\bar{p}H^-$ option. Presently this choice turns out to be useful also for reaching operating conditions at momenta below 100 MeV/c.

Cross sections for beam beam and intrabeam stripping have been evaluated using several different approximations[55,56). The cross section goes to zero for $\bar{p}H^-$ relative velocities smaller than $0.56\cdot 10^{-4}$ c, for which not enough kinetic energy is available to ionize the H^-, above this threshold it increases with the relative velocity and flattens to a value $\sigma_{st} \sim 10^{-14} \div 10^{-13} \text{cm}^2$ that depends on the model employed in the calculation. One may expect that the threshold velocity for intrabeam stripping is higher than for $\bar{p}$ stripping, as in the H^-H^- case the kinetic energy available in a collision can be shared to increase the internal energy of both H^- ions. The strip rate can be estimated similarly to the $p\bar{p}$ atom production rate.

The beam-beam stripping rate $R_{H^0}{}^{bb}$ is proportional to the number N_{H^-} of H^- and to the number $N_{\bar{p}}$ of antiprotons:

$$R_{H^0}{}^{bb} = N_{H^-} \frac{N_{\bar{p}}}{SC} \langle \sigma_{st} \Delta v \rangle \gamma^{-2} \qquad (16)$$

The intrabeam stripping rate $R_{H^0}{}^{ib}$ is proportional to the square of N_{H^-}

$$R_{H^0}{}^{ib} = N_{H^-} \frac{N_{H^-}}{SC} \langle 2\sigma_{st} \Delta v \rangle \gamma^{-2} \qquad (17)$$

(The factor 2 comes from the assumption that both H^- get stripped in one collision). The H^- beam lifetime due to intrabeam stripping is then inversely proportional to N_{H^-}. In the $\bar{p}$ and H^- beam conditions that provide atomic beams of 10^4 $\bar{p}p$ s^{-1} ($N_{\bar{p}}=10^9$, $N_{H^-}=10^9$, $\Delta p/p \sim 10^{-3}$, $S \sim 1cm^2$) the H^- lifetime due to beam-beam and intrabeam stripping is about 10 s. If one would have 10^{10} H^-, the lifetime would be only about 1 s due to intrabeam stripping. If one increases $N_{\bar{p}}$ to 10^{10}, the rate of formation of $p\bar{p}$ atoms increases but the H^- beam lifetime reduces to 1 s and frequent refillings become necessary.

Shortly before the Tignes Workshop the LEAR team realized the record of storing negative hydrogen ions in an accelerator[57]. The H^- beam lifetime was however shorter than expected from formula (14) and at the limit of operational feasibility of $\bar{p}H^-$ operation (~10 sec) already with an H^- coast away from critical intrabeam. Recent machine developments performed shortly before this school have verified experimentally H^- beam lifetimes in excess of 50 sec at 300 MeV/c momentum and with more than 10^9 H^- injected and stored into the ring[58]. Additional stripping effects due to radiation dissociating the H^- have been identified and suppressed to reach these achievements. Not only the feasibility of the $\bar{p}H^-$ scheme is ascertained, but quantitative measurements of the intrabeam stripping cross section are in scope[60] by studying the variation of the slope of the H^- lifetime. This measurement should be able to discriminate between theoretical models used to calculate σ_{st}[55,56].

Operation of LEAR with H^- is very important for machine studies in $\bar{p}$ polarity and for monitoring of the injection lines[53]. Moreover, when a H^- beam circulates in LEAR, the intense H^0 beams emerging from the straight sections can provide a good simple and external monitor of the beam properties and vacuum conditions for each straight section. This is valid independently from the complex operation of co-rotating beams[37]. H^0 beams with a minimum intensity guaranteed by gas stripping emerge from the straight sections continuously (intensity in the range $10^6 \div 10^8$ when more than 10^9 H^- are injected into LEAR).

H^0 bunches can be produced by pulsed radiation dissociating H^- in one straight section. This is foreseen in order to measure with high accuracy the H^0 beam velocity by resonant spectroscopy[39] in order to calibrate $p\bar{p}$ with pe^- spectroscopy[38]. Successful tests of this possibility have been performed[59] and lead also to new possibilities : beam sculpture and H^0 coded transmission. By tailoring in time the radiation beam crossing the H^- beam a distribution of holes ("antibunches") can be made in the coasting beam together with the H^0 bunches emerging from the straight section. The antibunches rotate in the ring and their evolution can be studied by the monitoring equipment. By tailoring the intersection of two radiation beams with the H^- coasting beam, scraping and even tailoring of the H^- beam in the transverse plane is conceivable. Beam holes have actually been observed together with the associated H^0 pulses in a recent MD[59].

Stripping from electrons in cases when electron cooling would be used occurs only for a velocity mismatch of the two beams larger than $\delta v = 17 \cdot 10^{-4}$ c due to longitudinal velocity mismatch or misalignment. The typical temperature of e^- in their rest frame is 0.2 eV, well below the 0.75 eV binding energy of the peripheral e^- in H^-. Electron cooling of the H^- beam will make sense only if the cooling time will compete with the internal cooling mechanism provided by intrabeam stripping.

COMMENTS ON THE FEASIBILITY OF IDEFIX

The IDEFIX programme of high resolution protonium spectroscoy requires:

A) A dedicated H^- source
B) LEAR with
   i) ultra high vacuum to store H^-
   ii) a long and accessible prolongation of one straight section not interfering with existing civil engineering, with beam transport and experimental equipment of the experimental area
   iii) short prolongations of the other LEAR straight sections for monitoring and H^0 measurements
   iv) dedicated $\bar{p}H^-$ co-rotating beams operation mode
C) An annihilation vertex detector on the $p\bar{p}$ neutral beam line
D) Large acceptance low background position sensitive X-ray detectors
E) A special beam chamber with thin Be windows
F) Power lasers for H^0 control spectroscopy
G) Radiation sources for $p\bar{p}$ resonant spectroscopy.

After the running in of the OBELIX experiment, that is in course of installation with the central detector on the axis of the $p\bar{p}$ beam line, and with the overall design of the LEAR experimental area planned for ACOL time[61)] conditions A, B(i)÷B(iii) and C will be fulfilled. Machine development will be necessary to implement condition B(iv) and to measure experimentally the production rate of $p\bar{p}$ atoms. Concerning points D to E nothing is trivial or cheap but the necessary technology and know-how exist. Further work will be necessary to define if and how to intervene on the $p\bar{p}$ atomic cascade before $p\bar{p}$ atoms reach the region useful to measure the annihilation vertex. It is not improbable that high resolution protonium spectroscopy starts within 15 years from the initial proposal.

ACKNOWLEDGEMENTS

On most of the topics covered in this paper I have profited from constructive criticism and discussions with several colleagues of the LEAR team, of the ASTERIX and OBELIX collaborations and of the IDEFIX study group.

REFERENCES AND FOOTNOTES

1. ASTERIX Collaboration, S. Ahmad et al, Phys.Lett. 157B:33 (1985).
2. U. Gastaldi, The X-ray Drift Chamber (XDC), N.I.M. 157:441 (1978).
3. U. Gastaldi, The Spiral Projection Chamber (SPC), N.I.M. 188:459 (1981).
4. ASTERIX Collab., presented by U. Schaefer, to be published in Proc. of 7th European Antiproton Symposium, Thessaloniki, 1-5 Sept. 1986
5. U. Gastaldi, $p\bar{p}$ experiments at very low energy using cooled antiprotons, in Proc. 4th European Symp. on Antiproton Interactions, Barr (Strasbourg) 1978, ed. A. Friedman (CNRS, Paris 1979), vol. 2, p. 607.

6. ASTERIX Collab., R. Armenteros et al, CERN/PSCC/80-101 (1980).
7. U. Gastaldi, M. Heel, H. Kalinowsky, E. Klempt, R. Landua, R. Schneider, O. Schreiber, R.W. Wodrich and M. Ziegler, Construction and operation of the Spiral Projection Chamber of the ASTERIX experiment, to be submitted to Nucl. Instrum. Methods.
8. OBELIX Collab., R. Armenteros et al, CERN/PSCC/86-4 (1986).
9. Crystal Barrel Collab., E. Aker et al, CERN/PSCC/85-56 (1985).
10. U. Gastaldi, OBELIX note CERN/11-84 (1984)
11. U. Gastaldi, OBELIX note CERN/01-85 (1985).
12. OBELIX study group, presented by U. Gastaldi, in proc. 3rd LEAR Workshop on "Physics with Antiprotons at LEAR in the ACOL Era", Tignes 1985, ed. U. Gastaldi, R. Klapisch, J.M. Richard and J. Tran Thanh Van (editions Frontières, 1985), p. 369.
13. U. Gastaldi, Antinucleon annihilations at low energies at LEAR, to be published in Proc. 8th Autum School, Lisbon 13-18 Oct. 1986, and CERN-EP/87-19 (1987).
14. Y. Onel, A. Penzo and R. Rossmanith,A spin splitter for antiprotons at LEAR, these proceedings.
15. PS173 Collab., presented by E. Steffens, same proc. as ref. 12, p. 245.
16. Notice that the motivations for these developments are mainly for formation experiments with stored $\bar{p}$ circulating in LEAR.
17. ASTERIX Collab., R. Armenteros et al, presented by U. Gastaldi, in Proc. 2nd LEAR Workshop on "Physic at LEAR with Low-Energy Cooled Antiprotons", Erice 1982, eds. U. Gastaldi and R. Klapisch (Plenum, New-York 1984), p. 109.
18. ASTERIX Collab., presented by F. Feld, in "Fundamental Interaction in low energy systems", eds. P. Dalpiaz, G. Fiorentini and G. Torelli (Plenum New-York 1985), p. 279.
19. ASTERIX Collab., presented by C. Amsler, in Proc. 7th European Sym. on Antiproton Interactions, Durham 1984, ed. M.R. Pennington (Institute of Physics Conference Series 73, Bristol, 1985), p 287. Notice that the conclusions presented in this paper concerning the S-P wave ratio are wrong and at variance with previous (ref. 18) and following work (ref. 20).
20. ASTERIX Collab., presented by C. Amsler, same proc. as ref. 12, p. 353.
21. E. Borie, same proc. as ref. 17, p. 561.
22. J.M. Richard and M.E. Sainio, Phys. Lett. 110B:349(1982).
23. B. Kerbikov, in Proc. 5th European Symp. on $N\bar{N}$ interactions, ed. M. Cresti (CLEUP, Padua 1980) p. 423.
24. W.B. Kaufmann, in Proc. Kaon Factory Workshop, ed. M.K. Craddock, (TRIUMF report, TRI-79-1, 1979), p. 160.
25. R. Armenteros and B. French, $N\bar{N}$ interactions, in "High-Energy Physics", ed. E.H.S. Burhop (Academic Press, New-York, 1969), vol. 4, p. 237.
26. U. Gastaldi et al, N.I.M. 176:99 (1980).
27. A.J.P.L. Policarpo et al, N.I.M. 102:337 (1972).
28. W.J.C. Okx et al, N.I.M. A252:605 (1986).
29. C.A. Baker et al, these proceedings.
30. R. Bacher et al, these proceedings.
31. T.P. Gorringe et al, Phys. Lett. 162B:71 (1985).
32. R.W. Schmieder and R. Marrus, N.I.M. 110:459 (1973).
33. U. Gastaldi, Lamb-shift type experiments on protonium, CERN-EP internal report 76-23 (1976).
34. U. Gastaldi, A possible new experimental approach to the study of the $p\bar{p}$ system at low energies, in "Exotic Atoms", Proc. 1st Int. School of Physics of Exotic Atoms, Erice 1977, eds. G. Fiorentini and G. Torelli (Servizio di documentazione dei Laboratori di Frascati, Rome, 1977), p. 205; and CERN $p\bar{p}$ Note 13 (1977).

35. U. Gastaldi, A scheme to maximize the $p\bar{p}$ atom formation rate, CERN $p\bar{p}$ Note 30 (1977).
36. U. Gastaldi, Protonium spectroscopy in flight at LEAR with the $\bar{p}H^-$ parallel beams approach, CERN/PSCC/79-67 (1979).
37. U. Gastaldi and D. Möhl, Co-rotating beams of antiprotons and H^- in LEAR and high resolution spectroscopy of $p\bar{p}$ atoms in flight, same proc. as ref. 17, p. 649.
38. U. Gastaldi, in Atomic Physics 9, eds. R.S. Van Dyck and E.N. Forston (World Scientific, Singapore 1984), p. 118, and references quoted therein.
39. A. Coc et al, same proceedings as ref. 12, p. 683.
40. I am indebted to L. Palffy for discussions on this point and for making me aware of ref. 32.
41. J.B. Donahue et al, IEEE Transactions NS-28:1203 (1981).
42. T. Bergeman et al, Shape resonances in the hydrogen stark effect in fields up to 3 MV/cm, in Atomic Physics 9 - Conference Abstracts, ed. R.S. Van Dyck and E.N. Forston (1984) Contribution B27.
43. H.C. Bryant et al, Phys. Rev. A27:2889 (1983).
44. R. Bellazzini et al, N.I.M. A252:453 (1986).
45. U. Gastaldi, E. Iacopini and R. Landua, CERN $\bar{p}$ LEAR-Note 21 (1979).
46. Design study of a facility for experiments with low energy antiprotons, CERN/PS/DL/80-7 (1980).
47. D. Möhl, Technical implications of possible future options for LEAR, same proc. as ref. 12, p. 65.
48. C. Habfast et al, Status and perspectives of the electron cooling device under construction at LEAR, same proc. as ref. 12, p. 129.
49. The electron cooling device in preparation for LEAR does not cover all the momentum span where $\bar{p}H^-$ operation is possible. It is however conceivable to cool at H^- injection energies and to have no cooling at data taking momentum.
50. Schemes to form exotic atoms in flight by radiative capture were considered early in 1970: D.L. Morgan and V.W. Hughes, Phys. Rev. D2:1389 (1970); V.W. Hughes and B. Maglic, Bull Am. Phys. Soc. 16:65 (1971); V. Hughes et al, Formation of atoms involving two unstable particles, (1971) unpublished. I am indebted to V.W. Hughes for having informed me of these references.
51. VAPP-NAP Group, Report CERN/77-08 (1977).
52. L. Bracci, G. Fiorentini and O. Pitzurra, Phys. Lett. 85B:280 (1979).
53. P. Lefèvre, LEAR present status, future developments, in same proc. as ref. 12, p. 33.
54. U. Gastaldi, Estimate of the lifetime of H^- ions in $\bar{p}H^-$ overlapping storage rings, CERN $p\bar{p}$ Note 32 (1977). In this note are compiled former references for motional electric stripping of H^- going through magnets.
55. G. Fiorentini and R. Tripiccione, Phys. Rev. A27:737 (1982).
56. J.S. Cohen and G. Fiorentini, Phys. Rev. A33:1590 (1985).
57. E. Asseo et al, Charge exchange injection and diagnostics with H^- in LEAR, same proc. as ref. 12, p. 99.
58. LEAR team, report in preparation.
59. LEAR team, H. Duong S. Liberman and J. Pinard, report in preparation.
60. I am indebted to R. Giannini, P. Lefèvre and D. Möhl for informations and discussions on these points.
61. D. Simon et al, The LEAR experimental areas, status report and possible developments, in same proc. as ref. 12, p. 47.

A CRITIQUE OF THE VARIOUS TECHNIQUES TO POLARIZE LOW ENERGY ANTIPROTONS

David B. Cline *)
CERN, Geneva, Switzerland
and
Physics Department
University of Wisconsin, Madison, Wisconsin

Abstract

The use of a polarized antiprotons would add a new technique to many low energy $p\bar{p}$ experiments and to the measurement of fundamental properties of antiprotons. Of the various ideas that have been presented the techniques known as the spinfilter, spinsplitter, polarization of $\overline{H}$ and polarization using strong interactions remain the most likely ones to work. We give a brief critique of each method.

1. Introduction

The possibility of polarizing antiprotons for use in $p\bar{p}$ collisions is of great interest in the study of high energy interactions. In addition to the production of (W,Z) particles in polarized $p\bar{p}$ collisions and thus searching for right handed currents, it would be possible to carry out new tests of QCD. There are also interesting low energy experiments that can be carried out with polarized $\bar{p}'s$. For these reasons there are now many new studies of the possible technique to polarize antiprotons. All of these techniques have substantial problems and uncertainty. Nevertheless there is little experimental evidence against any of these schemes. In this brief communication we will briefly introduce the different schemes and make some comments on the test of the idea and the prospects for useful polarization.

2. Spin Filter Technique

This idea uses the possibility that there exists substantially different $p\bar{p}$ total cross–section for the triplet and singlet spin–spin states in

$$(\sigma_{\uparrow\uparrow} \neq \sigma_{\uparrow\downarrow})$$

low energy scattering. A polarized proton target is inserted into a storage ring and the different cross-section for the different $\bar{p}$ spin states results in a net reduction of the amplitude for that state. After a certain period determined by the magnitude of the cross–section a net polarization builds up. The most important factor in this technique is the overall reduction of the anitproton intensity through the interaction with the target. The formalism for this technique is

$$\frac{dI \pm (t)}{dt} = -\sigma \pm (t) \tag{1}$$

where

$$I_{+}(u) = I_{-}(u) = \frac{1}{2} \quad \text{and} \quad \sigma_{+} = \sigma^{T}_{\uparrow\uparrow} \,; \sigma_{-} = \sigma^{T}_{\uparrow\downarrow} \,;$$

$$\sigma\pm = \sigma_0 \pm \sigma_1 \,;\, \sigma/\sigma_0 = -\frac{1}{2}(\Delta\sigma^{T}/\sigma)_{TOT}$$

The solution to this equation is

$$I(t) = I_o e^{-t/t_0} \, cosh(t/t_1)$$

$$P(t) = \frac{I_{+} - I_{-}}{I_{+} + I_{-}} = -\, tanh(t/t_1)$$

An example given by Steffen is $n = 10^{14}/cm^2$, $f = 10^6/sec$ E=30 MeV and $\sigma_1/\sigma_0 \sim$ 0.07 gives $I(t) = I_0/100$ P(t) = - $\frac{1}{3}$ and T = 48 hours.

Thus in this example the $\bar{p}$ polarization never exceeds 33% while the intensity has been reduced by a factor of 100. The key to this technique is to find an energy in $p\bar{p}$ (or $\bar{p}$ nucleus collisions where

$$\sigma^{T}_{\uparrow\uparrow} \neq \sigma^{T}_{\uparrow\downarrow}$$

by a large difference. In recent reports by C. Dover (private communication) this possibility seems unlikely due to the nature of the strong annihilation potential in low energy $p\bar{p}$ interactions. Unfortunately even in an optimal senario the flux of antiprotons is reduced by this method.

3. Spin Spletter Technique

This technique uses the possibility of separating spin by inserting a Stern–Gerlach element into a storage ring. The technique does not lead to a large reduction of antiprotons but is undoubtly at the edge of feasibility in present accelerators. The proponents of the scheme are Niinskoski, Onel, Penzo and Rossmanith. The effect of the field gradiant on the magnetic moment is extremely small compared to the effects due to the charge of the circulating particles, however, the choice of a possible coherence condition (or resonant condition) between the spin and orbit motion may allow a constructure build up of a spatial separation (in Betatron space) of the spin components. The concept and feasibility of this approach is widely debated these days. A specific proposal to test the scheme has been submitted for LEAR and detailed calculations of the effect in LEAR are in progress. (For a description see the paper by Onel, Penzo and Rossmanith in "Antimatter Physics at Low Energies", Fermilab Low Energy Antiproton Facility Workshop). This idea would be very interesting if it could be applied to a high energy machine like the $\bar{p}'s$ in TeVI. There are several types of criticisms of this scheme some of which are:

(a) The stability condition for the machine and the resonant condition are hard to achieve in practice (i.e. LEAR);

(b) the concept relies on linear conditions in the machine. Unknown non–linear effects could destroy the effect.

4. $\overline{H}$ Productional Hyperfine Interactions to Produce Polarized Antiprotons

This scheme uses the possibility of producing $\overline{H}$ by using e^+ beams to capture on the $\bar{p}(\bar{p}e^+)$ and to subsequently polarize the $\bar{p}$ using hyperfine interactions and laser initiated transitions. A brief description of the idea is given by K. Imai (Particles and Fields 145, AIP proceedings, p. 229). The technique relies on the production of intense $\overline{H}$ which is interesting in its own right.

This scheme, although very interesting is unlikely to result in significant sources of polarized antiprotons.

5. Polarization Using Strong Interactions

An idea that was discussed at the Bodega Bay workshop on polarized antiproton sources (D. Cline) that has been overlooked is to use the strong interaction between polarized protons and antiprotons to either filter out one spin state or to produced polarized antiprotons in the small angle scattering. These are well known spin dependent effects in very small angle scattering that might be used. No serious study of this possibility has been carried out. For example there are considerable spin effects in the Coulomb–Nucleon interference region (for $t < 0.01$) (the asymmetry was calculated by Lapidus et al. Sov. Jour. Nucl. Phys. 19, 114 (1977)). These effects might be used to polarize the $\bar{p}$ beam in the TeVI provided a beam cooling mechanism (electron cooling) was used to correct for the scattering effects at high energy.

6. Spin Transfer from Intense Cold e^- Beams

An idea that was discussed at the Bodega Bay meeting was the possibility of using polarized e^- beams in a spin transfer method. This idea has recently emerged as a posible technique due to the interest in high energy electron cooling at Fermilab. A new generation of high tune electron storage rings that could lead to very cold e^- beams for cooling the antiprotons in the TeV machine might also be used for a spin transfer polarizer. T. Niinikoski has recently looked into the possibility.

7. Summary

The interest in polarizing antiprotons for high energy and low energy $p\bar{p}$ collisions remains high. Unfortunately no simple believable scheme for polarizing antiprotons while maintaining high intensity has emerged. The spin spletter may be such a technique provided the very difficult machine conditions can be met.

Acknowledgements

I thank K. Steffen, T. Niinikoski, A. Penzo and A. Skrinsky for helpful discussions on antiproton polarizers.

POLARIZATION OF STORED ANTIPROTONS BY THE STERN-GERLACH EFFECT

T.O. Niinikoski

CERN

Geneva, Switzerland

We discuss the polarization of stored antiproton beams by spatially separating particles with opposite spin directions, using the Stern–Gerlach effect in alternating quadrupole fields.

INTRODUCTION

It has been recently proposed[1–5] that particle beams could be self-polarized in storage rings or colliders. This self-polarization is based on the Stern–Gerlach effect: in an inhomogeneous magnetic field, the particles with spins aligned parallel or antiparallel to the field are deflected in opposite directions and become spatially separated. We may immediately note the well-known fact that the Stern–Gerlach experiment, with a single pass through a quadrupole field, only works for neutral or ion beams having electronic magnetic moment, because the uncertainty principle together with the Lorentz force causes proton or electron beams to smear much more than their splitting. The uncertainty principle, however, becomes insignificant in our case where the beam particles receive repetitive kicks along their otherwise stable orbits.

A conventional storage ring has quadrupole magnets with inhomogeneous fields. Considering one single magnet, particles with different spin directions are deflected by different angles in the quadrupole field. Since this difference in the deflection angle is extremely weak, the next quadrupole with opposite polarity would almost exactly cancel the effect of the first. However, the bending magnets between the quadrupoles may rotate the spin in such a way that the kicks obtained in the quadrupoles add up, always following the variation of the vertical deflection angle. For a FODO cell, therefore, we must require that the vertical betatron phase advance from the first to the second quadrupole approximately equals the spin phase advance minus π over complete precession cycles in the dipole magnet between the quadrupoles. A more general additional requirement is that the spin and vertical betatron phase advances, modulo 2π, in a complete machine cycle are equal. This can be expressed as

$$a\gamma - n = \nu_y - m \, ,$$

where $a = (g - 2)/2$ ($= 1.793$ for protons), $\gamma = (1 - \beta^2)^{-1/2}$ is the Lorentz factor, n is the nearest integer below $a\gamma$ (the number of spin cycles in a machine turn), ν_y is the number of vertical betatron oscillations in a machine turn, and m is the nearest integer below ν_y. A mathematically rigorous treatment[4] leads to the condition

$$a\gamma \pm \nu_y = kP \, , \tag{1}$$

where k is an integer and P is the periodicity of the machine lattice. This corresponds to an intrinsic spin resonance, which quickly destroys any vertical polarization by dephasing the vertical spin component. The horizontal spin components, however, will remain synchronized by the resonant horizontal fields, and there is no horizontal depolarization in first order.

This situation can be visualized in analogy with magnetic resonance in solids, where transformation to rotating frame allows the main time dependence of the spin Hamiltonian to be removed. It is known experimentally that in solids no depolarization will occur as long as the oscillating transverse field is maintained, in spite of large randomly fluctuating dipolar fields.

The above arguments are valid for a perfectly linear machine and for the case where the synchrotron oscillations average out the differences in the energy dependences of the spin and betatron tunes. For a real machine a complicated spin matching must be performed. This paper focuses on the first-order beam-splitting speed; the problems for real machines are only briefly discussed.

SEPARATION OF PARTICLES WITH DIFFERENT SPIN DIRECTIONS

The field of a quadrupole magnet is described in first order by

$$B_x = -by\,,$$

$$B_y = bx\,,$$

in a coordinate system where x and y are the transverse horizontal and vertical axes, respectively, and z is tangent to the design orbit. The force on the magnetic dipole $\bar{\mu}$ is[6]

$$\delta\bar{F} = \nabla(\bar{\mu}\cdot\bar{B}) = b\nabla(-\mu_x y + \mu_y x) = b(\bar{i}\mu_y - \bar{j}\mu_x)\,, \tag{2}$$

and the additional deflections of the particle in the x–z and y–z planes are, respectively,

$$\delta x' = -b\mu_y \Delta L/E\,,$$

$$\delta y' = -b\mu_x \Delta L/E\,,$$

where μ_x and μ_y are the projections of the magnetic moment along the x and y directions, ΔL is the length of the quadrupole, and $E = \gamma m$ is the beam energy. We note that when the spin points in the direction of the beam, there is no deflection; when it points along x, the particle is deflected vertically; when it points up, the deflection is horizontal. The radial field gradient b is positive for quadrupoles focusing in the vertical plane and negative for those focusing in the horizontal plane.

Between the quadrupoles there are bending magnets with homogeneous field. The motion of the spin in an homogeneous field is described by the BMT equation:[7]

$$d\bar{\sigma}/dt = (q/m\gamma)\bar{\sigma} \times [(1+a)\bar{B}_{\parallel} + (1+a\gamma)\bar{B}_{\perp}]\,.$$

Here γ and a are as defined earlier, $\bar{\sigma}$ is a unit vector describing the spin direction in the laboratory frame, q is the charge of the particle, and m is the rest mass; $\bar{B}_{\parallel}$ and $\bar{B}_{\perp}$ are laboratory frame fields parallel and perpendicular to the motion of the particle. The required integrated field strength for a 90° spin rotation for protons is 2.7 T·m, independent of energy in the relativistic limit.

The vertical betatron motion can be described by an homogeneous differential equation

$$y'' + \omega_y^2(s)y = 0\,, \tag{3}$$

with the well-known solutions $y_1 = \sqrt{\beta_y(s)}\cos[\psi_y(s)]$ and $y_2 = \sqrt{\beta_y(s)}\sin[\psi_y(s)]$; here $\beta_y(s) = [d\psi(s)/ds]^{-1}$, and s is the distance along the design orbit. When the betatron motion is excited by the Stern–Gerlach effect, Eq. (3) becomes

$$y'' + \omega_y^2(s)y = -\mu/\gamma m\, b(s)\, e^{-i\Phi_x(s)} . \tag{4}$$

Here b(s) is the gradient of the field and $e^{-i\Phi_x(s)}$ is the projection of the spin on the x-axis.

The solution of this inhomogeneous differential equation has been presented[4]; we shall quote here the resulting vertical splitting rate of the beam halves with opposite μ_x:

$$\frac{dy}{dt} = \frac{\mu}{\gamma m 4\pi\nu_y f}\int_{s_0}^{s_0+L} \beta^{1/2}(s)e^{-i\psi_y(s)}b(s)\,e^{-i\Phi_x(s)}\,ds . \tag{5}$$

By inserting values of b/γ, ν_y, and f that are appropriate for existing or planned large colliders such as the CERN Super Proton Synchrotron (SPS), the Fermilab Tevatron, and the Superconducting Super Collider (SSC), Eq. (5) gives separation speeds below 1 mm/day, which makes the method somewhat impractical. It is interesting, however, to extrapolate toward a purpose-built machine, which is optimized for spin separation. Quadrupole magnets can be built with gradient $b = 100\ \mathrm{T\cdot m^{-1}}$. If we assume that alternating gradient magnets have this strength all over the ring with f = 50 kHz (such as the CERN SPS), and if we operate at proton energy $\gamma m_p = 100\ \mathrm{GeV/c^2}$ and $\nu_y \cong 10$, we find

$$(\Delta y/\Delta t)_{max} = 4.5\ \mathrm{mm/h} .$$

This rate of separation of the beam into polarized halves would seem to be much faster than the rate of blow-up or loss.

POLARIZING LOW-ENERGY ANTIPROTONS

Examining Eq. (5) we note that the splitting rate is inversely proportional to the beam energy γ. Accelerators usually operate at constant b/γ; however, if a machine lattice could be designed which had a very high value of b while conserving a low tune ν_y, a substantial increase in the splitting rate could be observed. This requirement could be achieved by placing the strong quadrupole pairs as close together as possible while maintaining a spin rotation by π in the interleaved dipole. Also, the machine should have a low cycling frequency f, which means a large perimeter. Summarizing, the machine would look like a large weak-focusing stretcher ring with very strong closely spaced quadrupole pairs.

The superperiods of the machine could also include reverse bends or vertical bends, in order to achieve a sufficient length of quadrupoles while maintaining a large diameter and low tune.

A recent proposal[5] explores the use of a solenoid spin rotator for the same purpose as a dipole magnet between the quadrupoles.

DISCUSSION

At least the following problems must be solved before our suggested scheme may be considered as a serious alternative for polarized beams:

1) Beam energy spread makes a spread both in the spin tune $a\gamma$ and the betatron tune ν_y. Can these be compensated so that all spins will maintain phase coherence with the betatron oscillations?

The energy dependence of the betatron tune is suppressed to a large extent by the averaging due to synchrotron oscillations at low energies. The spin motion associated with the acceleration and deceleration due to synchrotron oscillations is adiabatic and thus conserves the beam polarization in the horizontal plane. The adiabatic spin manipulation has been proven experimentally during acceleration and deceleration.[8]

The acceleration or deceleration far from the exact resonance $a\gamma = kP \pm \nu_y$ would turn the rotating horizontal polarization adiabatically in the vertical orientation, conserving the opposite alignments in the beam halves as long as the betatron frequency spread allows vertical coherence to be maintained.

2) The spin and betatron oscillations must be phase-locked over extended periods of time. Is this possible with existing beam-diagnostic devices?

The harmonics of the betatron frequency can be measured by specially built pick-ups and low-noise amplifiers. The additional motion due to the spin [Eq. (4)] might be observable as a sideband to a suitable high harmonic of the betatron frequency,[9] and allow the phase locking of the two oscillations. This technique might also allow the monitoring of the separation build-up, and hence polarization.

3) Higher-order effects may destroy (dephase) the rotating horizontal polarization. One such mechanism may be the coupling of the vertical and horizontal betatron oscillations due to uncompensated axial fields, for example.

Without attempting to find definite answers to this question, we first would like to point out that there are several techniques to make the spin tune energy-independent. These are generally based on spin rotation about the longitudinal or the transverse horizontal axis.[10] This can be accomplished with so-called 'Siberian Snakes',[11] which rotate the spin through an angle independent of energy. Other spin manipulation techniques might involve vertical-bend spin rotators and closed-orbit distortions, which need to be optimized only at the energy of storage if further acceleration or deceleration of the polarized beam is not required.

Bypassing all the complexities of spin resonances in accelerators, we would like to draw some analogies with spin resonances in solids in order to hint at a way towards complete spin matching for achieving efficient self-polarization. The snake spin-rotator in an accelerator is exactly analogous with the spin-echo scheme, where incoherence resulting from Larmor frequency spread can be almost indefinitely repaired by 180° spin-reversing pulses. Much more complex schemes do exist in solid-state NMR, which remove the effects of spin diffusion[12,13] and dipolar broadening;[14–16] their analogies could probably be easily implemented in accelerators by the introduction of various snakes, vertical bends, and orbit distortions.

If we suppose that the technical difficulties in tuning a collider accurately to a spin resonance during extended periods can be overcome, it may be worth while to discuss the possible ways by which experiments could use the polarization in the beam halves. Firstly, each beam bunch has zero average polarization in spite of the fact that the particles with opposite horizontal spin components are spatially separated. The obvious solution is to steer the bunches so that beam halves with definite spin states enter in collision.

Secondly, during the beam separation at intrinsic resonance $a\gamma \pm \nu_y = kP$, the spin phase cannot be fixed in the machine lattice, because this would require running at a strong instability condition $\nu_y =$ integer. In all experiments, however, spin reversal is necessary, and this could now be accomplished simply by correlating the spin phase with the scattering data. This would require the tagging of the events with the vertical betatron phases of the colliding particles. A possible way of doing this is to determine the vertical momentum of the individual events with sufficiently high accuracy, so that the sum of the vertical momenta of the beam particles, and therefore the difference of their polarization vectors can be determined.

Thirdly, selective slow extraction from the polarized beam halves should lead to a polarized extracted beam, again because the betatron and spin phases are correlated. This, however, would require the extraction to take place in a vertical plane, and the remaining circulating beam would not acquire overall polarization because of the spread of the vertical betatron and horizontal spin phases.

In order to get an overall polarization in the circulating beam, the upper or lower half of the beam would have to be extracted or destroyed in one machine turn, and the polarization in the remaining beam would have to be turned adiabatically into vertical alignment by accelerating or decelerating immediately (in a short time compared with the coherence time of vertical betatron oscillations) off the intrinsic spin resonance at which the beam separation is done. Alternatively, the machine tune could be adiabatically shifted off the resonance by changing the quadrupole magnet currents immediately after the fast extraction of one half of the beam.

Finally, if the polarization can be adiabatically turned into the vertical position, as discussed above, the experiments could operate with beams in pure spin states of arbitrary alignment by using spin rotators.

We conclude that self-polarization of protons in existing colliders and storage rings might be impractically slow. However, a purpose-built machine could provide a sufficiently fast polarization speed, so that it might be worth examining the outstanding problems related with the method.

Acknowledgements

The author would like to thank J. Bell, Y. Onel, A. Penzo and R. Rossmanith for many stimulating discussions on the self-polarization of antiprotons.

REFERENCES

1. T.O. Niinikoski and R. Rossmanith, preprint CERN-EP/85-46 (1985).
2. T.O. Niinikoski and R. Rossmanith, Proc. 6th Int. Symp. on Polarization Phenomena in Nuclear Physics, Osaka, 1985 [Suppl. J. Phys. Soc. Japan 55:1134 (1986)].
3. T.O. Niinikoski and R. Rossmanith, Proc. Int. Workshops on Polarized Beams at the SSC, Ann Arbor, 1985, and on Polarized Antiprotons, Bodega Bay, 1985, eds. A.D. Krisch, A.M.T. Lin and O. Chamberlain (AIP Conf. Proc. No. 145, New York, 1986), p. 244.
4. T.O. Niinikoski and R. Rossmanith, Self-polarization of protons in storage rings, Brookhaven Nat. Lab. Accelerator Division Internal Report 86-4 (1986), to be published in Nuclear Instruments and Methods in Physics Research.
5. Y. Onel, A. Penzo and R. Rossmanith, Polarized antiprotons at LEAR, Proc. Symp. on High-Energy Spin Physics, Protvino, 1986 (to be published);
   A. Penzo, these Proceedings.
6. R. Leighton, Principles of modern physics (McGraw-Hill, Inc., New York, 1959).
7. V. Bargmann, L. Michel and V.L. Telegdi, Phys. Rev. Lett. 2:435 (1959).
8. E. Grorud et al., Proc. Int. Symp. on High-Energy Spin Physics, Brookhaven, 1982 (AIP Conf. Proc. No. 95, New York, 1983), p. 407.
9. This suggestion is due to S. van der Meer (private communication).
10. E. Courant, J. Phys. 46:C2-713 (1985).
11. Ya.S. Derbenev and A. Kondratenko, Proc. Int. Symp. on High-Energy Physics with Polarized Beams and Targets, Argonne, 1978 (AIP Conf. Proc. No. 51, New York, 1979), p. 292.
12. H.Y. Carr and E.M. Purcell, Phys. Rev. 94:630 (1954).
13. S. Meiboom and D. Gill, Rev. Sci. Instrum. 29:6881 (1958).
14. E.D. Ostroff and J.S. Waugh, Phys. Rev. Lett. 16:1097 (1966).
15. P. Mansfield and D. Ware, Phys. Rev. 168:318 (1968).
16. L.M. Stacey, R.W. Vaughan and D.D. Elleman, Phys. Rev. Lett. 26:1153 (1971).

A SPIN SPLITTER FOR ANTIPROTONS IN LEAR

Y. Onel

DPNC, University of Geneva, Switzerland

A. Penzo

INFN and University of Trieste, Italy

R. Rossmanith

DESY, Hamburg, Fed. Rep. Germany

ABSTRACT

A method for obtaining polarized antiprotons in LEAR via the separation of opposite-spin states is suggested. The method can be tested with protons circulating in the same machine.

INTRODUCTION

Even with the advent at CERN of the Low-Energy Antiproton Ring (LEAR) and its antiproton beams of superior quality, the perspectives for obtaining a complete description of the $N\bar{N}$ amplitudes are limited, mainly because of the lack of polarized antiprotons. Several methods have been suggested for overcoming this deficiency. Most of these methods, such as scattering on hydrogen or nucleon targets,[1] differential absorption on a polarized target,[2] or polarization of antihydrogen,[3] are of limited practical value as they entail severe losses of intensity and result in only modest final beam polarization. A much more appealing scheme, which in principle does not involve a large reduction of flux in the course of polarization, has recently been proposed[4] for high-energy machines. This scheme is based on the spatial separation of circulating particles, with opposite spin directions, by means of the magnetic moment interaction with the field gradients of quadrupoles in the machine. Although the effect of the field gradient on the magnetic moment is extremely small compared with the effect on the charge of the circulating particle, the choice of specific coherence conditions between the orbit and spin motions allows a constructive build-up of separation over many revolutions, leading to a mechanism of self-polarization.

The field of a quadrupole magnet is described to first order by

$$B_x = b \cdot y \quad , \qquad B_y = b \cdot x \quad , \tag{1}$$

in a coordinate system where x and y are the horizontal and the vertical axis, respectively, transverse with regard to the axis z along the design orbit. The force acting on a magnetic dipole is

$$\mathbf{F} = \nabla(\mu \cdot \mathbf{B}) = b(\mu_x y + \mu_y x) \quad , \tag{2}$$

and the deflections of the particles in the x–z and y–z planes are, respectively,

$$\delta x' = b\mu_y L/p\beta \quad ,$$
$$\delta y' = b\mu_x L/p\beta \quad , \tag{3}$$

where μ_x and μ_y are the projections of the magnetic moment along the x and y directions, L is the length of the quadrupole, and p is the momentum of the beam. We note that there is no deflection when the spin points in the direction of the beam. When the spin points along the horizontal plane the particle is vertically deflected; when it points along the vertical direction the spin is horizontally deflected.

Since the difference in the deflection angle is extremely weak for particles with opposite spin, the next quadrupole in the ring, with opposite polarity, would almost exactly cancel the effect of the first. Elements between the quadrupoles have to rotate the spin in such a way that the kicks obtained in the quadrupoles add up.

In the original proposal[4] the rotation of the spin around the vertical axis is performed by the bending magnets between the quadrupoles, according to a specific machine lattice.

In order to achieve a build-up over many revolutions, the spin tune and betatron frequency must be, modulo 2π, identical. This can be expressed by

$$\nu_s = a \cdot \gamma = Q \pm n \quad , \tag{4}$$

where $a = (g - 2)/2$ (= 1.793 for protons); $\gamma = (1 - \beta^2)^{-1/2}$ is the Lorentz factor, n is an integer, ν_s is the spin tune, and Q is the frequency of betatron motion in a suitably chosen (vertical or horizontal) plane.

In the following we present a scheme for low-energy storage rings which, under certain circumstances, can be quite simple. In particular, we discuss a possible layout for LEAR.

THE LAYOUT FOR LEAR

For LEAR, a machine designed for low energies, the polarizing elements are not distributed around the machine, but are concentrated in a compact configuration: the so-called spin splitter.[5] This spin splitter consists of two quadrupoles with a solenoid between them (Fig. 1), placed in one straight section of the ring. The quadrupoles have opposite polarity. Between the quadrupoles the solenoid rotates the spin by 180° around the direction of particle motion.

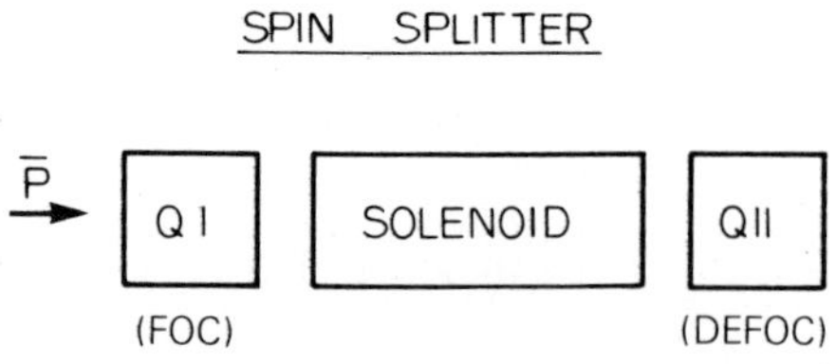

Fig. 1 The spin-splitter scheme

The characteristics of this configuration are as follows:

i) The existing machine lattice is not changed, and the spin-splitter configuration will not have any effect on the rest of the ring for the orbit motion (see end of this section). Furthermore, large separations between particles with opposite spin can be obtained by using strong (superconducting) quadrupoles.

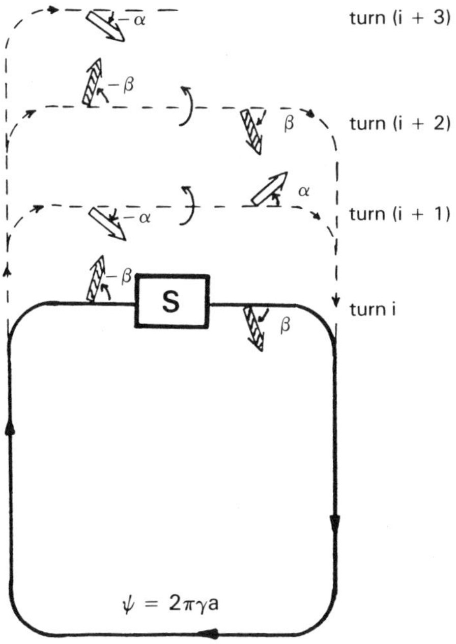

Fig. 2 The spin direction in the machine

ii) The spin tune is affected in a characteristic way by the spin splitter because the solenoid behaves like a 'Siberian snake' for the low-energy storage ring (Fig. 2). Each second revolution, the spin of each particle points in the same direction. The non-integer part of the spin tune is therefore 1/2 and is independent of the energy of the particle: $\nu_s = \frac{1}{2} + m$ (m integer). This has two consequences. First of all it significantly reduces the depolarization due to the finite energy spread. On the other hand, according to formula (4) the machine would run on a half-integer betatron tune: one possible way of avoiding this half-integer resonance is to make the rotating angle of the solenoid slightly smaller than π and allow vertical periodic spin motion. This increases the polarization time by a modest amount with respect to an ideal situation.

The second difficulty is that the Q-value has a certain spread, and only a small part of the beam particles could be excited. In order to obtain high degrees of polarization, one has to cure this problem by introducing a coupling betwen spin tune and vertical betatron motion. The technique for doing this can be the same as in electron machines.[6]

iii) The solenoid would introduce a coupling between the horizontal and the vertical plane, which would affect the beam behaviour in the machine. This effect can be cured by rotating the quadrupoles of the spin splitter by 45° (skew quadrupoles) and adjusting their fields with respect to the solenoid field in order to minimize such coupling. A possible residual coupling could, if necessary, be compensated by adding less strong elements. The only effect resulting from the use of skew quadrupoles is to exchange the planes of deflection $(x,z) \rightleftarrows (y,z)$ for vertical and horizontal spins with respect to the previous situation with normal quadrupoles. Such an arrangement has also another advantage: it can introduce a controlled coupling between spin tune and betatron motion to correct for Q-spread. This is done by using the coupling between horizontal and vertical betatron motion introduced by the spin-splitter configuration, and locking the betatron motion and the spin motion together over a sufficiently wide Q-range.

THE BUILD-UP TIME FOR POLARIZATION

The separation time can be deduced from formula (3); μ is 1.41×10^{-23} erg/G. Going to low energies, the separation speed increases. If we have a momentum of 200 MeV/c and a gradient of

20 T·m^{-1}, 1 m long, the deflection per turn is 42×10^{-15} rad, and for one second (7.8×10^5 turns) 3.35×10^{-8} rad. Particles with opposite spin direction are separated with a speed of 2.5 mm/hour (a quarter betatron wavelength of 10 m was assumed). As the beam in LEAR has vertical diemensions of this order, this separation speed seems to be quite adequate. This energy for applying the build-up of polarization has to be chosen as a compromise between separation rate and emittance growth; at 600 MeV/c, for example, the separation speed is three quarters of the speed at 200 MeV/c, but the emittance is a factor of 2 smaller.

TEST OF THE PROPOSED METHOD WITH PROTONS

Before any further consideration of the possible implementation of the scheme described here, it would seem of great interest to perform a test with protons stored in LEAR, in order to establish the feasibility of the proposed method.[5]

As a prototype spin splitter for a test with protons, a configuration consisting of existing elements would require very little investment for the test. Two of the superconducting quadrupoles previously used for the high luminosity ISR low-beta section would be adequate to provide the necessary field gradient: they have a physical length of 1.1 m and an internal diameter of 173 mm, with a gradient of 39.9 T·m^{-1}.

The solenoid between the two quadrupoles could be the superconducting one previously used for the EMC large polarized target assembly with a 25 kG field, 1.9 m overall physical length, and 150 mm internal diameter. The array of three elements would fit in an overall length of 5 m (Fig. 3), compatible with LEAR straight sections, and their operating conditions would be comfortably within the specified limits.

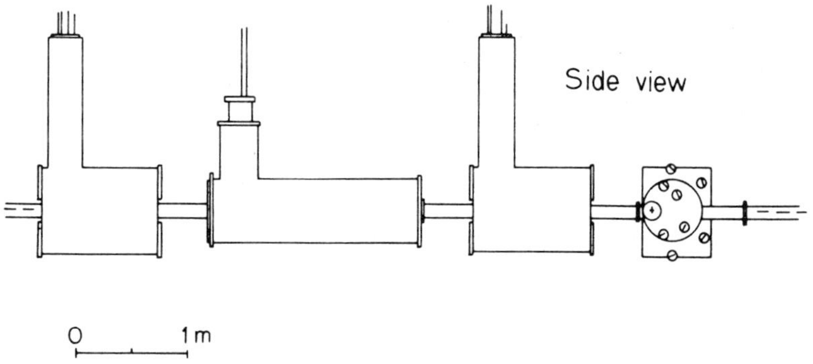

Fig. 3 The superconducting elements for the spin splitter and the polarimeter set-up

THE POLARIMETER

The measurement of the polarization space separation in the circulating beam can be performed by using a simple proton polarimeter measuring the pC or pp elastic scattering asymmetry from a thin carbon or CH_2 wire target scanning the stored beam.

With 10^8 circulating protons, the pp coincidence rate from a 10 μm diameter wire would be a few hundred counts per second at specific correlated laboratory angles corresponding to c.m. angles where maximum analysing power is obtained. Therefore, almost instantaneous real-time measurements of the polarization across the circulating beam profile would be made available by sweeping the target wire across the beam and measuring the polarimeter rates gated in phase with the betatron frequency. Normalization is given by left–right counters. The analysing power of pp scattering is adequate above 500 MeV/c;[7] at energies in the lower range of LEAR, the pC scattering should be used, as it provides greater analysing power[8] (Fig. 4). The same set-up could also be employed for antiprotons by detecting $\bar{p}p$ scattering with the correlated telescopes.

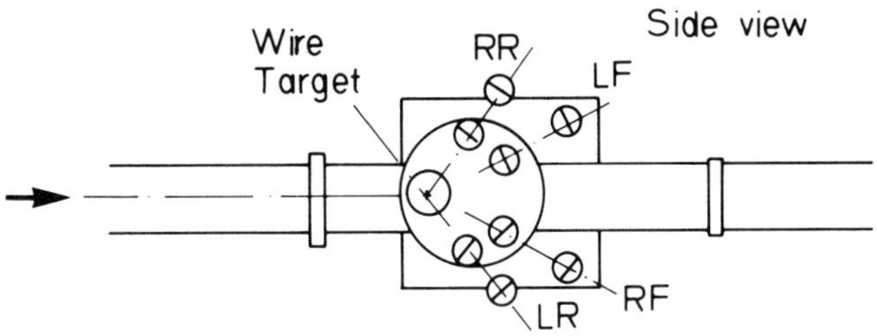

a) The polarimeter for the measurement of pp and pC elastic scattering

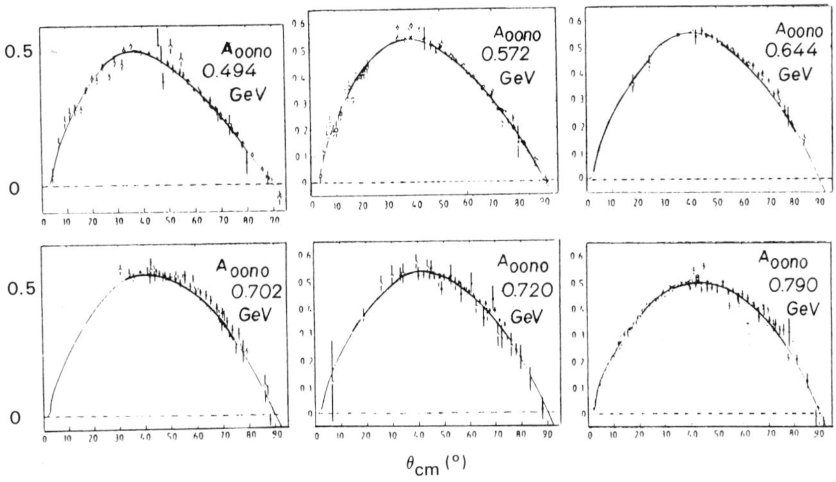

b) Analysing power for pp scattering above 400 MeV

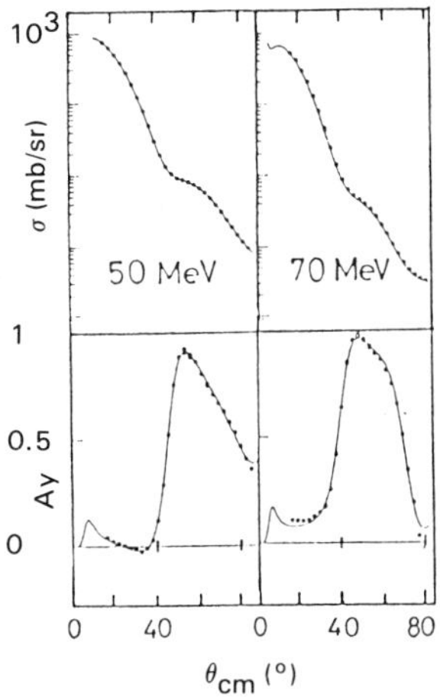

c) Analysing power for pC scattering below 100 MeV

Fig. 4

Currently available results for p̄p elastic scattering show an appreciable analysing power (≈ 20%) between 900 and 2000 MeV/c at sufficiently large angles[9] (Fig. 5): these data will be improved by the results from PS172 at LEAR, between 400 and 1500 MeV/c, in such a way that a precise normalization of the polarimeter will be possible over an extended energy range.

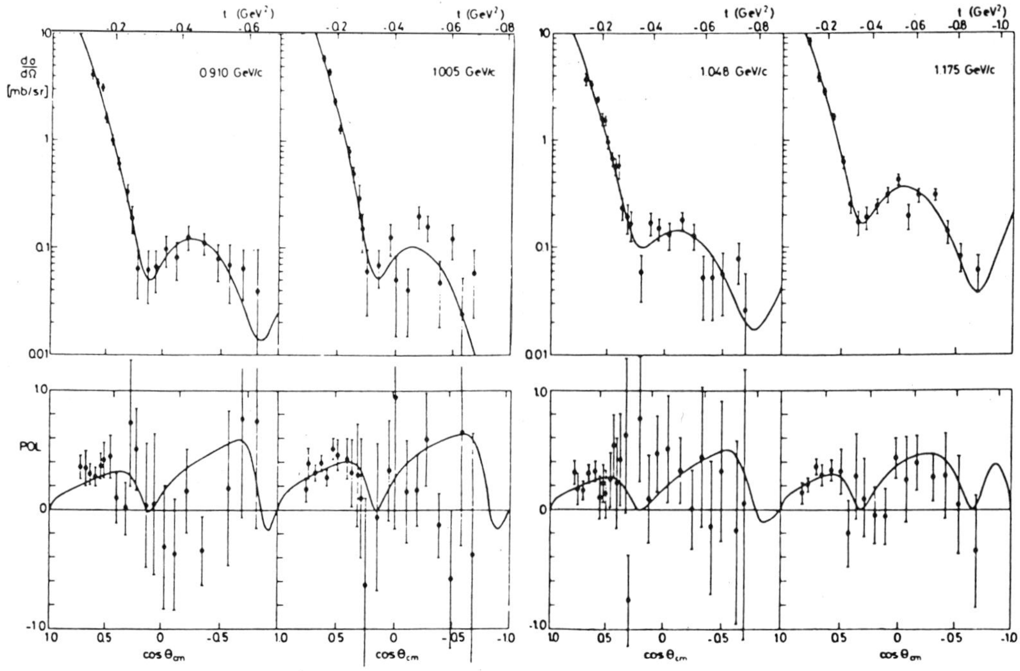

Fig. 5 p̄p elastic scattering (cross-sections and polarization)

CONCLUSIONS

It is unlikely that many of the problems concerning the $N\bar{N}$ interaction mechanisms and the spectroscopy of new exotic states under investigation at LEAR will receive a clear answer without polarization measurements. Polarized antiprotons are essential for these studies, and the method proposed for obtaining them has definite advantages over others. It does not require special hardware developments and can be fully tested with protons in LEAR, with confidence that, if successful, it would be safely applicable to antiprotons—it does not imply a severe loss of the stored beam, and it is advantageous from the point of view of antiproton economy. The demonstration of the validity of this method might have far-reaching consequences for other storage-colliding machines also at much higher energies.

Acknowledgements

We would like to thank P. Lefèvre and D. Möhl for very fruitful discussions regarding the compatibility of the spin splitter with the LEAR machine. Advice from and encouragement by J. Bell, D. Cline and T. Niinikoski are gratefully acknowledged.

REFERENCES

1. R. Birsa et al., Phys. Lett. 155B:437 (1985).
2. H. Dobbeling et al., Proposal P92, CERN/PSCC/85-80 (1985).
3. K. Imai, Kyoto Univ. preprint KUNS 801, July 1985.
4. T.O. Niinikoski and R. Rossmanith, preprint CERN-EP/85-46 (1985).
   T.O. Niinikoski and R. Rossmanith, Proc. Workshop on Polarized Antiprotons, Bodega Bay, 1985, eds. A.D. Krisch, A.M.T. Lin and O. Chamberlain (AIP Conf. Proc. No. 145, New York, 1986), p. 244.
   T.O. Niinikoski and R. Rossmanith, Proc. 5th Int. Workshop on Polarized Sources and Targets, Montana, 1986, eds. S. Jaccard and S. Mango, Helv. Phys. Acta 59:710 (1986).
5. Y. Onel, A. Penzo and R. Rossmanith, CERN/PSCC/86-22, PSCC/M258 (1986).
6. R. Rossmanith and R. Schmidt, Nucl. Instrum. Methods A236:231 (1985).
7. J. Bystricky et al., Nucl. Instrum. Methods A239:131 (1985); also Saclay preprint DPhPE 82-12 (1982).
8. M. Ieiri et al., Proc. 6th Int. Symposium on Polarization Phenomena in Nuclear Physics, Osaka, 1985, eds. M. Kondo, S. Kobayashi, M. Tanifugi, T. Yamazaki, K.-I. Kubo and N. Onishi, Suppl. J. Phys. Soc. Japan 55:1106 (1986).
9. M.G. Albrow et al., Nucl. Phys. B37:349 (1972).

POLARIZED ANTIPROTONS FROM ANTIHYDROGEN

H. Poth

Kernforschungszentrum Karlsruhe
Institut für Kernphysik
Karlsruhe, Fed. Rep. Germany

INTRODUCTION

Beams of polarized antiprotons are continuously requested by intermediate and high-energy physicists. As in the nucleon–nucleon system, spin observables play a key role also for the antinucleon–nucleon system. In the latter case the study of these observables is even more important since the Pauli principle, which reduces the number of independent amplitudes in the NN system, does not apply here. An urgent search for a method of polarizing antiprotons is going on.

Several possibilities were discussed recently during a workshop on antiproton polarization.[1] The most promising ones that do not require antihydrogen formation are the spin-filter[2] and the spin-splitter[3] method. In the following we shall only present ideas based on the interim formation of antihydrogen.[4] The details of the antihydrogen formation are discussed elsewhere.[5]

POLARIZATION THROUGH OPTICAL PUMPING

Fast beams of antihydrogen are formed in the capture of positrons by antiprotons circulating in a storage ring. In spontaneous radiative capture mainly the ground state is populated. Capture into higher atomic states is followed by de-excitation, so that after a sufficiently long flight path all atoms are in the ground state. Once in the ground state the antihydrogen atoms can be manipulated with light in order to achieve polarization by populating a specific magnetic substate. Zelenskiy et al.[6] proposed a method of polarizing protons that is based on the optical pumping of a fast neutralized H^- beam with circularly-polarized laser light. This idea was then applied, by Imai,[7] to a fast atomic beam of antihydrogen. In the following the method is briefly described and illustrated by Fig. 1a.

Atoms are excited from the ground state to the 2p level with circularly-polarized laser light. Owing to selection rules only transitions can take place that change the magnetic quantum number by one unit (e.g. +1). Then all those atoms that are in the $m_F = -1$ ($m_F = 0$) ground state are moved to the $m_F = 0$ ($m_F = 1$) substate of the 2p levels. Spontaneous de-excitation leads to a population of the $m_F = -1$, $m_F = 0$, and $m_F = 1$ ($m_F = 0$, $m_F = 1$) substates of the ground state. Induced transitions populate only the $m_F = -1$ state. It is evident that a repetition of this excitation and de-excitation process several times accumulates population in the $m_F = 1$ substate of the ground state. With a laser of 100 $kW \cdot cm^{-2}$ a cycle of 10 excitations and de-excitations would be sufficient to achieve a polarization of 90%.[6] A relativistic hydrogen beam has been chosen in order to make use of the Doppler effect, which boosts the laser wavelength from the

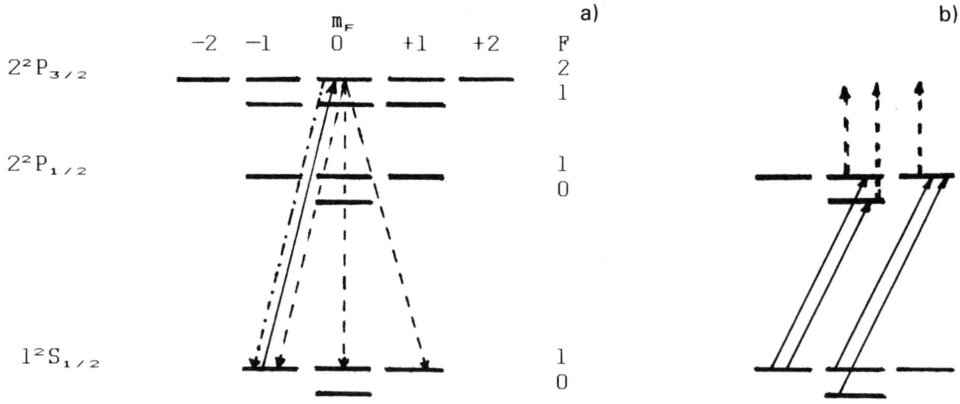

Fig. 1 Hyperfine structure of the 1S and 2P levels of the (anti)hydrogen atom; a) illustrating the optical pumping to achieve polarization in the ground state; b) allowed transitions between the $1^2S_{1/2}$ and the $2^2P_{1/2}$ levels for circularly-polarized light.

visible to the vacuum-ultraviolet (VUV) of the Lyman-alpha line (1s → 2p). This helps to find a sufficiently powerful laser. At the end of the process the atom is ionized to produce the polarized (anti)proton.

RESONANT IONIZATION

As shown elsewhere,[4] selective excitation to the $2^2P_{1/2}$ states is possible if the stored beam can be cooled to a small momentum spread, so that the smearing out of the laser wavelength due to the Doppler effect is small compared to energy difference between the $2^2P_{1/2}$ and the $2^2P_{3/2}$ states. As can be seen from Fig. 1b, the remaining allowed transitions populate only the $m_F = 0, +1$ magnetic substates of the $2^2P_{1/2}$ (F = 1) level and the $m_F = 0$ state of the $2^2P_{1/2}$ (F = 0) level. These atoms can be ionized with another laser. The remaining atoms populate only one magnetic substate, i.e. $^2S_{1/2}$ (F = 1, $m_F = +1$), and the nucleus is therefore 100% polarized; they can be separated from the primary antiproton beam and ionized. It might be possible to feed back into the storage ring those antiprotons which originated from the two-step ionization. They are also polarized to a certain extent. Hence keeping track of their polarization in the storage ring and repeating the process many times may build up a polarization in the stored beam. The method of polarizing protons by a two-step photo-ionization was proposed and shown to work with cold hydrogen gas by Rottke and Zacharias.[8] It depends very much on the momentum spread which can be achieved in the antihydrogen beam whether this method can be applied successfully also to a relativistic beam of antihydrogen.

LAMB-SHIFT SPIN FILTER

The classical Lamb-shift source for producing polarized hydrogen and deuterium requires atoms in the metastable 2s state that are partially quenched with the 2p state in a magnetic field to filter out one of the magnetic substates of the 2s level (Fig. 2). This method is also applicable to a fast beam of antihydrogen atoms.[4]

Antihydrogen atoms can be formed in the n = 2 levels through induced capture. Those reaching the 2p levels will decay rapidly to the ground state. By forming the atoms in a magnetic field close to 574 G and a small electric field, a part of the substates are mixed with $2^2P_{1/2}$ substates and decay rapidly to the ground state. Only atoms in the $2^2S_{1/2}$ ($m_F = 0,1$) state are left over. An adiabatic decrease of the magnetic field then produces a 50% polarization of the antiprotons. Instead of that, one of the magnetic substates can be depleted by feeding in an additional radiofrequency field that induces transitions to the $m_F = 0$ or -1 substate, which is mixed with the $2^2P_{1/2}$ level. This leads to fully nuclear polarized atoms in the $2^2S_{1/2}$ level. These

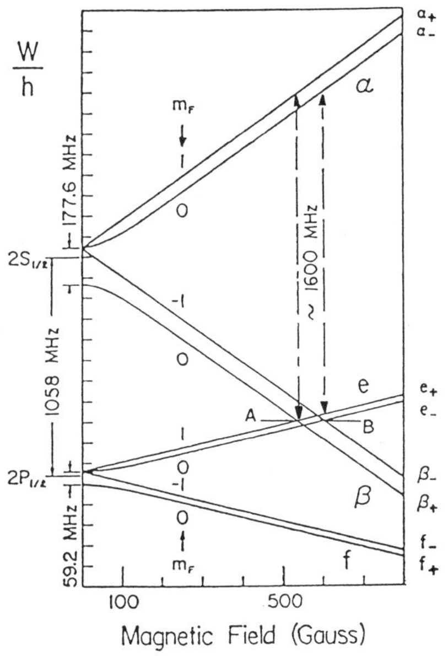

Fig. 2 Energy-level diagram for the principal quantum number n = 2 in a magnetic field.

atoms are re-ionized by laser light or a combination of excitation and stripping, and thus fully polarized antiprotons are obtained. Instead of depleting one of the magnetic substates by feeding in radiofrequency one might be able to apply the diabatic field-reversal method (Sona method),[9] where the external magnetic field is adiabatically decreased and then flipped in sign fast enough compared to the Larmor precession. In this case both magnetic substates contribute to the antiproton polarization. It increases the polarization efficiency by a factor of 2.

Instead of forming the antihydrogen atoms in the n = 2 state by induced capture one can use all the atoms which have eventually reached the ground state and excite them to the 2s state. Whichever procedure is more advantageous depends on the details of the arrangement and the use of polarized antiprotons.

POLARIZED POSITRONS

Antiproton polarization can also be achieved by starting with polarized positrons.[10] It was pointed out that polarized positrons can be obtained from radioactive sources and that their polarization can be maintained during thermalization in a moderator.[10] Positron beams needed for the formation of antihydrogen can be produced using such cold positron sources. Acceleration, if done carefully, conserves the polarization. Spontaneous capture of the positrons by antiprotons does not change the orientation of the positron spin.

Let us consider capture of polarized positrons into the ground state. If it takes place in a magnetic field the substates with $m_j = +\frac{1}{2}\,(-\frac{1}{2})$ corresponding to the states a_+ and a_- of Fig. 2 (β_+ and β_-) are populated. An adiabatic decrease of the magnetic field leads to a 50% polarization of the antiprotons originating from atoms formed in the 1s state. By applying the Sona method they can probably be polarized to 100%. For capture into the 2s state the procedure described in the last section can be used. Concerning capture into other atomic states the situation is more complicated and one has to examine these cases in more detail. However, the orientation of the positron spin during the atomic cascade is hardly changed. This is known from muonic atoms.

Inducing the capture by illumination with circularly-polarized light would help to raise the initial polarization since only the magnetic substates of the p-levels in the continuum with $m_F > 0$

(for positron spin and magnetic field parallel) are populated so that only transitions $m_F = 0$ to $m_F = +1$ are allowed for left-circularly-polarized light. Induced capture of the ground state would then immediately produce a high antiproton polarization.

CONCLUSION

It was emphasized that antiprotons can be polarized passing through intermediate formation of antihydrogen. There are several methods which allow for the achievement of a high degree of polarization and polarization efficiency. The crucial obstacle so far seemed to be the low formation rate of antihydrogen atoms. However, by putting together the present know-how in beam cooling, thermalization of positrons, and laser technology, one could produce at least several tens of thousands of antihydrogen atoms per second[11] in a few years from now. A comparable number of polarized antiprotons could then be obtained. These antiprotons could be accumulated or used directly in an experiment.

There are other methods proposed to polarize stored antiprotons that are also very attractive. The methods described in this paper have the advantage that a) they do remove only those antiprotons from the beam which are used for the subsequent experiment with polarized antiprotons (great economy), b) they do not depend on the stability of polarization (and polarization build-up) in the storage ring, c) they implement beam extraction.

REFERENCES

1. Workshop on Polarized Antiprotons, Bodega Bay, 1985, A.D. Krisch, A.M.T. Lin and O. Chamberlain, eds., AIP Conference Proc. No. 145, New York (1986).
2. E. Steffens et al., Proposal for measurement of spin dependence of $p\bar{p}$ interaction at low momenta, *in* Proc. Third LEAR Workshop on Physics with Cooled Low-Energy Antiprotons in the ACOL Era, Tignes, 1985, U. Gastaldi, R. Klapisch, J.-M. Richard and J. Tran Thanh Van, eds., Editions Frontières, Gif-sur-Yvette (1984), p. 245.
3. Y. Onel, A. Penzo and R. Rossmanith, A spin-splitter for antiprotons in low energy storage rings, *in* Proc. Conf. on the Intersection between Particle and Nuclear Physics, Lake Louise, 1986, D.F. Geesaman, ed., AIP Conf. Proc. No. 150, New York (1986), p. 1229.
4. H. Poth and A. Wolf, Antiproton polarization through antihydrogen, KfK 4098 (1986).
5. R. Neumann, Possible experiments with antihydrogen, these proceedings.
6. A.N. Zelenskiy et al., Nucl. Instrum. Methods 227:429 (1984).
7. K. Imai, *in* Proc. 6th Int. Symposium on Polarization Phenomena in Nuclear Physics, Osaka, 1985, M. Kudo et al., eds., Suppl. J. Phys. Soc. Japan, 55:302 (1986).
8. H. Rottke and H. Zacharias, Phys. Rev. A33:736 (1986).
9. P. Sona, Energia Nucleare 14:295 (1967);
   T.B. Clegg, G.R. Plattner and W. Haeberli, Nucl. Instrum. Methods 62:343 (1968).
10. R.S. Conti and A. Rich, The status of high intensity, low energy positron sources for anti-hydrogen production, presented at the Workshop on Antimatter Facility, Madison, Wisconsin (1985).
11. H. Poth, Physics with antihydrogen, *in* Proc. Conf. on the Intersection between Particle and Nuclear Physics, Lake Louise, 1986, D.F. Geesaman, ed., AIP Conf. Proc. No. 150, New York (1986), p. 580.

LARGE POLARIZATION ASYMMETRIES IN NUCLEON-NUCLEON SCATTERING

A.O. Barut*

International Centre for Theoretical Physics

P.O. Box 586, Trieste, Italy

When planning antiproton polarization experiments, as discussed at this Meeting, it may be good to remember the large polarization effects discovered in recent years in proton-proton and neutron-protron experiments[1]. Similar effects should be expected in the antiproton case. The large spin effects came unexpectedly, not only because one generally thought the spin effects in strong interactions to be small, but also the perturbative QCD predicted no such spin asymmetries. This is due to a theorem that for γ_μ-type interactions at higher energies the helicity is conserved for arbitrary tree-Feynman diagrams [2]. It was no surprise to other models in which one expected large spin (rather magnetic moment) effects at short distances[3]. At any rate these experiments test more crucially the perturbative QCD than unpolarized experiments, and actually show that PQCD is not sufficient to understand these phenomena.

There have been a number of attempts to fit these new experimental polarization asymmetries[4].

I present in this note a simple idea which accounts for the spin effects in a most direct way, namely via the anomalous magnetic moment of the nucleons.

The five (standard) helicity amplitudes in N-N scattering are

$$\phi_1 = H_{++++}; \quad \phi_2 = H_{++--}; \quad \phi_3 = H_{+-+-}$$
$$\phi_4 = H_{+--+} \quad \text{and} \quad \phi_5 = H_{+++-} .$$

In terms of which the cross-section σ_0 and three of the many possible polarization parameters which are measured are given by

$$\begin{aligned}
\sigma_0 &= \tfrac{1}{2}\left(|\phi_1|^2 + |\phi_2|^2 + |\phi_3|^2 + |\phi_4|^2 + 4|\phi_5|^2\right) \\
\sigma_0 A &= \mathrm{Im}\left[\left(\phi_1 + \phi_2 + \phi_3 - \phi_4\right)^* \phi_5\right] \\
\sigma_0 A_{NN} &= \mathrm{Re}\left(\phi_1\phi_2^* - \phi_3\phi_4^* + 2|\phi_5|\right)^2 \\
\sigma_0 A_{SL} &= \mathrm{Re}\left[\left(\phi_1 + \phi_2 - \phi_3 + \phi_4\right)\phi_5^*\right] \qquad (1)
\end{aligned}$$

* Permanent address: Department of Physics, University of Colorado, Boulder, CO 80309, USA.

We have calculated all the amplitudes and asymmetry parameters for the scattering of two fermions assuming the couplings

$$e\gamma_\mu + a\sigma_{\mu\nu}q^\nu \quad , \qquad (2)$$

in the Born approximation, thus with direct and exchange diagrams[5]. [γ_5 couplings have also been treated which we shall omit here.]

The effect of the Pauli-coupling on the asymmetries is very remarkable. They are not only very large, but also highly energy dependent. I shall now indicate that they are qualitatively of the same nature as found experimentally for pp-scatterings.

In lowest order calculations the coupling $e\gamma_\mu + a\sigma_{\mu\nu}q^\nu$ is equivalent to a coupling $f\gamma_\mu = a(p+p')_\mu$ with $f = e + 2ma$, that is a vector and a scalar coupling. Secondly, the magnetic a-coupling gives an effect proportional to energy E so that the relevant parameter in the asymmetries is $V = \frac{aE}{f}$, relative to the pure γ_μ-coupling. At energies $E > 1$ GeV the magnetic term dominates. With this model we are able to reproduce the angle and energy dependence of the polarization asymmetry[6] A_{NN}. For example, at $\vartheta = 90^\circ$ in the centre of mass system, we find[6]

$$A_{NN} = \frac{2(B^2 - AC)}{A^2 + 2B^2 + C^2} \quad , \qquad (3)$$

where

$$A = 4 + x - 4v(2 + x) + 2v^2(3 + 2x)$$
$$B = 1 - 2v - v^2(3 + 2x)/x$$
$$C = x - 4v(1 + x) + 2v^2(3 + 2x)\left(1 + \frac{1}{x}\right) \quad ,$$

with $x = m^2/p^2$ and $v = ma/f$ (p = centre of mass momentum).

The Born amplitudes are real, hence the polarization parameter A in eq.(1) is zero in this approximation. Experimentally also A is rather small except two points at large $p_\perp^2$ ($\gtrsim 6$ (GeV)^2). Of course, the amplitudes will have some imaginary parts. They may be seen to arise due to diffractive scattering or Regge-type amplitudes, or due to the fact that our coupling (3) applies to the constituents of the nucleon (as well as to the nucleon as a whole), in which case there will be cross diagrams leading to an imaginary part of the amplitudes. At this stage the simplest method of parametrizing is to assume a relative phase between the amplitude ϕ_5 and the others[6]. Then we see from eq.(1) that A_{NN} does not change, and A becomes different from zero. We also as a check suggest the measurement of A_{SL} which is essentially a counterpart of A (real and imaginary parts).

Independent of the scattering experiments, the spin effects can also be studied in bound state problems. The theory must account for and calculate both in a consistent way. The anomalous magnetic spin forces have been found to be quite important and according to some models even dominant[3,7] in hadron spectroscopy.

REFERENCES

1. D.G. Crabb et al., Phys. Rev. Lett. 41:1257 (1978);
E.A. Crosbie et al., Phys. Rev. D23:600 (1981);
K.A. Brown et al., Phys. Rev. D31:3017 (1981);
for the latest experiment and further references see G.R. Court et al., Phys. Rev. Lett. 57:507 (1986).

2. See e.g. the reviews: C. Bourrely, E. Leader and J. Soffer, Phys. Reports 59:95 (1980);
   N.S. Craigie, K. Hidaka, M. Jacob and F.M. Renard, Phys. Reports 99:69 (1983);
   Proc. 6th Intern. Symposium on High Energy Spin Effects, Marseille 1984, published in Journ. de Phys. 46 C2 (1985).
3. A.O. Barut, in Proc. Orbis Scientiae 1982, "Field Theory and Elementary Particles",B. Kursunoglu et al., ed., Plenum Press, 1983, p.323.
4. G. Preparata and J. Soffer, Phys. Lett. 86B:304 (1979); 93B:187 (1980);
   M. Anselmino and E. Predazzi, Z. f. Phys. C28:303 (1985) and references therein;
   S.M. Troshin and N.E. Tyurin, Phys. Lett. 144B:260 (1984); Journ. de Phys. 46 C2:235 (1985);
   C. Avilez, G. Cocho and M. Moreno, Phys. Rev. D24:634 (1981).
5. T. Anders, A.O. Barut and W. Jachman, to be published.
6. T. Anders, A.O. Barut and W. Jachman, to be published.
7. A.O. Barut, Spin forces in hadron spectroscopy, in Procs. of the Kaziemierz Conference on Elementary Particle Physics, University of Warsaw Press, 1986.

LECTURERS

Dr. J.S. Bell	CERN - TH CH-1211 Geneva 23 Switzerland
Dr. W. Buchmüller	Institut für Theor. Physik Universität Hannover Appelstrasse 2 D-3000 Hannover 1 FRG
Professor F. Calaprice	Physics Department Princeton University P.O. Box 708 Princeton New Jersey 08544 USA
Dr. G. Gabrielse	Dept. of Physics, FM-15 The University of Washington Seattle Washington 98195 USA
Professor V. Hughes	Physics Department Yale University New Haven Connecticut 06520 USA
Professor I. Mannelli	Departimento di fisica Università di Pisa Piazza Torricelli 2 I-56100 Pisa Italy
Professor G. Morpurgo	I N F N Via Dodecaneso 33 I-16146 Genova Italy
Professor R. Neumann	Physikalisches Institut Universität Heidelberg Philosophenweg 12 D-6900 Heidelberg 1 FRG

Professor L. Wolfenstein

Physics Department
Carnegie-Mellon University
Pittsburgh
Pennsylvania 15213
USA

PARTICIPANTS

L. Adiels	Research Institute of Physics Freskativ. 24 S-104 05 Stockholm 50 Sweden
R. Bacher	S I N CH-5234 Villigen Switzerland
A.O. Barut	International Centre for Theoretical Physics P.O.Box 586 - Miramare I-34100 Trieste Italy
D. Cline	Physics Department University of Wisconsin Madison Wisconsin 53706 USA
Ph. Bloch	CERN - EP CH-1211 Geneva 23 Switzerland
R. Decker	Institut für Theoretische Kernphysik Universität Karlsruhe Postfach 6980 D-7500 Karlsruhe 1 FRG
H.T. Duong	Laboratoire Aimé Cotton Bâtiment 505 F-91405 Orsay France
K. Elsener	CERN - EP CH-1211 Geneva 23 Switzerland
U. Gastaldi	CERN - EP CH-1211 Geneva 23 Switzerland
Th. Geralis	NRC "Democritos" Aghia Paraskevi Athens Greece

P.J. Hayman	Physics Department University of Liverpool P.O. Box 147 Liverpool L69 3BX England
D. Holtkamp	Group P-3, MS D449 Los Alamos National Laboratory P.O. Box 1663 Los Alamos New Mexico 87545 USA
R.J. Hughes	Group T-8, MS-B285 Los Alamos National Laboratory P.O. Box 1663 Los Alamos New Mexico 87545 USA
E. Iacopini	Scuola Normale Superiore Piazza dei Cavalieri I-56100 Pisa Italy
R. Klapisch	CERN - DG CH-1211 Geneva 23 Switzerland
H.-J. Kluge	CERN - EP CH-1211 Geneva 23 Switzerland
R. Landua	CERN - EP CH-1211 Geneva 23 Switzerland
R. Le Gac	Laboratoire René Bernas Bâtiment 108 F-91406 Orsay France
U. Mall	Institut für Physik Universität Basel Klingelbergstrasse 82 CH-4056 Basel Switzerland
V. Matsinos	NRC "Demokritos" Aghia Paraskevi Athens Greece
G. Mezzorani	Dipartimento di Scienze Fisiche Università di Cagliari Via Ospedale, 72 I-09100 Cagliari Italy

N. Nägele
Austrian Academy of Science
Boltzmanngasse 3
A-1090 Wien
Austria

T.O. Niinikoski
CERN - EP
CH-1211 Geneva 23
Switzerland

P. Pavlopoulos
CERN - EP
CH-1211 Geneva 23
Switzerland

A. Penzo
Dipartimento di fisica
Università di Trieste
Via A. Valerio 2
I-34127 Trieste
Italy

I. Picek
D E S Y
Notkestrasse 85
D-2000 Hamburg 52
FRG

H. Poth
Institut für Kernphysik (KfK)
Postfach 3640
D-7500 Karlsruhe 1
FRG

R. Rickenbach
Institut für Physik
Universität Basel
Klingelbergstrasse 82
CH-4056 Basel
Switzerland

A. Schopper
Institut für Physik
Universität Basel
Klingelbergstrasse 82
CH-4056 Basel
Switzerland

F. Scuri
INFN - Sezione di Pisa
I-56010 San Piero a Grado
Italy

L. Simons
S I N
CH-5234 Villigen
Switzerland

P. Staroba
Institute of Physics
Czechoslovak Academy of Sciences
Na Slovance 2
CZ-180 40 Prague 8-Liben
Czechoslovakia

N.W. Tanner
Nuclear Physics Laboratory
Oxford University
Keble Road
Oxford OX1 3RH
England

C. Thibault
Laboratoire René Bernas
Bâtiment 108
F-92406 Orsay
France

G. Torelli
Dipartimento di fisica
Università di Pisa
Piazzale Torricelli 2
I-56100 Pisa
Italy

F. Touchard
Laboratoire René Bernas
Bâtiment 108
F-91506 Orsay
France

INDEX